Environmental Challenges and Solutions

The Environmental Challenges and Solutions series aims to improve our understanding of the Earth's most important environmental challenges, and how we might more effectively solve or at least mitigate these challenges. Books in this series focus on environmental challenges and solutions in particular geographic regions ranging from small to large spatial scales. These books provide multidisciplinary (technical, socioeconomic, political, etc.) analyses of their environmental challenges and the effectiveness of past and present efforts to address them. They conclude by offering holistic recommendations for more effectively solving these challenges now and into the future.

We are welcoming proposals specifically on the following topics:

- Environmental sciences and social aspects
- Environmental sciences and human conflicts
- Boosting de-carbonization of energy system to face climate and economic emergency
- Climate emergency and health: sources, science-based analysis and global solutions
- Indoor and outdoor chemicals, environmental quality and public health in a changing world
- Demographic, climate and environmental change: evidence and recommendation for global sustainability
- Technological, industrial and corporate change for supporting environmental and social sustainability
- Nanotechnology applied to water decontamination/purification processes
- Sorbents and other materials in environmental protection
- Raw materials supply and circular economy
- Strategies to boost clean mobility

All books are written in a concise and readable style, making them suitable for both specialists and non-specialists starting at first year graduate level. Proposals for the book series can be sent to the Publishing Editor, Éva Lörinczi eva.loerinczi@springer.com.

Abhishek Guldhe • Jaya Lakkakula
Nilesh Shirish Wagh
Editors

Biomaterials for Environmental Applications: Detection and Remediation

Springer

Editors
Abhishek Guldhe
Amity Institute of Biotechnology
Amity University Maharashtra
Mumbai, Maharashtra, India

Jaya Lakkakula
Amity Institute of Biotechnology
Amity University Maharashtra
Mumbai, Maharashtra, India

Nilesh Shirish Wagh
Amity Institute of Biotechnology
Amity University Maharashtra
Mumbai, Maharashtra, India

ISSN 2214-2827 ISSN 2214-2835 (electronic)
Environmental Challenges and Solutions
ISBN 978-3-032-12700-6 ISBN 978-3-032-12701-3 (eBook)
https://doi.org/10.1007/978-3-032-12701-3

This Springer imprint is published by the registered company Springer Nature Switzerland AG
The registered company address is: Gewerbestrasse 11, 6330 Cham, Switzerland

Preface

Environmental pollutants pose a serious threat to ecological balance, human health, and the global climate. Early detection and effective remediation of these pollutants are critical to prevent further damage. Over the years, both natural and synthetic materials have been explored and applied for pollutant detection and remediation. This book, *Biomaterials for Environmental Applications: Detection and Remediation*, presents current techniques, critically evaluates their efficiency, and highlights the future prospects of various biomaterials in addressing environmental pollution.

Biomaterials have gained increasing importance in tackling environmental challenges due to their unique advantages such as biocompatibility, specificity, ease of fabrication, biodegradability, cost-effectiveness, and eco-friendly nature. In recent years, a wide range of biomaterials has been developed for potential applications including dye removal, heavy metal detection, and use as pharmaceutical excipients. Moreover, the integration of biomaterials with other advanced materials has paved the way for ultrasensitive and highly selective pollutant detection and remediation strategies. This book also explores the advancements in biomolecules employed in biosensors for pollutant detection. It covers both the fundamental concepts and the latest developments in bioremediation using biomolecules such as chitosan, alginate, pectin, enzymes, and biochar. Additionally, the synthesis and applications of advanced nanomaterials derived from biological sources such as bacteria, fungi, algae, and plants are discussed in detail.

The book is designed to serve a wide audience including researchers, engineers, technicians, and students from diverse fields such as biochemistry, microbiology, biotechnology, and environmental sciences. Each chapter provides insights into advanced conversion techniques for synthesizing biomaterials, application strategies for pollutant detection and remediation, associated challenges, as well as their economic and environmental impacts.

We are deeply grateful to the contributors from leading institutions worldwide, whose expertise and commitment have greatly enriched the content of this book.

We hope that this volume will serve as a valuable reference and inspire further innovation in the development of biomaterials for environmental applications.

Mumbai, Maharashtra, India Abhishek Guldhe
Mumbai, Maharashtra, India Jaya Lakkakula
Mumbai, Maharashtra, India Nilesh Shirish Wagh

Contents

Chapter 1
Enzyme-Based Biosensors for Detection of Environmental Pollutants

Jugal Patil, Arya Algikar, and Shankar Khade

Abstract The modern world provides simple access to all areas necessary for a comfortable living and interpersonal interactions. The world is gradually becoming more exposed to the most dangerous aspect of existing environmental pollution while providing such a resource. If the action in question is not handled seriously, it will have a negative impact on environmental security and pose serious issues for all of the living things in this ecosystem. Environmental pollutants are compounds that are introduced into an open environment and disrupt living ecosystems. Some of these substances may naturally occur, but most are introduced by human activities such as energy generation, agriculture, and industry. Numerous scientifically based procedures and initiatives have been launched to identify and regulate dangerous pollutants. However, these measures alone are insufficient to bring about a significant shift, and the problem will remain difficult to fix. Simple, quick, inexpensive, and sensitive procedures are required for the development of the initiative to address global pollution and to detect pollutants. To create biosensors, methods like enzyme immobilization could be applied. Biosensors are among the items listed above. Enzyme, whole cell, antibody, aptamer, and DNA-based approaches for the detection of environmental contaminants have been popular recently. This chapter reviews and discusses types of contaminants, development hurdles, and techniques for detecting environmental pollutants using enzyme-based biosensors.

Keywords Biosensors · Enzymes · Enzyme immobilization · Pollutants

---

J. Patil · A. Algikar · S. Khade (✉)
Department of Biotechnology Engineering, School of Engineering, Ajeenkya DY Patil University, Pune, Maharashtra, India

A. Guldhe et al. (eds.), *Biomaterials for Environmental Applications: Detection and Remediation*, Environmental Challenges and Solutions,
https://doi.org/10.1007/978-3-032-12701-3_1

1.1 Introduction

One of the main objectives of sustainable development is the detection and remediation of environmental toxins, which is a modern necessity. One of the factors contributing to rising carbon emissions is an increase in human activity related to daily commuting. On the other hand, synthetic chemicals are found in cosmetics, insecticides, fast-paced industrialization (Shindhal et al. 2021), medications, cleaning supplies, etc. Things do not stop at this point; one of the major reasons is also uncontrolled population growth and human-induced activities that have caused the uncontrolled discharge of different toxic pollutants like mercury (Hg), lead (Pb), cadmium (Cd), copper (Cu), chrome (Cr), nickel (Ni), selenium (Se), and arsenic (As) (Coronado-Apodaca et al. 2023), radioactive elements (radium "Ra" and uranium "U"), organic compounds like acrylamide and polyacrylamide, nitride and nitrates, benzene phenols, etc. Methods are available for detection of the above-mentioned pollutants, like the quantum dot method with fluorescence, hydride generation atomic absorption spectroscopy, graphite furnace AAS spectrometry, X-ray fluorescence spectrometry, and energy dispersive X-ray fluorescence. Heavy metals, including arsenic, in drinking water samples and other complex matrices, the ICP-MS method is the most common method used (González-González et al. 2023), etc. On the other hand, standard methods for the detection of organic compounds in water bodies are gas chromatography and mass spectrometry. However, despite the many advantages of this technology, the weakness of the methods is that sophisticated instruments are required with proper knowledge, which is necessary for handling such equipment. Also, due to the expensive and high-demand technique, this system has failed in terms of bioavailability. To overcome certain gaps, a complementary system is developed that is made from living systems and is used for detection as well as analysis. The examples for those various systems are bacteria, fungi, blue-green algae, and enzymes, which are used in this type of analysis, including single-celled microorganisms, especially bacteria. The reason for consideration is their rapid growth and low maintenance (Nigam and Shukla 2015). Over this period, the proteins present in these cells are also applied to specific pollutants and analytes (Acha et al. 2009). Therefore, it is necessary to create a simple, specific, sensitive, rapid, and portable method to monitor and evaluate environmental safety threats; thus, biosensors have become a screening tool option.

Mechanism of Biosensor

A biosensor is defined by the IUPAC as a device that detects chemical molecules, usually through electrical, thermal, or optical signals, by means of certain biochemical reactions mediated by separated enzymes, proteins, antibodies, tissues, organelles, or entire cells (Inshyna et al. 2020). A normal biosensor is represented in Fig 1.1.

The components of a normal biosensor are as follows:

Analyte: A substance or material that is of interest and also requires detection. The most common example of an analyte is glucose.

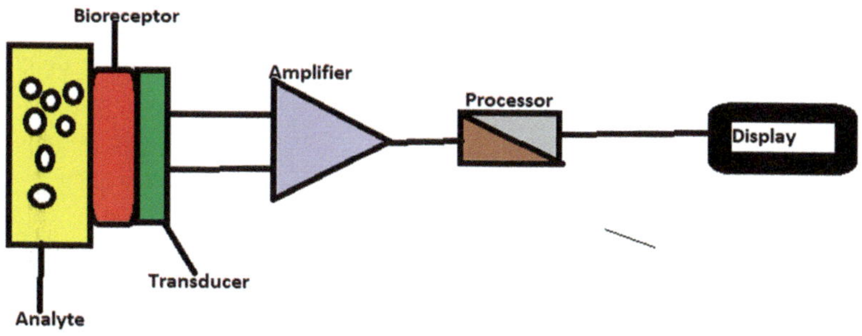

Fig 1.1 Schematic diagram of a biosensor (Kan et al. 2004)

Bioreceptor: A bioreceptor is a molecule that analyses or identifies the analyte specifically, e.g. enzymes, DNA, etc.

Transducer: A transducer could be simply described as a device that detects or observes change, may it be physical or chemical, into a measurable signal. The signals may be optical or electrical, which are proportional to the quantity of analyte-bioreceptor interactions.

Electronics: This part is solely responsible for processing the transduced signals for display. It mainly consists of a circuit that amplifies and converts the signal from analog to digital and further transfers it to the display.

Display: The display is mainly made of liquid crystal, which is a direct computer or printer that makes numbers or curves that are understandable by the user. The output could be graphic, numeric, image, or tabular.

1.2 The Role of Enzymes in Biosensors

In biosensors, enzymes could be used as essential part for recognition element. The enzyme used is immobilized within its supported matrix on the surface of transducer for maintaining the enzymatic activity. Therefore, the advantage of using enzyme as a part of biosensor is their high specificity, high turnover rates, etc. The principle of enzymes as biosensing is to determine changes in the proton (H+) concentration, the release and uptake of gases (CO_2, NH_3, O_2, etc.), light emission, absorption or reflection, and heat emission in order to identify the presence of analytes (Nguyen et al. 2019). The basic principle of enzyme-based biosensors is the absorption of enzymes, i.e., their byproducts can be measured by following substrate degradation on electrodes. By closely observing a biocatalytic process, the electrode monitors the changes either amperometrically or potentiometrically. An enzyme is applied to the electrode surface to make enzyme biosensors. An enzyme helps a substrate break down, and the end products are measurable. These systems usually include the catalytic process of redox reactions in which an electrical charge is present in either the substrate or the product. Many different enzyme-based biosensors are used for detecting and monitoring environmental pollutants, for example, pollutants like parathion, nitrate, and formaldehyde can be detected by sulphite parathion hydrolase, nitrate reductase, and formaldehyde dehydrogenase (Cosnier et al. 1994; Liu et al. 1995; Nader et al. 1990).

1.3 Detection of Environmental Pollutants with Enzyme-Based Biosensors

Heavy metals, hormones, organophosphates, phenolic compounds, antibacterials, and dyes are among the environmental toxins that are currently deserving of attention because of their toxicity to the environment and potential harm to humans and other creatures (Hart et al. 2018). Enzymes are a class of complex proteins that function in living cells as catalysts. Similar to a catalyst, an enzyme quickens the pace of a chemical reaction. An enzyme changes one or more compounds, often referred to as substrates, into another class of substances, known as products, during a chemical process.

Enzymes can be divided into six main classes: oxidoreductase, transferase, hydrolase, isomerase, lyase, and ligase (Nelson et al. 2008). Of these several categories of enzymes, two primary categories—oxidoreductase and hydrolase—are primarily employed effectively as biosensors for the identification of numerous dangerous metalloids and heavy metals. The measurement of the immobilized system's enzyme activity involves monitoring variations in conductivity, heat release, current, mass, voltage, and ionic strength as well as optical characteristics like reactance or reflectance. The immobilization process can be done directly or indirectly using the transducer system. Acquired in either of the aforementioned forms, the signals are then processed once again via an advanced microprocessor system to enhance the signal intensity, enabling accurate detection of even the smallest changes during an enzyme activity. Through indirect blockage of the active sites by other substances, these enzymatic processes either cause an inhibition of enzyme activity or cause the enzyme to become activated as a direct enzyme-substrate complex form. In the first situation, the concentration of substrates that interact with enzymes is measured, whereas the concentration of substrates that inhibit enzyme activity is measured (Table 1.1).

1.4 Principle of Enzyme-Based Sensor

Enzyme-based biosensors are devices that, in essence, convert data from changes in the catalytic domain during chemical reactions between enzymes and substrates into quantifiable signals in a variety of forms, such as pH, temperature, ions, the absorption efficiency, charges, molecular weight, and atomic mass utilizing a biosensing as well as transducer system as depicted in Fig. 1.2.

Table 1.1 Specificity of enzymes in detecting the pollutants

Pollutants/analytes	Enzyme	Method for enzyme immobilization	Type of pollution	References
Mercury, cadmium, and arsenic	Urease enzyme	Entrapment, adsorption, covalent attachment	*WP*	Pal et al. (2009)
Phenols	Horseradish peroxidase	Covalent attachment to PEG-coated magnetite nanoparticles	*WP*	Yadav et al. (2021)
Organic micropollutants	Melon peel peroxidase	Entrapment in alginate beads	*WP*	Al-Madhagi et al. (2023)
Nitrogen oxides (NOx) and sulphur oxides (SOx)	Nitrate reductase and sulphite oxidase	Entrapment in silica sol-gel matrices or polyacrylamide gels	*AP*	Sohail and Adeloju (2016)
Volatile organic compounds (VOCs)	Alcohol dehydrogenase and aldehyde dehydrogenase	Covalent binding to chitosan beads or silica gel particles	*AP*	Sohail and Adeloju (2016)
Petroleum hydrocarbons (TPH)	Bacterial enzymes (e.g., alkane hydroxylase)	Encapsulation in calcium alginate beads or polyvinyl alcohol gel beads	*SP*	Li et al. (2020)
Pesticides	Acetylcholinesterase	Entrapment	*WP, SP, AP*	Ali and Singh (2020)
Dyes	Peroxidase	Covalent attachment	*WP, SP*	Garg and Vemaraju (2024)

AP Air pollution, *WP* Water pollution, *SP* Soil pollution

The biosensing enzyme in addition to a transducing unit (electrodes, photodetector, thermistor, or piezoelectric) makes up the majority of an enzyme-based biosensor. The transducer and biosensing enzyme are typically connected by a suitable immobilization mechanism (Rajangam et al. 2018).

1.5 Development of Enzyme-Based Biosensors: Design and Fabrication

The design of an enzyme-based biosensor consists of different components as mentioned in the following:

1. *Recognition Element*: An enzyme acts as a recognition element which specifically interacts with the target analyte. The choice of the enzyme is important for ensuring that it is selective and specific towards the targeted analyte.
2. *Immobilization Technique*: This involves attaching the enzyme to a support material like a transducer or electrode. This improves enzyme stability, reusability, and orientation on the sensor surface.

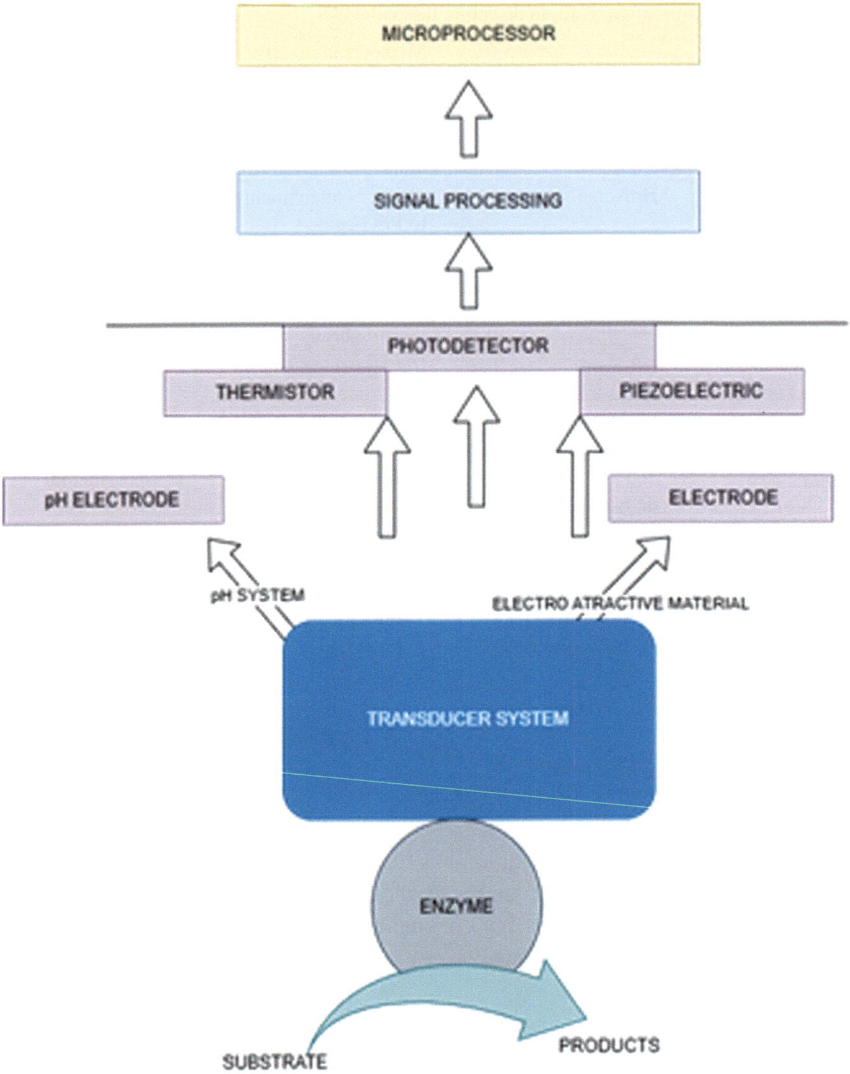

Fig. 1.2 Typical diagram for enzymatic biosensor (Sarkar et al. 2019)

3. *Transducer*: It changes the biochemical signal generated from the interaction between enzyme and analyte to a measurable chemical or physical signal. Common transducers are electrochemical, optical, and piezoelectric transducers.
4. *Signal Processing Unit*: Signal processing unit amplifies, filters, and processes the transducer's output signal in order to generate quantitative readout.

1.6 Fabrication of Enzyme-Based Biosensors

In general, fabrication of enzyme-based biosensors involves the following steps:

Preparation of the Substrate: The substrate may be a conductive electrode or a microfluidic channel which is prepared to provide a suitable surface for enzyme immobilization.

Immobilization of Enzyme: Different methods such as physical adsorption, entrapment, crosslinking, and self-assembly are used to immobilize selected enzymes onto the chosen substrate. Enzyme activity and orientation must be preserved during immobilization.

Integration of Transducer: The transducer is integrated in the sensor design to ensure proper electrical contact or optical coupling with the immobilized enzyme.

Packaging of Sensor: To ensure stability, avoid contamination, and ease handling, a sensor is packaged in protective housing.

1.7 Recent Advancements and Applications of Enzyme-Based Biosensors for Pollutant Detection

(a) *Nanomaterials-Based Enzyme Immobilization*

Incorporation of nanomaterials into enzyme immobilization techniques has improved enzyme stability, orientation, and biocompatibility. Carbon nanotubes, gold nanoparticles, and silica nanoparticles among other nanomaterials possess high surface area for enzymes attachment hence improving enzyme loading and reducing enzyme aggregation. Besides, the presence of these materials around the enzyme helps to control its microenvironment thereby making it more stable and active in different conditions (Liang et al. 2021).

(b) *Enzyme Engineering for Enhanced Selectivity*

Enzymes with enhanced selectivity towards specific target pollutants are being engineered using directed evolution and rational design strategies. These methods involve introducing mutations into an enzyme's protein sequence or structure so that it better recognizes the target pollutant while minimizing its interaction with interfering substances. Such targeted engineering of enzymes has enabled development of biosensors that have increased selectivity but decreased false negatives (Korendovych and DeGrado 2020; Porter et al. 2016).

(c) *Microfluidic and Miniaturized Biosensors*

Miniaturization is a key trend in biosensor development, enabling the creation of portable, user-friendly devices for on-site pollutant detection. Microfluidic technology plays a crucial role in this miniaturization, allowing for the precise manipulation of small volumes of fluids and the integration of multiple sensor components onto a single chip. Miniaturized enzyme-based biosensors offer the potential for real-time, in-field monitoring of pollutants in environmental and industrial settings (Fattahi and Hasanzadeh 2022; Karunakaran et al. 2025).

(d) *Microbial Pollution Detection*

Enzyme-based biosensors have been utilized for the identification of microbial pollutants, such as *E. coli* and Salmonella, in water samples. These biosensors employ enzymes like β-galactosidase and alkaline phosphatase, which are immobilized on nanomaterials including quantum dots and graphene. The integration of these nanomaterials has significantly enhanced the sensitivity and selectivity of the biosensors, making them suitable for real-time monitoring of microbial contamination (Kucherenko et al. 2019).

(e) *Detection of Phenolic Compounds by Oxidoreductases:*

Enzymes like tyrosinase and laccase, classified as oxidoreductases, are utilized in biosensors to identify phenolic compounds present in water samples. They catalyse the oxidation of phenolic substances, leading to observable changes in electrochemical signals. Recent investigations have focused on the use of nanomaterials, including carbon nanotubes and magnetic nanoparticles, for enzyme immobilization, which enhances the functionality of these biosensors (Lzaod and Dutta 2022).

(f) *Hydrolases in the Detection of Organophosphates:*

Enzymes such as organophosphorus hydrolase (OPH) are employed in biosensors for the identification of organophosphate pesticides. These hydrolases facilitate the hydrolysis of organophosphates, resulting in detectable alterations in electrochemical signals. Recent research has concentrated on the application of sophisticated immobilization methods, including sol-gel encapsulation and cross-linking, to improve the efficacy of these biosensors (Ali and Singh 2020).

(g) *Transferases in the Detection of Pharmaceutical Compounds:*

Transferases are employed in biosensors to identify pharmaceutical compounds in water samples. These enzymes catalyse the transfer of specific functional groups, resulting in measurable variations in electrochemical signals. Recent studies have examined the use of nanomaterials, such as gold nanoparticles and magnetic nanoparticles, for the immobilization of enzymes, which improves the performance of these biosensors (Kucherenko et al. 2019).

1.8 Analysis of Viability of Enzyme-Based Biosensors

Even though a number of biosensors have already been created for the purpose of identifying and tracking both organic and inorganic environmental contaminants, many of them—such as hormones, endocrine disruptors, poisons, heavy metals, and pharmaceuticals—remain undetected readily because of their intricate features. The significant advantages of biosensor use over the compactness of environmental contaminants are one of the traditional analytical techniques in terms of portability, size, flexibility to use a smaller sample volume, and in situ monitoring aptitude (Nigam and Shukla 2015). Nonetheless, a number of factors limit the true uses of biosensors. The primary issues with environmental samples are their complexity, heterogeneity, and decrease of signal as a result of external interference. "Component-based failure" or "biocompatibility-based failure" are the two main causes of inefficiencies in enzyme-based biosensors (Rocchitta et al. 2016). Proteins, cellular detritus, or living cells cling to the electrode area and outside biosensor surface during biofouling, blocking the diffusion process and lowering the sensor signal. Other chemicals in environmental samples also cause a decrease in signal intensity by forming a polymeric coating on transducer surfaces. An enzyme-based biosensor's sustained activity is the single factor that affects how long it lasts, and it might get less over time. Good immobilization methods can aid to maintain enzyme activity, which in turn increases the stability of the biosensor. Examples of these methods include adsorption, the sol-gel method, the use of aqueous-insoluble coupling agents, and polymeric films (Sassolas et al. 2013).

1.9 Challenges and Limitations of Enzyme-Based Biosensors in Environmental Detection

1. Enzyme Stability and Activity:
   Enzymes are proteins that are susceptible to denaturation under various environmental conditions, such as changes in temperature, pH, and ionic strength. This instability can lead to a loss of enzyme activity and reduced biosensor performance (Adamian et al. 2021; Sheldon 2007).
2. Enzyme Selectivity:
   While enzymes exhibit high specificity towards their target analytes, they may also interact with non-target molecules, leading to false positives or reduced sensitivity. This issue is particularly challenging in complex biological or environmental samples (Liang et al. 2021).
3. Sensitivity:
   The sensitivity of enzyme-based biosensors is crucial for detecting low concentrations of analytes. However, the sensitivity of these biosensors can be limited by factors such as the enzyme's activity, the transducer's efficiency, and the background signal (Bucur et al. 2018).
4. Miniaturization and Integration:
   Miniaturization of enzyme-based biosensors is essential for portability and point-of-care applications. However, miniaturization can pose challenges in maintaining enzyme stability, integrating multiple sensor components, and ensuring reliable signal transduction (Liao et al. 2018).
5. Cost-Effectiveness:
   Enzyme-based biosensors can be relatively expensive to produce due to the use of purified enzymes and complex fabrication processes. Reducing costs is essential for making these biosensors more accessible and widely applicable (Ispas et al. 2012).

1.10 Case Studies

Heavy Metal Detection in Water Bodies: Nigam and Shukla's study showed how to employ biosensors based on enzymes to find heavy metals in water bodies. Water samples were subjected to a study that employed a microbiological biosensor which identified heavy metals like lead, cadmium, and mercury (Nigam and Shukla 2015).

Monitoring Air Quality in Urban Areas: Second study showed how air quality could be monitored in urban areas using enzyme-based biosensors. Nitrogen dioxide (NO2) could be found in air samples using an amperometry biosensor (Nigam and Shukla 2015).

Finding Soil Contamination: Another study showed how biosensors could be used for analysing soil sample for detecting soil pollutants which could turn adverse if such cultivation is pumped in markets (Kucherenko et al. 2019).

1.11 Future Aspects and Consideration

Enzyme biosensors have demonstrated great promise in supplanting traditional, expensive equipment-based analytical methods. However, a number of issues covered in Sect. 1.5 limit the practical uses of enzyme biosensors. Future studies on biosensors based on enzymes for use in the real-world detecting, identifying, and measuring of emerging pollutants should concentrate on the following areas:

1. Increasing stability and sensitivity
2. Increasing the level of selectivity (among developing contaminants)
3. Lab-on-a-chip (LiP) miniaturization of biosensors based on enzyme activities
4. Enzyme-based biosensors going into business

To increase the sensitivity of enzyme biosensors, researchers have recently investigated the efficient use of nanomaterials and enzyme-analyte reaction optimization. It has also been demonstrated that using nanoparticles in the biosensor fabrication procedure increases stability.

1.12 Conclusion

Enzyme-based biosensors have emerged as powerful tools for the detection and control of environmental pollutants due to their high specificity, sensitivity, and quick response times. These biosensors also provide different advantages over traditional analytical methods, including portability, cost-effectiveness, and reusability. Recent advancements in enzyme immobilization techniques, enzyme engineering, and biosensor design have further enhanced the capabilities and applications of enzyme-based biosensors.

Recent advancements in enzyme immobilization techniques, enzyme engineering, and biosensor design have significantly enhanced the capabilities of enzyme-based biosensors. Novel immobilization methods, such as nanomaterial-based immobilization, have improved enzyme stability and facilitated miniaturization. Directed evolution and rational design strategies have enabled the development of highly selective and robust enzymes for target pollutants. Additionally, innovative biosensor designs, such as microfluidic and integrated biosensors, have expanded the applications of these biosensors.

Despite these advancements, several challenges remain in the development and widespread adoption of enzyme-based biosensors for environmental monitoring. Ensuring long-term enzyme stability under harsh environmental conditions is crucial for reliable and continuous monitoring. Overcoming interference from complex sample matrices is essential for accurate and precise measurements. Miniaturizing biosensors without compromising performance is critical for portability and point-of-care applications. Additionally, reducing the cost of biosensor fabrication and materials is essential for broader accessibility and adoption.

As research continues to address these challenges, enzyme-based biosensors are poised to revolutionize environmental monitoring and pollution control. These biosensors will play a pivotal role in early detection of environmental contaminants, enabling timely interventions to prevent environmental damage and protect human health. With further advancements, enzyme-based biosensors have the potential to transform environmental monitoring from a reactive to a proactive approach, paving the way for a cleaner and more sustainable future.

References

Acha V et al (2009) Tissue-based biosensors. In: Recognition receptors in biosensors. Springer, pp 365–381

Adamian Y et al (2021) Recent developments in the immobilization of laccase on carbonaceous supports for environmental applications-a critical review. Front Bioeng Biotechnol 9:778239

Ali R, Singh DV (2020) Biosensors: a biotechnological tool for monitoring environmental pollution. In: Bioremediation and biotechnology, vol 3: Persistent and recalcitrant toxic substances. Springer International Publishing, Cham, pp 331–348

Al-Madhagi H, Yazbik V, Abdelwahed W (2023) Food wastes clean water wastes: melon peel peroxidase isolation and immobilization onto magnetite nanoparticles for phenol removal. Chem Biol Technol Agric 10(1):121

Bucur B et al (2018) Advances in enzyme-based biosensors for pesticide detection. Biosensors 8(2):27

Coronado-Apodaca KG et al (2023) Immobilized enzyme-based novel biosensing system for recognition of toxic elements in the aqueous environment. Top Catal 66(9):606–624

Cosnier S, Innocent C, Jouanneau Y (1994) Amperometric detection of nitrate via a nitrate reductase immobilized and electrically wired at the electrode surface. Anal Chem 66(19):3198–3201

Fattahi Z, Hasanzadeh M (2022) Nanotechnology-assisted microfluidic systems for chemical sensing, biosensing, and bioanalysis. TrAC Trends Anal Chem 152:116637

Garg A, Vemaraju S (2024) Effect of green logistics management on biodegradability studies of lignin-based materials with sustainable environment. Libr Prog 44(3). https://doi.org/10.48165/bapas.2024.44.2.1
González-González RB et al (2023) Pharmaceutical pollution fingerprinting and waterbodies remediation using waste-derived carbon dots as sustainable advanced nanomaterials. Environ Res 238:117180
Hart JE et al (2018) Exposure to hazardous air pollutants and risk of incident breast cancer in the nurses' health study II. Environ Health 17:1–9
Inshyna NM et al (2020) Biosensors: design, classification and application. IntechOpen
Ispas CR, Crivat G, Andreescu S (2012) Recent developments in enzyme-based biosensors for biomedical analysis. Anal Lett 45(2–3):168–186
Kan J, Pan X, Chen C (2004) Polyaniline–uricase biosensor prepared with template process. Biosens Bioelectron 19(12):1635–1640
Karunakaran R et al (2025) An overview of microfluidic-assisted strategies for synthesis and applications of molecularly imprinted polymers. Anal Sens 5(3):e202400029
Korendovych IV, DeGrado WF (2020) De novo protein design, a retrospective. Q Rev Biophys 53:e3
Kucherenko I et al (2019) Advances in nanomaterial application in enzyme-based electrochemical biosensors: a review. Nanoscale Adv 1(12):4560–4577
Li Y, Pan J, Ma Y (2020) Elucidation of multiple alkane hydroxylase systems in biodegradation of crude oil n-alkane pollution by Pseudomonas aeruginosa DN1. J Appl Microbiol 128(1):151–160
Liang X et al (2021) Immobilized enzymes in inorganic hybrid nanoflowers for biocatalytic and biosensing applications. J Mater Chem B 9(37):7597–7607
Liao Z et al (2018) Recent advances in microfluidic chip integrated electronic biosensors for multiplexed detection. Biosens Bioelectron 121:272–280
Liu TZ et al (1995) Determination of organonitriles using enzyme-based selectivity mechanisms. 2. A nitrilase-modified glassy carbon microelectrode sensor for benzonitrile. Anal Chem 67(10):1679–1683
Lzaod S, Dutta T (2022) Recent advances in the development of oxidoreductase-based biosensors for detection of phenolic antioxidants in food and beverages. ACS omega 7(51):47434–47448
Nader PA, Vives SS, Mottola HA (1990) Studies with a sulfite oxidase-modified carbon paste electrode for detection/determination of sulfite ion and SO2 (g) in continuous-flow systems. J Electroanal Chem Interfacial Electrochem 284(2):323–333
Nelson DL, Lehninger AL, Cox MM (2008) Lehninger principles of biochemistry. Macmillan
Nguyen HH et al (2019) Immobilized enzymes in biosensor applications. Materials 12(1):121
Nigam VK, Shukla P (2015) Enzyme based biosensors for detection of environmental pollutants-a review. J Microbiol Biotechnol 25(11):1773–1781
Pal P et al (2009) The detection of mercury, cadium, and arsenic by the deactivation of urease on rhodinized carbon. Environ Eng Sci 26(1):25–32
Porter JL, Rusli RA, Ollis DL (2016) Directed evolution of enzymes for industrial biocatalysis. Chembiochem 17(3):197–203
Rajangam B, Daniel DK, Krastanov AI (2018) Progress in enzyme inhibition based detection of pesticides. Eng Life Sci 18(1):4–19
Rocchitta G et al (2016) Enzyme biosensors for biomedical applications: strategies for safeguarding analytical performances in biological fluids. Sensors 16(6):780
Sarkar A et al (2019) An overview of enzyme-based biosensors for environmental monitoring. In: Tools, techniques and protocols for monitoring environmental contaminants. Elsevier, Amsterdam, pp 307–329
Sassolas A, Hayat A, Marty J-L (2013) Enzyme immobilization by entrapment within a gel network. In: Immobilization of enzymes and cells, 3rd edn. Humana Press, Springer, New York, pp 229–239
Sheldon RA (2007) Enzyme immobilization: the quest for optimum performance. Adv Synth Catal 349(8–9):1289–1307

Shindhal T et al (2021) A critical review on advances in the practices and perspectives for the treatment of dye industry wastewater. Bioengineered 12(1):70–87

Sohail M, Adeloju SB (2016) Nitrate biosensors and biological methods for nitrate determination. Talanta 153:83–98

Yadav N et al (2021) Detection and remediation of pollutants to maintain ecosustainability employing nanotechnology: a review. Chemosphere 280:130792

Chapter 2
Nano-biomaterials in the Detection of Environmental Pollutants

Parvindar M. Sah, Harshala S. Naik, Aniket Gade, and Rajesh W. Raut

Abstract This book chapter highlights the critical role of nano-biomaterials in detecting environmental pollutants. The chapter covers various types of nano-biomaterials, such as carbon-based, metal-based, and polymeric nano-biomaterials, and the properties that make them suitable for detecting different types of pollutants. Additionally, the chapter describes different detection techniques that employ nano-biomaterials, including colorimetry, fluorescence-based, and electrochemical methods.

Furthermore, the chapter deals with nano-biomaterials' applications in detecting heavy metals, pesticides, organic pollutants, and biological pollutants. In this section, the chapter outlines the advantages of nano-biomaterials over traditional detection methods and discusses how the sensitivity and selectivity of these materials can aid in pollutant detection.

The chapter also examines the challenges and future prospects of using nano-biomaterials in pollutant detection. Some of the challenges discussed include nano-biomaterials' toxicity and the difficulties associated with their large-scale production. However, the chapter highlights nano-biomaterials' potential in advancing pollutant detection technology and their potential for real-world applications.

In conclusion, the chapter emphasizes the importance of nano-biomaterials in detecting environmental pollutants and their potential to revolutionize the field of pollutant detection. Overall, this book chapter serves as an informative resource for researchers and scientists interested in the applications of nano-biomaterials in detecting environmental pollutants.

P. M. Sah · H. S. Naik · R. W. Raut (✉)
The Institute of Science, Constituent and Lead Institute of Dr. Homi Bhabha State University, Mumbai, Maharashtra, India

A. Gade
Department of Biological Sciences and Biotechnology, Institute of Chemical Technology, Mumbai, India

A. Guldhe et al. (eds.), *Biomaterials for Environmental Applications: Detection and Remediation*, Environmental Challenges and Solutions,
https://doi.org/10.1007/978-3-032-12701-3_2

Keywords Nano-biomaterials · Environmental pollutants · Detection techniques · Heavy metals · Pesticides · Prospects

2.1 Introduction

Environmental pollution is a major issue that threatens the health of living organisms and ecosystems. To combat this issue, there is an increasing need for effective techniques to detect pollutants in environmental samples. Nano-biomaterials have recently emerged as a promising tool for detecting environmental pollutants due to their unique physical and chemical properties. With high surface area-to-volume ratio, biocompatibility, and functionality, nano-biomaterials have proven to be useful in various detection methods, such as adsorption, catalysis, and sensing (Ibrahim et al. 2016).

By using nano-biomaterials, it is possible to overcome the limitations of traditional detection methods, such as low sensitivity and specificity. Nano-biomaterials have the ability to detect pollutants at trace levels, with high sensitivity and selectivity, even in complex mixtures. Furthermore, they can be regenerated and reused, making them an economical and sustainable solution for pollutant detection (Lofrano et al. 2017).

This chapter will provide an overview of the latest research on the use of nano-biomaterials in detecting environmental pollutants. It will cover the different types of nano-biomaterials used in pollutant detection, their physical and chemical properties, and their roles in detection methods. The advantages and limitations of using nano-biomaterials for detecting environmental pollutants will also be discussed, along with the potential environmental and health risks associated with their use. This chapter will be useful for researchers, students, and professionals in the field of environmental science and engineering, providing valuable insights into the use of nano-biomaterials in detecting environmental pollutants.

2.1.1 Definition of Nano-biomaterials

Nano-biomaterials are materials that have dimensions at the nanoscale and are used in biological applications. They are a rapidly growing field of research that combines the principles of nanotechnology with biomaterials science. Nano-biomaterials have unique properties that make them highly suitable for various biological and medical applications such as drug delivery, tissue engineering, and biosensors (Yadav et al. 2017a).

The properties of nano-biomaterials depend on their size, shape, surface area, and surface chemistry. At the nanoscale, materials exhibit different physical and chemical properties compared to their bulk counterparts. For example, nano-biomaterials have a large surface area-to-volume ratio, which increases their reactivity and allows for more efficient interactions with biological systems. They also exhibit unique optical, electronic, and magnetic properties that can be exploited for various biomedical applications (Koul and Taak 2018).

Nano-biomaterials can be classified into three main categories: organic, inorganic, and hybrid. Organic nano-biomaterials include polymers, lipids, and proteins. Inorganic nano-biomaterials include metals, metal oxides, and quantum dots. Hybrid nano-biomaterials are a combination of both organic and inorganic materials (Chong et al. 2022).

2.1.2 Importance of Detecting Environmental Pollutants

Ensuring the environmental quality and safeguarding public health heavily rely on the detection of pollutants. Pollution can infiltrate the environment via diverse channels, including industrial operations, agricultural practices, and transportation systems. The repercussions of these pollutants can range from minor discomfort to severe health complications and even fatalities, exerting adverse impacts on both the environment and human well-being. Consequently, it becomes imperative to identify and measure these pollutants accurately in order to evaluate the risks they pose. This information serves as a foundation for formulating suitable strategies to prevent or alleviate their detrimental consequences (Gumpu et al. 2015).

The detection of pollutants is crucial for several reasons. Firstly, it ensures compliance with environmental regulations by monitoring and identifying sources of non-compliance. This helps enforce environmental laws and hold industries and individuals accountable. Secondly, pollutant detection helps assess the impact of pollutants on the environment and human health, such as ozone depletion and eutrophication. It enables the identification of sources and the development of measures to prevent or mitigate harmful effects. Lastly, pollutant detection is important for monitoring emerging pollutants, which are substances not yet regulated but recognized as potential hazards. Detecting emerging pollutants helps identify risks and develop appropriate prevention or mitigation strategies (Yan et al. 2018; Liu et al. 2020a; Liu and Lu 2003).

2.1.3 Role of Nano-biomaterials in Detection of Environmental Pollutants

Nano-biomaterials are highly promising for detecting environmental pollutants due to their unique physicochemical properties. A nanomaterial-based sensors offer significant advantages in terms of sensitivity, selectivity, rapid response, miniaturization, multi-parameter detection, cost-effectiveness, and reusability. These features make them indispensable tools for detecting and monitoring environmental pollutants, aiding in pollution control, risk assessment, and the development of effective mitigation strategies (Lofrano et al. 2017).

In addition, these versatile materials offer efficient solutions for pollutant detection in air, water, and soil samples, enabling enhanced sensitivity and reliable identification of pollutants. Nano-biomaterials play a crucial role in addressing environmental pollution by providing effective tools for monitoring and mitigating the impact of pollutants on ecosystems and human health (Yadav et al. 2017b).

2.2 Types of Nano-biomaterials Used in Detecting Environmental Pollutants

Nano-biomaterials are increasingly being used for detecting environmental pollutants because of their unique properties, including high surface area, selective binding, and strong adsorption. Carbon nanotubes (Lam et al. 2014), quantum dots (Yu et al. 2015), metal oxide nanoparticles (Gaviria-Arroyave et al. 2020), magnetic nanoparticles (Parida et al. 2021), and nanoporous materials (Lee et al. 2019) are some of the most used types of nano-biomaterials for this purpose. By modifying and functionalizing these materials, it is possible to enhance their sensitivity and selectivity for detecting pollutants, making them a promising approach for environmental monitoring (Jain et al. 2021).

2.2.1 Carbon-Based Nano-biomaterials

Carbon-based nano-biomaterials such as carbon nanotubes and graphene are used for detecting environmental pollutants due to their unique properties including high surface area, electrical conductivity, and biocompatibility. These materials can be functionalized with specific biomolecules to selectively bind pollutants and detect them with high sensitivity. They have been used for detecting heavy metals, pesticides, PAHs, nitroaromatics, and other pollutants in water and soil samples. Carbon-based nano-biomaterials offer a promising approach for environmental pollution detection due to their unique properties that can be tailored for specific pollutant detection (Abdelbasir and Shalan 2019).

2.2.1.1 Carbon Nanotubes

In the field of environmental pollution detection, CNTs have been extensively used as sensing platforms due to their high sensitivity and selectivity toward specific pollutants. A wide array of detection techniques has been utilized for the identification of heavy metal ions following their sorption and preconcentration using nanoscale carbon-based materials. Notably, electrochemical methods stand out due to their ability to employ carbon-based electrode materials as both sorbents and detectors for multiple heavy metal ions concurrently. Prominent examples of these electrochemical techniques include voltametric stripping analysis, potentiometric stripping analysis, field effect transistor configuration, and electrical impedance. CNTs can be functionalized with specific biomolecules, such as antibodies or aptamers, to selectively bind to target pollutants and detect them with high sensitivity (Wanekaya 2011). Nanoscale carbon-based materials can be modified onto electrode surfaces either by casting colloidal dispersions or by direct growth using methods such as chemical vapor deposition (CVD), enabling enhanced properties and functionalities for various applications. For example, functionalized CNTs have been used for the detection of heavy metals, pesticides, and organic compounds in water and soil samples (Ibrahim et al. 2016).

Here are a few more examples of how carbon nanotubes (CNTs) have been used for detecting environmental pollutants: Polyaromatic hydrocarbons (PAHs) are organic compounds composed of fused aromatic rings. They are formed during incomplete combustion processes and are commonly found in various environmental sources such as industrial emissions, vehicle exhaust, and natural sources like forest fires. PAHs are of significant interest due to their potential harmful effects on human health and the environment. Polyaromatic hydrocarbons (PAHs) encompass a range of compounds with fused aromatic rings, including naphthalene, anthracene, phenanthrene, and pyrene. Carbon nanotubes (CNTs) offer promise in PAH detection due to their high surface area, electrical conductivity, and sensitivity to analytes. Detection methods involving CNTs include functionalizing CNT-based field-effect transistors (FETs) with selective receptors for PAH binding, utilizing CNT-coated quartz crystal microbalances (QCMs) to measure changes in resonant frequency upon PAH adsorption, and developing optical sensors using functionalized CNTs to induce changes in optical properties for PAH detection. Ongoing research in this field aims to refine and expand the capabilities of CNT-based sensors for PAH detection (Akinpelu et al. 2019).

A composite material consisting of multi-walled carbon nanotubes incorporated into a porous monolith was used for the extraction and detection of polycyclic aromatic hydrocarbons (PAHs). The method achieved efficient extraction of PAHs with good recovery values and low detection limits (ranging from 0.02 to 0.22μg/L). The presence of carbon nanotubes enhanced the extraction performance of the solid phase adsorbent. This composite material demonstrated advantages such as precision, low detection limits, reproducibility, and good recovery values for the detection of PAHs in environmental samples (Al-Rifai et al. 2018).

A Pd-CuO/MWCNTs modified glassy carbon electrode served as an electrochemical nano-sensor for the detection of ametridione pesticide in water samples. The method achieved a limit of detection of 0.0796μg/mL^{-1}, a limit of quantification of 0.5560μg/mL^{-1}, and a concentration range of 0.1–10.0μg/mL^{-1}. The sensor demonstrated successful detection of ametridione in tap water, agricultural run-off, and river water samples, with mean recoveries exceeding 98% (Bebi et al. 2021).

A DNA-based electrochemical method has been developed to detect Cd (II) ions using ethyl green and multi-walled carbon nanotubes. By immobilizing double-stranded DNA on a glassy carbon electrode with the help of MWCNT, researchers employed differential pulse voltammetry (DPV) analysis. Cd (II) ions cause the unwinding of dsDNA, allowing ethyl green to bind to the resulting single-stranded DNA and produce a higher reduction peak current. The method achieves a linear detection range of 2 nM to 10.0 nM, with a sensitivity of approximately 5 nM and a limit of detection of 2 nM. This study expands the application of MWCNT-modified electrodes as biosensors and presents potential for the detection of other heavy metals in environmental samples (Sreekanth et al. 2021).

2.2.1.2 Graphene

The use of graphene-based nano-biomaterials for the detection of environmental pollutants is a promising area of research. Graphene's high surface area, electrical conductivity, and mechanical properties make it an ideal material for biosensors. Some examples of graphene-based biosensors include the detection of heavy metals, polycyclic aromatic hydrocarbons, and pesticides. In each case, the graphene-based sensor exhibited a high sensitivity and low detection limit (Anku et al. 2018). Figure 2.1 shows the types of graphene-based nanomaterial.

Here are some detailed examples of how graphene-based nano-biomaterials are being used for the detection of environmental pollutants: A cysteine-functionalized graphene oxide (cGO) was synthesized, and a nanocomposite film of cGO/polypyrrole (PPy) was grown on the working electrode surface of a screen-printed electrode (SPE) via controlled one-step electrochemical deposition. The cGO/PPy-SPE was used to detect lead ions (Pb^{2+}) in water by depositing Pb^{2+} on the working electrode surface and anodic stripping by differential pulse voltammetry (DPV). The sensor has a measurable detection limit of 0.07 ppb and can effectively adsorb and remove Pb^{2+} in water samples (Seenivasan et al. 2015).

The study found that DNA sequences containing polycytosine can be strongly adsorbed onto the surface of graphene oxide to enhance the physical adsorption between graphene oxide and DNA. The researchers developed two biosensors for detecting Hg^{2+} ions using fluorescence resonance energy transfer. The first biosensor had a limit of detection of 1.11 nM, while the second had a limit of detection of 0.84 nM. To improve the detection signal, the researchers covered an ultra-thin silica shell with metal nanomaterial Ag. This improved the stability and sensitivity of the biosensors, resulting in lower limits of detection of 0.36 nM and 0.15 nM for the two sensors, respectively (Lv et al. 2022).

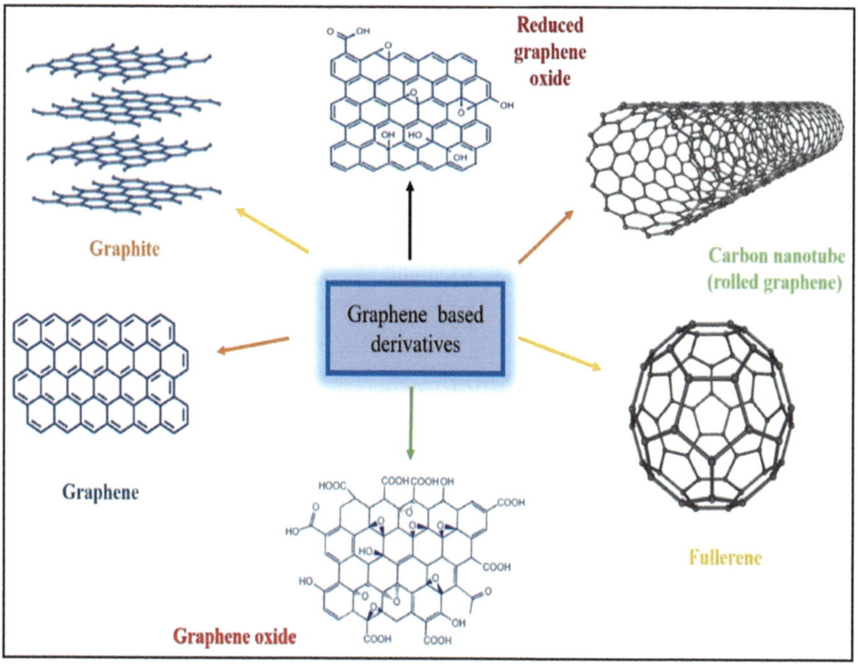

Fig. 2.1 Graphene-based nanomaterials for the detection of environmental pollutant. (Image adapted from (Kumar et al. 2023) with permission)

In one study, a cost-effective approach for the detection of polycyclic aromatic hydrocarbons (PAHs) using graphene quantum dots (GQDs) as fluorescence sensors was developed. GQDs were synthesized from graphene oxide (GO) and citric acid (CA), and their properties were analyzed. The GQDs were then applied to detect pyrene in water samples using a "turn-off-on" mechanism. Both GO-GQDs and CA-GQDs showed a linear response within a concentration range of 2×10^{-6} mol L^{-1}, with detection limits of $0.325 \times 10–6$ mol L^{-1} and 0.242×10^{-6} mol L^{-1}, respectively. The sensor was also tested with lake water, and satisfactory recoveries were achieved. These findings suggest the potential of GQDs as an affordable and eco-friendly approach for environmental monitoring (Nsibande and Forbes 2020).

The study proposes a non-enzymatic electrochemical-sensing electrode modified with electrochemically reduced graphene oxide (ERGO) for detecting the presence of organophosphate-based pesticides such as parathion (PT) in environmental samples including soil, water, vegetables, and cereals. The ERGO sensor showed high sensitivity, selectivity, and stability for interference-free detection of PT residues in actual samples. The electrochemical nano-sensor has a wide dynamic range and a lower detection limit, making it suitable for point-of-care detection of PT. The nanosensor's performance was validated through absorption spectroscopy after adding PT to natural samples. The results suggest the design of easy-to-use nano-sensor-based analytical tools for rapidly monitoring other environmental samples (Sen et al. 2022).

The study synthesized a nanocomposite of reduced graphene oxide (rGO) and gold nanoparticles (AuNPs) using a microwave-assisted hydrothermal method for detecting sulfamethazine (SMZ) in swine effluent. The rGO-AuNPs composite was characterized and modified with the AuNPs. A glassy carbon (GC) electrode was used with the composite to detect SMZ, and the GC/r GO-AuNPs electrode achieved detection limits of 0.1μ mol L^{-1}. The sensor was successfully applied to detect SMZ in synthetic swine effluent samples (Silva and Cesarino 2019).

2.2.2 *Metal-Based Nano-biomaterials*

Nano-biomaterials based on gold, silver, and iron oxide have shown promise in the detection of environmental pollutants. These nanomaterials can be functionalized with specific molecules that bind to the target pollutants, allowing for sensitive and selective detection. Gold nanoparticles, for example, have been used to detect heavy metals such as lead and mercury, while silver nanoparticles have been used to detect pesticides and herbicides. Iron oxide nanoparticles have also been used for the detection of heavy metals and organic pollutants. These metal-based nano-biomaterials have the potential to provide rapid and cost-effective methods for environmental monitoring and can contribute to the development of sustainable and greener environmental practices.

2.2.2.1 Gold Nanoparticles

Gold nanoparticles (AuNPs) have unique physicochemical properties, such as a high surface area to volume ratio, which allows for efficient interaction with target analytes. They also exhibit unique optical properties, including surface plasmon resonance, which leads to intense and tunable absorbance and scattering of light. These properties make AuNPs attractive for use in various sensing applications, including the detection of environmental pollutants (Wang and Yu 2013; Gutierrez et al. 2017; Sung and Wu 2014; Valera et al. 2008; Priyadarshini and Pradhan 2017).

Here are some examples of gold nano-biomaterials for environmental pollutant detection taken from a review article (Wang and Yu 2013):

According to Liu and Lu (2007a, b), a colorimetric sensor was developed for the detection of Pb^{2+} with a detection limit of 3 nm. Lee et al. (2009) introduced a cation-specific functionalized sensor capable of detecting Hg^{2+} with a detection limit of 0.2 ppb. Kim et al. (2001) proposed a direct electrostatic aggregation sensor for divalent heavy metal ions, but the specific detection limit was not mentioned. Gooding et al. presented an electrochemical sensor for Cu^{2+} with a detection limit of less than 1 pm. In terms of aromatic compounds, Dasary et al. (2009) reported a cysteine-modified AuNP-based label-free sensor that could detect

2,4,6-trinitrotoluene (TNT) with a detection limit of 2 ppm. Liu et al. (2007a, b) developed an electrochemical sensor for nitrobenzene with a detection limit of 0.3 mg/l (Wang and Yu 2013).

For the detection of inorganic pollutants, Daniel and Astruc (2009) described a crosslinking colorimetric sensor for nitrite ions with a sub-ppm detection limit. Xiao and Yu (2010) introduced a non-crosslinking colorimetric sensor for nitrite ions with a detection limit below 1 ppm. Zhang and Yi (2009) presented an electrochemical sensor for nitrite ions with a detection limit of 0.1μM. Tsai et al. (2011) developed an electrochemical sensor for hypochlorite with a detection limit of 1μM. Rassaei et al. (2008) reported an electrochemical sensor for arsenic with a detection limit of 30 ppb (Wang and Yu 2013).

Newman et al. (2007) presented an optical sensor for organophosphates and phosphonates (OPPs) with a detection limit of 0.5μM. Chauhan et al. (2011) developed an electrochemical sensor for OPP compounds/pesticides with a detection limit of 0.1 nM. Regarding toxin detection, Uzawa et al. (2008) used gold nanoparticles (AuNPs) conjugated with a lateral flow strip assay to detect ricin with a detection limit of 10 pg/ml. Bo et al. (2007), Liu et al. (2007a, b), and Shim et al. (2009) developed immunochromatographic strip assays for the detection of Ochratoxin A (OTA), zearalenone (ZEA), and aflatoxin B1 with detection limits of 0.32 ng/ml and 2.5 ng/ml (Wang and Yu 2013).

Gold nanoparticles (GNPs) can be used as colorimetric probes by modifying their surface through a process called functionalization. This involves conjugating GNPs with ligand molecules that have a specific affinity for the target analyte or molecule of interest. For example, maleic acid-functionalized GNPs are employed for the detection of Pb^{2+} ions (Wei et al. 2008). Various methods for the colorimetric detection of As^{3+} ions have been demonstrated. Wu et al. used a specific aptamer and a cationic polymer (PDDA) to induce aggregation of gold nanoparticles (GNPs) in the presence of As^{3+} ions, resulting in a color change (Le Thao Nguyen et al. 2018). Zhan et al. achieved stabilization of GNPs using an aptamer alone, and the presence of As^{3+} ions led to GNPs destabilization and aggregation. Another approach involved the use of poly (allylamine hydrochloride) (PAH)-conjugated GNPs, where the binding of As^{3+} ions to amine groups caused a measurable shift in the surface plasmon resonance (SPR) peak. Additionally, laser-induced desorption/ionization mass spectrometry (LDI-MS) coupled with GNPs-modified membranes enabled the detection of As^{3+} ions through aggregation-induced color change. These methods provide diverse strategies for sensitive and specific colorimetric detection of As^{3+} ions (Wu et al. 2012).

Colorimetric sensors for the detection of As^{3+} ions are predominantly based on the strong binding affinity of As^{3+} to sulfur-containing compounds. Kalluri et al. demonstrated the use of glutathione (GSH), dithiothreitol (DTT), and cysteine (Cys) functionalized gold nanoparticles (GNPs) for the colorimetric detection of As^{3+} ions. The minimum detection limits (MDL) for the GNP probes were reported as 5 ppb, 20 ppb, and 25 ppb for DTT, GSH, and Cys-conjugated GNPs, respectively. To enhance the detection limit, dynamic light scattering (DLS) analysis was employed, achieving a sensitivity of 10 ppt, indicating DLS as a more sensitive detection method (Kalluri et al. 2009).

DTT facilitated the conversion of As^{5+} to As^{3+}, which then bound to the GNPs, resulting in aggregation, color change, and a shift in the surface plasmon resonance (SPR) peak. GSH-DTT-Cys-PDCA functionalized GNPs exhibited better performance in detecting As^{3+} compared to GSH-DTT and Cys functionalized GNPs. The MDL of GSH-DTT-Cys-PDCA functionalized GNPs was 2.5 g/L, and the minimum quantification limit was 8.4 g/L (Domínguez-González et al. 2014).

Another metal like mercury (Hg) is a highly toxic heavy metal ion that contaminates water bodies. One effective method for detecting Hg^{2+} ions is using gold nanoparticles (GNPs) functionalized with various molecules. For example, GNPs functionalized with amino acids, peptide moieties, and quaternary ammonium groups can efficiently detect Hg^{2+} ions by causing the aggregation of the functionalized GNPs. By adding Hg^{2+} ions, the charge on the GNP surface is destabilized, leading to aggregation and a shift in the surface plasmon resonance (SPR) peak (Wang and Yu 2013). Other functional molecules, such as 3-mercaptopropionate acid, adenosine monophosphate, mercaptopropionic acid (MPA), homocysteine, l-cysteine, and papain (a cysteine protein), have also been used to functionalize GNPs for Hg^{2+} detection. These molecules enable the aggregation of GNPs in the presence of Hg^{2+} ions, resulting in a change in color or plasmon shift (You et al. 2013). Oligonucleotide-functionalized GNPs and DNA-functionalized GNPs have shown high sensitivity and selectivity for detecting Hg^{2+} ions. By conjugating GNPs with specific DNA sequences, the presence of Hg^{2+} ions can induce conformational changes in the DNA, leading to differential GNP aggregation and changes in optical properties (Tiwari et al. 2011).

As like lead and chromium, various colorimetric methods have been employed for the detection of copper ions using gold nanoparticles (GNPs) with different surface functionalizations. One approach involved the use of citrate-capped GNPs, where the negatively charged citrate groups on the GNP surface prevented aggregation by mutual electrostatic repulsion. However, interaction with Cu^{2+} ions destabilized the surface charge, leading to aggregation and a red shift in the surface plasmon resonance (SPR) peak (Salcedo and Sevilla 2013). Similarly, cysteamine-functionalized GNPs underwent aggregation in the presence of Cu^{2+} ions by reacting with a Schiff base formed on the GNP surface, resulting in a color change from red to blue (Li et al. 2014).

Another method, where calixarene-functionalized GNPs formed an alloy with Cu^{2+} ions, caused a change in color from dark brownish to green and a red shift in the SPR peak. Core-satellite GNPs, functionalized with l-Cysteine, showed cross-linking and the formation of core-satellite structures upon addition of Cu^{2+} ions, resulting in a color change from red to blue. Catalytic etching of GNPs using Cu as a catalyst was another detection method, where the presence of Cu^{2+} ions enhanced the etching process, leading to a decrease in the intensity of the SPR peak and a slight red shift. Catalytic leaching of gold nanorods by thiosulfate or H_2O_2 in the presence of Cu^{2+} ions caused changes in color and SPR peak intensity (Gunupuru et al. 2014).

A gold nanoparticles (AuNPs) probe was developed for detecting Copper (II) ions (Cu2+) without the need for labelling. The probe was fabricated using plant-derived, non-toxic polysaccharide carboxymethyl gum karaya (CMGK), which acted as a reducing, stabilizing, and functionalizing agent. The system showed a linear correlation in the range of 10–1000 nM of copper, with a limit of detection of 10 nM. The method was successfully applied to detect copper in real water samples, as well as biological samples like human plasma and urine (Gangapuram et al. 2017).

The detection of lead (Pb^{2+}) ions using gold nanoparticles (GNPs) with different surface functionalizations. One approach involves the use of glutathione-conjugated GNPs, maleic acid-functionalized GNPs, or calixarene-functionalized (CALNN) GNPs, which undergo aggregation in the presence of Pb^{2+} ions, leading to a color change and a red shift in the surface plasmon resonance (SPR) peak. Ion-dependent chelation-based GNP aggregation using a pentapeptide (CALNN) has also been utilized for Pb^{2+} detection, resulting in visible color changes and SPR peak shifts (Chai et al. 2010).

Additionally, DNAzyme-based methods have been explored, where DNAzyme-modified GNPs are used for visual sensing of Pb^{2+} ions. Cleavage of the DNAzyme-substrate complex by Pb^{2+} ions prevents GNP aggregation, resulting in a red color, while the absence of Pb^{2+} ions leads to aggregation and a color change. Different variations of DNAzyme structures and GNP sizes have been studied to enhance the sensitivity and selectivity of the colorimetric sensor. Moreover, the combination of DNAzyme-modified GNPs with graphene oxide has been demonstrated to achieve highly sensitive detection of Pb^{2+} ions with excellent selectivity (Wei et al. 2008).

Gold nanoparticles (GNPs) have been utilized for the efficient detection and monitoring of cadmium (Cd^{2+}) ions in water samples. One approach involves the conjugation of citrate-synthesized GNPs with di-(1H-pyrrol-2-yl) methanation, which exhibits a high affinity for Cd^{2+} ions. The chelation of Cd^{2+} ions by the functionalized GNPs results in a plasmonic shift to longer wavelengths, accompanied by a change in color from red to blue. Another method utilizes GNPs modified with 4-amino-3-hydrazino-5-mercapto-1,2,4-triazole (AHMT), which serves as an excellent chelator for Cd^{2+} ions. Chelation leads to inter-plasmon coupling and a distinct color change from red to blue (Sung and Wu 2014).

A label-free colorimetric detection method for Cd^{2+} ions involves unmodified GNPs, which tend to aggregate at high ionic strength concentrations. However, the presence of glutathione (GSH) prevents aggregation. When Cd^{2+} ions are added to the solution, they coordinate with GSH molecules, leading to a decrease in the amount of free GSH and resulting in diminished stability and aggregation in high salt concentrations. This method is highly specific for Cd^{2+} ions and has been successfully applied to detect Cd^{2+} ions in rice sample (Guo et al. 2014).

2.2.2.2 Silver Nanoparticles

Silver nanoparticles have also been used for the detection of environmental pollutants due to their unique optical and electronic properties. Silver biosensors were prepared using different methods, such as drop casting or sticking, and utilized for the detection of various pollutants. The pollutants detected include acetamiprid, acifluorfen, amikacin sulfate, atrazine, carbendazim, catechol, chloramphenicol, ciprofloxacin, clenbuterol, diene sterol, doxorubicin, diethylstilbesterol (DES), eriochrome black T, 5-fluorouracil, hydroquinone, lindane, malathion, methyl parathion, methylene blue, monocrotophos, 4-nitrophenol, 4-nitrotoluene, nitrofurazone, ochratoxin A, orange II, oxazepam, paraoxon, pentachlorophenol, phoxim, rhodamine B, rilpivirine, thiodicarb, thiourea, trichloroacetic acid, and chlorpyrifos. The biosensors demonstrated different limits of detection for each pollutant, indicating their sensitivity to the target analytes (Wang and Yu 2013).

Silver nanoparticles (AgNPs) and their functionalization with mercaptoundecanoic acid (11-MUA) (Fig. 2.2). The AgNPs showed a surface plasmon absorption band (SPAB) at 393 nm, which shifted to 417 nm after coating with 11-MUA. The nanoparticles selectively detected micromolar levels of Ni^{2+} ions, even in the presence of other cations such as manganese (Mn^{2+}), cobalt (Co^{2+}), cadmium (Cd^{2+}), copper (Cu^{2+}), zinc (Zn^{2+}), iron (Fe^{2+}), mercury (Hg^{2+}), lead (Pb^{2+}), and chromium (Cr^{3+}) (Rossi et al. 2021).

Figure 2.2 shows that a colorimetric sensor using green-synthesized silver nanoparticles (AgNPs) derived from the roots extract of *Bistorta amplexicaulis* is reported for selective and cost-effective detection of Hg^{2+} and Pb^{2+} ions. The AgNPs were characterized using various spectroscopic techniques, and their photophysical

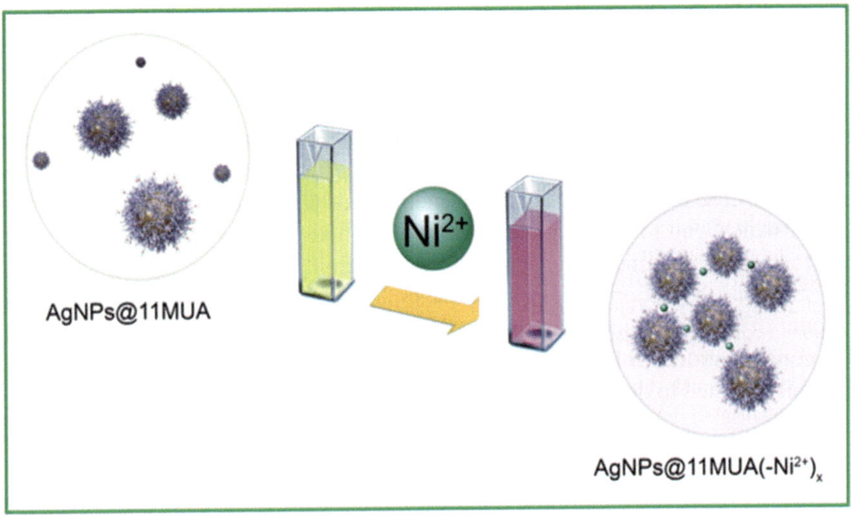

Fig. 2.2 Detection of Ni metal in the presence of silver nanoparticle using colorimetric sensor. (Image adapted from (Rossi et al. 2021) which comes under creative common attribution CC BY 4)

potential for metal cation recognition was explored through absorption spectroscopy and colorimetric assay. The AgNPs exhibited a hypochromic shift in the surface plasmon resonance (SPR) band, allowing visual detection of Hg^{2+} ions through a color change from dark brown to light yellow. Mixing AgNPs with Pb^{2+} ions led to a reduction in absorbance. The colorimetric sensor showed high sensitivity, with limits of detection (LOD) of 8.0×10^{-7} M and 2.0×10^{-7} M for Hg^{2+} and Pb^{2+} respectively (Ahmed et al. 2020).

Cysteine-coated silver nanoparticles (Cys-AgNPs) were investigated for their potential as label-free sensors for detecting heavy metals, specifically lead (Pb^{2+}) and mercury (Hg^{2+}). The Cys-AgNPs were characterized using various techniques, and their photophysical properties toward Hg^{2+} and Pb^{2+} ions were studied through absorption spectroscopy and colorimetric tests. The limit of detection (LOD) was determined to be 45.39×10^{-9} M for Hg^{2+} ions and 49.39×10^{-9} M for Pb^{2+} ions. Furthermore, the Cys-AgNPs demonstrated the ability to simultaneously detect Hg^{2+} and Pb^{2+} ions in water samples, with LODs of 43×10^{-9} M for constant Hg^{2+} and varying Pb^{2+} ions, and 37.95×10^{-9} M for constant Pb^{2+} and varying Hg^{2+} ions. These findings highlight the potential application of Cys-AgNPs as a quick and effective colorimetric sensor for the detection of individual or mixed Hg^{2+} and Pb^{2+} ions in water samples (Samuel and Rao 2023).

2.2.2.3 Iron Oxide Nanoparticles

Iron oxide nanoparticles (IONPs) have gained significant attention in recent years due to their unique magnetic and optical properties, making them useful in various applications, including environmental sensing and detection of pollutants.

Using Fe_3O_4 MNPs offers significant advantages, including their low cost, high efficiency, recyclability, regenerability, and magnetic controllability. These features make them an attractive option for developing a biosensor system to detect H_2O_2 in rainwater, which is important for monitoring the quality of water resources. Therefore, Fe_3O_4 MNPs have the potential to be a promising alternative to HRP and other existing biosensors for detecting H_2O_2 in rainwater (Zhuang et al. 2008).

2.2.3 Polymeric Nano-biomaterials

Dendrimers and nanogels are two types of polymer nanomaterials that have shown great promise for the detection of environmental pollutants. Both materials have unique properties that make them well-suited for sensing applications.

2.2.3.1 Dendrimers

Dendrimers are highly branched, three-dimensional polymers that can be synthesized to have a precisely controlled size, shape, and functionality. This makes them ideal for use as sensors, as they can be designed to selectively bind to specific pollutants. For example, dendrimers can be functionalized with ligands or receptors that bind to heavy metals or organic compounds, and can be used as fluorescent or colorimetric sensors for the detection of these pollutants in water or soil samples (Han et al. 2022).

2.2.3.2 Nanogels

Nanogels are a type of hydrogel nanoparticle that can swell or shrink in response to changes in their environment, such as changes in pH, temperature, or ionic strength. This property makes them useful for sensing applications, as they can be designed to undergo a change in size or shape in response to the presence of a pollutant. For example, nanogels can be functionalized with specific molecules that bind to a target pollutant, causing the nanogel to swell or collapse, and thereby indicating the presence of the pollutant (Han et al. 2022).

2.3 Detection Techniques Using Nano-biomaterials

Colorimetry, fluorescence, and electrochemical methods are used for detecting environmental pollutants. Colorimetry involves measuring the absorbance or transmission of light by a sample (Liu et al. 2020a), fluorescence involves measuring the emission of light by a sample after it has been excited, and electrochemical methods involve measuring the electrical properties of a sample. The choice of method will depend on the specific pollutant being detected and the required sensitivity.

2.3.1 Colorimetric Detection

2.3.1.1 Principle of Colorimetric Detection

The principle of colorimetry detection used in environmental pollutants is based on the measurement of the absorbance or transmission of light by a sample at a specific wavelength. In colorimetry, a reagent is added to the sample that reacts with the pollutant to produce a colored product. The intensity of the color is directly proportional to the concentration of the pollutant in the sample (Mortimer 2016; Gilchrist and Nobbs 2016; Alberti et al. 2020).

For example, a common colorimetric method for detecting ammonia in water involves adding a reagent called Nessler's reagent, which reacts with ammonia to produce a yellow-brown color. The intensity of the color is then measured using a spectrophotometer at a specific wavelength (Duarte et al. 2015).

The principle of colorimetry is based on the Beer-Lambert law, which states that the absorbance of light by a sample is directly proportional to the concentration of the absorbing species in the sample. The absorbance is defined as the logarithm of the ratio of the intensity of the incident light to the intensity of the transmitted light (Mortimer 2016).

2.3.1.2 Application of Colorimetric Detection Using Nano-biomaterials

Kong et al. developed a microfluidic paper-based colorimetric sensor for detecting bisphenol A (BPA), an environmental endocrine disruptor found in plastic products. The sensor combined molecular imprint polymer (MIP) membranes and $ZnFe_2O_4$ to achieve high selectivity. It exhibited a detection limit of 6.18 nM and a linear range of 10–1000 nM for BPA sensing, making it suitable for detecting BPA in food and environmental samples (Liu et al. 2020b).

In another study, Zhang et al. prepared BN nanosheet/CuS nanocomposites, which acted as mimetic peroxidase for the highly sensitive colorimetric determination of cholesterol. This sensor demonstrated a detection limit of 2.9μM and a linear range of 10–100μM for cholesterol detection (Liu et al. 2020b).

Ye et al. developed an angle-independent colorimetric sensor based on photonic crystals. The sensor used a photonic film with an inverse opal structure and a target-responsive aptamer hydrogel scaffold. It exhibited a detection limit of 1 nM and a wide detection range of 1 nM to 10μM for Hg^{2+}(Liu et al. 2020b).

2.3.2 Fluorescence-Based Detection

The increasing interest in nanomaterials that possess advanced optoelectronic properties has driven the progress of biosensors with diverse applications. In the realm of environmental studies, there is a pressing need for swift identification of pollutants in water sources, which has led to the emergence of fluorescent biosensors. These biosensors are characterized by high sensitivity and ease of use. In order to develop effective fluorescent detection systems, various nanomaterials such as carbon dots (CD), quantum dots (QD) (Pooja et al. 2019), gold nanoparticles, nanoclusters, graphene (G), graphene oxide (GO), transition metal dialcogens (TMDC), and organometallic frameworks (MOF) have been employed. These materials typically operate on the quenching principle of Förster resonance energy transfer (FRET). To achieve specific detection of the target pollutant, nanomaterials can be bioconjugated with molecules such as antibodies, enzymes, or aptamers. This allows for specific detection of the target pollutant, even in complex real-life samples (Gong et al. 2016).

Fluorescent biosensors use a fluorescent molecule as the detection element, which emits light when excited by a specific wavelength of light. This allows for very sensitive and specific detection of environmental pollutants. However, fluorescent biosensors require specialized equipment for measurement and can be expensive to produce (Bose et al. 2022).

2.3.2.1 Principle of Fluorescence-Based Detection

The principle of fluorescent-based detection techniques used in detecting environmental pollutants involves the use of a fluorescent molecule as a detection element that emits light when excited by a specific wavelength of light. The fluorescent molecule is coupled to a recognition element such as an enzyme or antibody that can specifically bind to the target pollutant (Cheng et al. 2018).

When the recognition element binds to the target pollutant, it causes a change in the fluorescence of the biosensor, which can be measured using specialized equipment. The change in fluorescence can be due to a variety of factors, including changes in the physical or chemical properties of the fluorescent molecule, changes in the distance between the fluorescent molecule and a quencher molecule, or changes in the energy transfer efficiency between the fluorescent molecule and a nearby chromophore (Gaviria-Arroyave et al. 2020).

The sensitivity and specificity of fluorescent-based detection techniques depend on several factors, including the choice of fluorescent molecule, the design of the biosensor, and the optimization of the experimental conditions. By carefully selecting the fluorescent molecule and the recognition element, it is possible to develop fluorescent biosensors that are highly sensitive and specific for detecting a wide range of environmental pollutants (Gaviria-Arroyave et al. 2020).

2.3.2.2 Application of Fluorescence-Based Detection Using Nano-biomaterials

A simple and sensitive detection system was developed for detecting chromium using carbon dots (CDs) based on fluorescence inhibition. The CDs were synthesized from papaya pulp residues through pyrolysis and demonstrated a strong fluorescent response in the blue spectrum. To achieve selective detection of chromium, the CDs were functionalized using ethylenediaminetetraacetic acid (EDTA) via carbodiimide chemistry. The method showed high sensitivity with a detection limit of 0.708 ppb (Pooja et al. 2019).

A different study also aimed to detect Cr (VI), using cellulose-synthesized CDs combined with isophorone dicyanate (IPDI) and β-cyclodextrin (β-CD) as stabilizing agents. The system exhibited decreased fluorescence upon excitation at 370 nm in the presence of the analyte, attributed to the chelation of metal ions by carbamate groups and hydrogen bonds. However, this sensor was designed for ion detection in digested soil, which requires extensive sample processing (Wang et al. 2017).

Acetylcholinesterase (AChE) has found extensive use in detecting pesticides enzymatically. AChE-dependent methods rely on the hydrolysis of acetylthiocholine (ATCh) to produce thiocholine, which contains a thiol group that can react chemically with metal cations, nanomaterials, and fluorophores. As a result, a measurable fluorescent or colorimetric signal can be generated (Yan et al. 2018).

Paraoxon is a highly toxic organophosphate compound that is often used as a pesticide. It is an acetylcholinesterase inhibitor, which means that it disrupts the normal functioning of the nervous system by preventing the breakdown of acetylcholine. The detection of paraoxon using acetylcholinesterase (AChE) can be achieved through a specific system that utilizes carbon dots (CDs) doped with nitrogen. CDs are a type of nanomaterial with unique optical properties that have been studied for various applications, including biosensing. Nitrogen doping is a common technique employed to enhance the quantum performance of CDs, which results in a higher fluorescence quantum yield. Moreover, nitrogen doping can generate a greater diversity of surface chemical groups, which makes it easier to functionalize CDs and tailor them for specific applications (Gong et al. 2016).

A biosensor has been developed that uses a nano beacon, which consists of gold nanoparticles functionalized with DNA aptamers and FAM as a fluorophore. This biosensor has achieved isocarbophos detection limits of 10μg/L. Initially, the fluorescence of FAM is inhibited due to its proximity to the nanoparticle, but it recovers upon binding of the aptamer and isocarbophos. When the pesticide is detected, a portion of the aptamer binds to it while another part binds to the nanoparticle, leading to a decrease in the initial fluorescence. This detection method corresponds to "turn off", where the nanoparticles act as an anchor for the beacon system as well as a quenching agent (Dou et al. 2015).

A new fluorescent biosensor has been developed by researchers, which uses aptamers and combines Quantum Dots and gold nanoparticles. The biosensor mentioned above has been tested with chlorpyrifos, diazinon, and malathion as model molecules. This biosensor has a detection limit of 0.73μg/L, with a "turn on" strategy used for smaller molecules. Negative samples are mixed with the AuNPs-Aptamer conjugates and rise toward the test zone by capillarity. If hybridization occurs, the fluorescence of the QDs is inhibited by the FRET phenomenon with the AgNPs, resulting in a dark line. Positive samples will not inhibit hybridization, allowing a positive fluorescence signal to be obtained. This biosensor has better detection limits and works on immobilized lateral flow type systems, making it an improvement over other aptamer-based system (Cheng et al. 2018).

2.3.3 *Electrochemical Detection*

Electrochemical biosensors are a popular method for detecting pollutants due to their high sensitivity, selectivity, and fast response times. These biosensors work by detecting changes in electrical properties caused by the interaction between the pollutant and a specific biological recognition element, such as an enzyme or an aptamer.

2.3.3.1 Principle of Electrochemical Detection

An electrochemical biosensor operates on the principle of utilizing a biological recognition element, such as an enzyme or an aptamer, which specifically targets the analyte or pollutant of interest. The biological recognition element is immobilized on an electrode surface, either directly or through a transducer layer, and acts as a sensing element.

When the sample is introduced to the biosensor, the target analyte or pollutant in the sample binds to the sensing element. This binding event causes a change in the electrochemical properties of the electrode, which is then measured by a signal transduction system. The signal transduction system typically involves an electrochemical reaction, such as the oxidation or reduction of the analyte, which generates a measurable signal, such as a change in current or voltage.

The change in the electrochemical properties of the electrode is directly proportional to the concentration of the analyte or pollutant in the sample. By measuring this change, the electrochemical biosensor can determine the concentration of the target analyte or pollutant in the sample. Thus, electrochemical biosensors offer high accuracy and sensitivity for the detection of target analytes or pollutants.

2.3.3.2 Application of Electrochemical Detection Using Nano-biomaterials

A novel detection approach has been introduced that utilizes a three-dimensional graphene framework combined with bismuth nanoparticles for simultaneous monitoring of multiple heavy metals. The sensor was able to detect cadmium and lead in a wide range of concentrations, with low detection limits. It also detected zinc, but with a smaller range of concentrations. The improved performance of the sensor was attributed to the increased binding sites for metal cations, better charge mobility, and improved conductivity due to the combination of bismuth and graphene (Shi et al. 2017).

A biosensor was created that can detect heavy metal cations through glucose oxidase enzymatic inhibition. It was built on carbon film electrode supports with immobilized glucose oxidase cross-linked with glutaraldehyde. A poly (neutral red) film was added through electro polymerization to act as a redox mediator. The biosensor was able to detect cadmium, copper, lead, and zinc in the presence of glucose, with detection limits of 1μ g L^{-1} for cadmium, 6μ g L^{-1} for copper, 3μ g L^{-1} for lead, and 9μ g L^{-1} for zinc (Ghica and Brett 2008).

Another study presents a biosensor that employs self-assembled gold nanoparticles to detect mercury ions using a potentiometric urease inhibition-based approach. The biosensor incorporates a PVC-NH_2 matrix membrane pH electrode surface with N, N-dodecyl amino methyl benzene (DAMAB) as a neutral carrier. The gold nanoparticles are chemically adsorbed onto the electrode surface, and urease is immobilized on the nanoparticles. The study also investigated the response characteristics of the DAMAB/PVC-NH_2 pH-sensitive membrane and the effect of nanoparticle size. The biosensor has a low detection limit of 0.05μ mol L^{-1} and a linear range for determining Hg^{2+} of 0.09–1.99μmol L^{-1}. The self-assembled immobilization method has several advantages, including a rapid response time and easy regeneration (Yang et al. 2006).

A study developed an amperometric biosensor to detect paraoxon, a pesticide that inhibits acetylcholinesterase (AChE). The biosensor was designed using a zinc oxide matrix to immobilize AChE, which retained its activity for up to 3 months when stored in phosphate-buffered saline (pH 7.4) at 4 °C. The biosensor was able to detect paraoxon in the concentration range of 0.035–1.38 ppm and can also be used to detect other AChE-inhibiting organophosphate pesticides (Sinha et al. 2010).

2.4 Applications of Nano-biomaterials in Detecting Environmental Pollutants

2.4.1 Detection of Heavy Metals

Heavy metal pollution is a major environmental concern triggered by industrial, mining, and human activities. Industrial, mining, and human activities are responsible for heavy metal pollution, with notable examples including lead, mercury, cadmium, chromium, arsenic, copper, zinc, and nickel. These heavy metals contaminate soil, water, and air, giving rise to significant risks for both human health and ecosystems. They have the potential to cause diverse detrimental effects on plants, animals, and microorganisms, while their accumulation in the environment can disrupt delicate ecosystems. It is imperative to prioritize the control and reduction of heavy metal pollution in order to protect and preserve the environment and promote human well-being (Gumpu et al. 2015; Vonnie et al. 2022; Baig et al. 2021; Wen et al. 2019).

Nano-biomaterials, such as nanoparticles, carbon-based nanomaterials, quantum dots, and nanocomposites, are being utilized for the detection of heavy metals in the environment. These materials offer unique properties at the nanoscale that enable sensitive and selective detection of heavy metal contaminants. By functionalizing these biomaterials with specific receptors or ligands, they can selectively bind to target heavy metals, resulting in measurable changes in their optical, electrical, or magnetic properties. This allows for the rapid and accurate detection of heavy metal ions in environmental samples. Nano-biomaterial-based detection systems offer advantages such as high sensitivity, selectivity, and the ability to detect multiple heavy metals simultaneously. They can be incorporated into portable devices for on-site monitoring, aiding in the mitigation of risks associated with heavy metal pollution in the environment (Gumpu et al. 2015; Baig et al. 2021; Wen et al. 2019).

Various heavy metals, including Arsenic (III) (Boruah et al. 2019), Arsenate (Liu et al. 2015), Cobalt (Annadhasan et al. 2015), Nickel (Annadhasan et al. 2015), Lead (Antunović et al. 2019), Cadmium (Antunović et al. 2019), Mercury (May and Oluwafemi 2016), Iron (May and Oluwafemi 2016), Copper (Cheon and Park 2016), and Lead (Cheon and Park 2016), are detected using biosensors. Detection methods include colorimetric analysis using functionalized nanoparticles, fluorescence-based mechanisms, and electrochemical techniques. These biosensors offer selective and sensitive detection of heavy metals in diverse samples. Table 2.1 shows detection of various metals using different kinds of biosensor.

2.4.2 Detection of Pesticides

Various biosensors have been developed for the detection of different pesticides, including glyphosate (Zúñiga et al. 2022a), atrazine (Supraja et al. 2019), chlorpyrifos (Talan et al. 2018), DDT (Alam et al. 2021), and organochlorine pesticides (Goel and Arora 2018). These biosensors utilize different detection principles such as competitive immune reactions, electrochemical sensing, and colorimetric responses. The biosensors offer high sensitivity, wide detection ranges, and low detection limits, making them suitable for precise and real-time detection of pesticides in environmental samples. These advancements in biosensor technology contribute to the development of effective tools for monitoring and ensuring the safety of agricultural and environmental systems. Table 2.2 represents detection of various pesticides by different kinds of biosensor.

IgG-CDs were employed for selective glyphosate detection via a competitive immune reaction. By separating magnetic antigen beads using a magnetic field, glyphosate attached to IgG-CDs could be quantified, exhibiting a linear range of 0.06–473.18μM and a detection limit of 0.05μM (Zúñiga et al. 2022b).

A novel colorimetric method using cysteamine-stabilized gold nanoparticles (CS-AuNPs) was developed for rapid, sensitive, and low-cost detection of glyphosate, with a detection limit of 5.88×10^{-8} M (Zheng et al. 2013).

A sensitive electrochemical sensor was developed to detect glyphosate by creating cavities through electropolymerization of gold nanoparticles in the presence of glyphosate as a template molecule. The sensor exhibited a broad linear range of 1 pg/L to 1μg/L and a quantification limit of 0.8 pg./L, while demonstrating excellent selectivity toward glyphosate compared to its metabolite, aminomethylphosphonic acid (Do et al. 2015).

Table 2.1 Detection of heavy metal using nanoparticle-based sensor

Sr. no	Name of heavy metal	Method for detection	Limit of detection	References
1.	Arsenic (III)	Glucose functionalized gold nanoparticles	0.53 ppb	Boruah et al. (2019)
		A novel cobalt oxyhydroxide (CoOOH) nanoflakes	3.72 ppb	Wen et al. (2019)
		A novel oligonucleotide functionalized CuInS2 quantum dots@magnetic Fe_3O4 nanocomposite	0.13 nmol L^{-1}	Liu et al. (2015)
2.	Mercury	Quantum dot with conjugation of DNA and gold nanoparticle system	1.22pbb	Li et al. (2013)
		A tyrosine-based radiometric fluorescent probe	22.65 nM	Shuai et al. (2020)
		A coumarin-based fluorescent probe incorporating a vinyl ether group as the reactive site, exhibiting a "turn-on" fluorescence response	0.12μM	
		A novel colorimetric fluorescent probe, where dichlororesorcinol acts as the fluorophore and the thiocarbon group serves as the recognition site	0.49 nM	
		Nitrogen-doped carbon quantum dots	19 nM	
		A carbon quantum dot derived from muskmelon (*Cucumis melo*) fruit through acid oxidation	0.33μM	
		A carbon quantum dot probe doped (N, S-CD) with both nitrogen and sulfur atoms was developed from citric acid and cysteine	0.15μM	
		Boron-doped sulfur quantum dots (B–MoS_2 QDs)	1.8 nM	
3.	Cobalt and nickel	a green synthesis method utilizing natural sunlight irradiation to prepare gold nanoparticles (AuNPs) using N-cholyl-l-valine	–	Annadhasan et al. (2015)
4.	Lead and cadmium	Porous $MnCo_2O_4$ nanoparticles	Pb-8.06 nmol/dm^3 Cd-7.02 nmol/dm^3	Antunović et al. (2019)
5.	Mercury and iron	Dextrose and maltose-reduced silver nanoparticles (Ag-NPs)	10–12 M	May and Oluwafemi (2016)

(continued)

Table 2.1 (continued)

Sr. no	Name of heavy metal	Method for detection	Limit of detection	References
6.	Mercury and iron	Green and cost-effective fluorescent carbon dots	5.5 nM for Hg^{2+} and 0.075μM for Fe^{3+}	Yu et al. (2015)
7.	Copper and lead	A simple and eco-friendly method for preparing uniform silver nanoparticles (AgNPs) using poly DOPA as a reducing and stabilizing agent.	Pb^{2+} -9.4×10^{-5}μM and Cu^{2+} $-8.1 \times 10–5$μM	Cheon and Park (2016)
8.	Lead	Gold Nps and CdSe/ZnS quantum dots	1.79 pg mL^{-1}	Khawashi et al. (2023)
		Gold-modified graphene nanocomposite	1.67 pmol/L	
		Fe3O4@PDA@MnO2 core-shell nanocomposites	0.03μg L^{-1}	
		β-cyclodextrin functionalized gold nanoparticle	83.3μmol/L	
		Gold nanoparticle graphene selenocysteine composite	0.05 ppb	
		Ag@Pt core-shell NPs	0.8 nM	
9.	Cobalt	Ag–Au bimetallic nanoparticles	0.02 mM	Xu et al. (2017)
10.	Nickel	Dimethylglyoxime-doped GQDs	20.0μg L^{-1}	Pimsin et al. (2021)
		Aggregation of citrate-stabilized silver nanoparticles	0.75 mM	Almaquer et al. (2019)

Table 2.2 Detection of pesticide using nanoparticle-based sensor

Sr. no	Name of pesticide	Biosensor used in detection	Detection limit	References
1.	Glyphosate	Antibody-tagged carbon dot (IgG-CDs)	0.05μM	Zúñiga et al. (2022a)
		Cysteamine-stabilized gold nanoparticles (CS-AuNPs)	5.88 × 10–8 M	Zheng et al. (2013)
		Electro polymerization of gold nanoparticles	8 pg./L	Zúñiga et al. (2022b)
		A fluorescence carbon dot/silver nanoparticle (CDs/AgNPs)	12 ng mL^{-1}	Wang et al. (2016)
		Cu-BTC MOF	1.4×10^{-13}	Cao et al. (2019)
		A nano-porous copper film	4 nmol L^{-1}	Regiart et al. (2020)
		Multi-walled carbon nanotube (MWCNTS)-ionic liquid (IL) composite and copper oxide nanoparticle (CuO)	1.3 nM	Gholivand et al. (2018)
		Bipyramid-shaped gold nanoparticle	5.92μM	Nafisah et al. (2020)
		Luminol-gold-L-cysteine-Cu (II) nanoparticle composites (Lu-Au-L cys-Cu (II))	0.5 nM	Liu et al. (2020c)
2.	Atrazine	Manganese oxide nanofibers	0.22×10^{-21} g/mL	Supraja et al. (2019)
		Platinum nanoparticles/carbon nitride nanotubes	1.5 × 10–13 M	Yola and Atar (2017)
		G-protein-functionalized magnetic nanoparticles (MNPs@protG)	0.05–1 ng/mL	Yola and Atar (2017)
		Multiwalled carbon nanotube embedded with ZnO nanofiber	5.368 nM	Supraja et al. (2020)
		Antibodies labelled with gold nanoparticles	0.1–1gL	Valera et al. (2008)
3.	Chlorpyrifos	Fluorine-doped tin-oxide (FTO) with gold nanoparticles (AuNPs) and anti-chlorpyrifos antibodies (chl-Ab)	10 fM	Talan et al. (2018)
		Photoelectrochemical sensor based on the noble-metal-free $AgBr/Ti_3C_2$ Schottky interface in fruit and vegetables.	0.33 pg L^{-1}	Du et al. (2022)
		$MnCo_2O_4$ decorated (2D/2D) rGO/g-C_3N_4-based non-enzymatic sensor	0.14×10^{-12} M	Maji et al. (2022)
		Gold nanoparticles-silk fibroin (SF-AuNPs) dispersion, label-free colorimetric	10 ppb	Mane et al. (2020)

(continued)

Table 2.2 (continued)

Sr. no	Name of pesticide	Biosensor used in detection	Detection limit	References
4.	DDT	The gold nanostructure	0.750μ M	Alam et al. (2021)
		void@AuNPs@SiO2 microporous capsules	–	Moldovan et al. (2021)
		AgNPs prepared by self-assembly/in situ growing method	–	
		AgNPs@ α, ω-aliphatic diamines	13.7 nM	
5.	Malathion	Gold nanoparticle prepared by using citrate	1.5×10^{-7} mol/L	Kohzadi and Roushani (2016)
6.	Dieldrin	AuNPs/AgNPs@alkyl dithiols	0.82μM	Moldovan et al. (2021)

A fluorescence system (CDs/AgNPs) was developed for the detection of glyphosate, where the fluorescence intensity of the system showed a linear relationship with glyphosate concentration in the range of 0.025–2.5μg mL^{-1}, achieving a low detection limit of 12 ng mL^{-1} (Wang et al. 2016).

A novel electrochemical sensor utilizing Cu-BTC MOF material has been developed for the detection of non-electroactive glyphosate, demonstrating a large specific surface area that enhances the electrode reaction site and improves the detection performance. Cyclic voltammetry (CV) evaluated the electrochemical behavior of Cu-BTC frameworks for glyphosate detection, and under optimal conditions, differential pulse stripping voltammetry (DPV) was employed, revealing an ultralow detection limit of 1.4×10^{-13} mol L^{-1} and a wide detection range of 1.0×10^{-12} to 1.0×10^{-9} mol L^{-1} and 1.0×10^{-9} to 1.0×10^{-5} mol L^{-1} (Cao et al. 2019).

A nano-porous copper (NPC) film was electrodeposited on a copper microelectrode for electrochemical sensing of glyphosate (Glyp). The NPC film exhibited a highly pure and crystalline structure. The optimized NPC-modified Cu microelectrode showed high sensitivity (14 nA $nmol^{-1}$ L), a low detection limit (4 nmol L^{-1}), excellent reproducibility, and selectivity for Glyp. The sensor successfully detected Glyp in river water samples (Regiart et al. 2020).

A highly sensitive colorimetric method is introduced for the determination of malathion using gold nanoparticles (AuNPs). The stability of the synthesized AuNPs solution is ensured by the presence of citrate anions, which prevent aggregation of the AuNPs by exerting repulsive forces. The surface of the AuNPs is capped by malathion, leading to their aggregation in Britton-Robinson buffer solution. The progress of the reaction is monitored spectrophotometrically by measuring the decrease in the plasmon resonance band of the AuNPs at 527 nm after 9 min. The impact of various reaction variables on the sensitivity of the reaction is investigated, and the interference from common ions is effectively minimized. The calibration curve demonstrates a linear relationship over the concentration range of 3.3×10^{-7} to 3.3×10^{-6} mol/L of malathion, exhibiting good precision and accuracy.

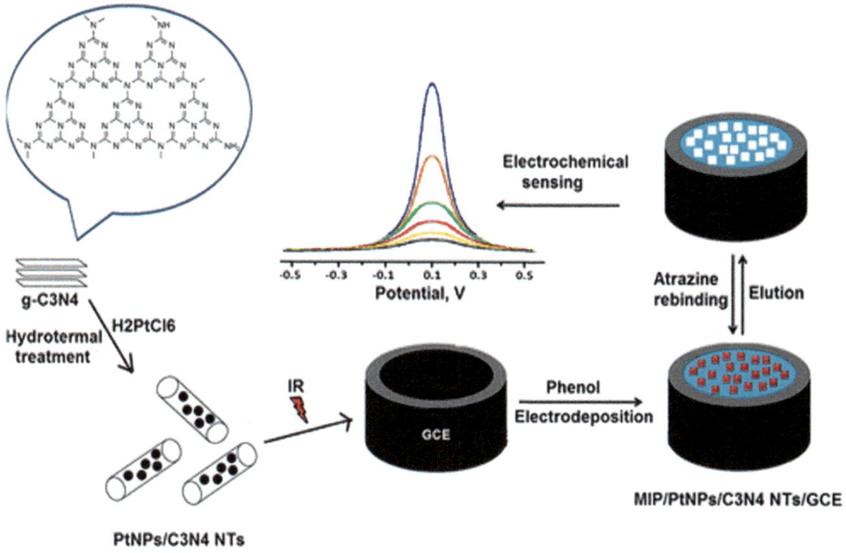

Fig. 2.3 Preparation of platinum nanoparticles/carbon nitride nanotubes for detection of atrazine. (Image adapted from (Yola and Atar 2017) with permission)

Furthermore, the detection limit of the method is as low as 1.5×10^{-7} mol/L (Kohzadi and Roushani 2016).

An ultrasensitive, label-free electrochemical immunosensor developed for detecting atrazine, a toxic pesticide. Utilizing electrospun manganese oxide nanofibers, the method achieves a low limit of detection of 0.22×10^{-21} g/mL and demonstrates superior performance compared to previous sensors, making it suitable for real-time applications (Supraja et al. 2019) (Fig. 2.3).

A highly sensitive and stable electrochemical nano-sensor utilizing fluorine-doped tin-oxide (FTO) with gold nanoparticles (AuNPs) and anti-chlorpyrifos antibodies (chl-Ab) was developed for the detection of chlorpyrifos, demonstrating a wide detection range and low limit of detection, making it suitable for on-site detection of chlorpyrifos traces in real samples.

The gold nanostructured electrochemical sensor achieved a low detection limit of 0.750μM for thiols, demonstrating its high sensitivity and suitability for precise detection.

Gold nanoparticles were used as a colorimetric sensor to detect and quantify the organochlorine endosulfan pesticide, with the nanoparticles' color changing from wine red to various shades of blue in the presence of different concentrations of the pesticide.

2.5 Challenges and Future Prospects

2.5.1 Challenges in Using Nano-biomaterials for Detection of Environmental Pollutants

Nano-biomaterials have shown immense potential in the detection of environmental pollutants. However, several challenges need to be addressed before these materials can be widely used for environmental monitoring.

One of the main challenges is the potential toxicity of the nanoparticles themselves. While nanoparticles have unique properties that make them useful for detection, these properties can also lead to toxicity concerns. For instance, the small size of nanoparticles can lead to cellular uptake and accumulation, which can cause cellular damage and inflammation. In addition, the surface chemistry of nanoparticles can also contribute to their toxicity. Therefore, it is important to thoroughly investigate the potential toxicity of nano-biomaterials and to develop strategies to mitigate their negative effects.

The selectivity and sensitivity are also challenges in biosensor development. Environmental samples can contain a range of different pollutants, some of which may have similar properties. Therefore, it is crucial to develop methods that can selectively detect the target pollutant of interest. In addition, the sensitivity of the detection method is important for accurately quantifying the pollutant concentration in the sample. Several strategies have been developed to enhance the selectivity and sensitivity of nano-biomaterial-based detection methods, such as functionalization with specific ligands and the use of amplification techniques.

Another challenge is the reproducibility and reliability of the detection methods. The synthesis and characterization of nano-biomaterials can be complex, and even minor variations in the synthesis conditions can lead to variations in the properties of the material. Therefore, it is important to establish standardized protocols for the synthesis, characterization, and application of nano-biomaterials. Furthermore, the robustness of the detection method should be thoroughly tested to ensure reliable and reproducible results.

Finally, the practicality and cost-effectiveness of the detection methods are also important considerations. The detection methods should be user-friendly, easy to operate, and require minimal sample preparation. Furthermore, the cost of the materials and equipment required for detection should be reasonable to enable widespread adoption of these methods.

2.5.2 *Prospects and Potential of Nano-biomaterials in Detecting Environmental Pollutants*

Nano-biomaterials have promising potential in detecting environmental pollutants, and the field has several prospects, including increasing sensitivity by improving surface area, shape, and size, developing new materials like metal-organic frameworks, miniaturization and portability for on-site detection, multi-analyte detection for comprehensive monitoring, and environmental remediation. Additionally, nano-biomaterials can reduce the environmental impact of traditional detection methods and enable real-time monitoring in industrial settings. Continued research and development in this field will lead to more efficient and effective monitoring and remediation methods to address the impact of environmental pollution on human health and the planet.

2.6 Conclusion

2.6.1 *Summary of Key Points*

The book chapter discusses the use of nano-biomaterials in the detection of environmental pollutants. It highlights the advantages of using nano-biomaterials, such as their high sensitivity, specificity, and selectivity, in environmental monitoring. The chapter provides various examples of nano-biomaterials, including gold nanoparticles, silver nanoparticles, and iron oxide nanoparticles, used for the detection of environmental pollutants, such as heavy metals, pesticides, and polycyclic aromatic hydrocarbons (PAHs). The challenges associated with the use of nano-biomaterials for environmental monitoring are also discussed, including their potential toxicity and the need for standardized protocols. The chapter concludes by outlining the potential future applications of nano-biomaterials in environmental monitoring and highlighting the importance of continued research in this area. Overall, the chapter provides a comprehensive overview of the use of nano-biomaterials in environmental monitoring and the potential impact they can have on improving our ability to detect and mitigate environmental pollutants.

2.6.2 *Importance of Nano-biomaterials in Detecting Environmental Pollutants*

Nano-biomaterials have shown great potential in detecting environmental pollutants due to their unique physicochemical properties, including large surface area, high reactivity, and strong affinity for certain pollutants. They can be designed to selectively target specific pollutants and provide highly sensitive and accurate detection with low limits of detection. Furthermore, the use of nano-biomaterials can greatly reduce the amount of sample needed for analysis, making them more environmentally friendly and cost-effective. The field of nano-biomaterials-based pollutant detection is constantly expanding, with new materials being developed and optimized for specific applications. The continued research and development of nano-biomaterials in this field have the potential to revolutionize the way environmental monitoring is performed, ultimately leading to a safer and healthier environment for all.

Author Contribution The first author contributed to the book chapter by drawing tables and obtaining permission to use photographs from respected journals. The second author of the book chapter was responsible for writing the manuscript. The corresponding author reviewed the manuscript, provided valuable feedback, and made necessary revisions to enhance the overall quality and coherence of the content.

Acknowledgments The first author expressed gratitude for RGST for providing financial support for research work. The second author expressed gratitude to the CSIR-UGC (Council of Scientific and Industrial Research-University Grants Commission) for providing a junior researcher scholarship (JUNE18-359427) that supported their PhD work.

References

Abdelbasir SM, Shalan AE (2019) An overview of nanomaterials for industrial wastewater treatment. Korean J Chem Eng 36(8):1209–1225. https://doi.org/10.1007/s11814-019-0306-y

Ahmed F, Kabir H, Xiong H (2020) Dual colorimetric sensor for Hg2+/Pb2+ and an efficient catalyst based on silver nanoparticles mediating by the root extract of Bistorta amplexicaulis. Front Chem 8(October):1–15. https://doi.org/10.3389/fchem.2020.591958

Akinpelu AA, Ali ME, Johan MR, Saidur R, Qurban MA, Saleh TA (2019) Polycyclic aromatic hydrocarbons extraction and removal from wastewater by carbon nanotubes: a review of the current technologies, challenges and prospects. Process Saf Environ Prot 122:68–82. https://doi.org/10.1016/j.psep.2018.11.006

Alam S, Augustine S, Narayan T, Luong JHT, Malhotra BD, Khare SK (2021) A chemosensor based on gold nanoparticles and dithiothreitol (Dtt) for acrylamide electroanalysis. Nanomaterials 11(10):2610. https://doi.org/10.3390/nano11102610

Alberti G, Zanoni C, Magnaghi LR, Biesuz R (2020) Disposable and low-cost colorimetric sensors for environmental analysis. Int J Environ Res Public Health 17(22). Multidisciplinary Digital Publishing Institute (MDPI):1–23. https://doi.org/10.3390/ijerph17228331

Almaquer FEP, Ricacho JSY, Ronquillo RLG (2019) Simple and rapid colorimetric sensing of Ni(II) ions in tap water based on aggregation of citrate-stabilized silver nanoparticles. Sustain Environ Res 1(1):1–10. https://doi.org/10.1186/s42834-019-0026-3

Al-Rifai A, Aqel A, Al Wahibi L, ALOthman ZA, Badjah-Hadj-Ahmed AY (2018) Carbon nanotube-based benzyl polymethacrylate composite monolith as a solid phase extraction adsorbent and a stationary phase material for simultaneous extraction and analysis of polycyclic aromatic hydrocarbon in water. J Chromatogr A 1535:17–26. https://doi.org/10.1016/j.chroma.2018.01.011

Anku WW, Kiarii EM, Sharma R, Joshi GM, Shukla SK, Govender PP (2018) Photocatalytic degradation of pharmaceuticals using graphene based materials. In: A new generation material graphene: applications in water technology. Springer International Publishing, pp 187–208

Annadhasan M, Kasthuri J, Rajendiran N (2015) Green synthesis of gold nanoparticles under sunlight irradiation and their colorimetric detection of Ni2+ and Co2+ ions. RSC Adv 5(15):11458–11468. https://doi.org/10.1039/c4ra14034f

Antunović V, Ilić M, Baošić R, Jelić D, Lolić A (2019) Synthesis of MnCo2O4 nanoparticles as modifiers for simultaneous determination of Pb(II) and Cd(II). PLoS One 14(2). https://doi.org/10.1371/journal.pone.0210904

Baig N, Kammakakam I, Falath W, Kammakakam I (2021) Nanomaterials: a review of synthesis methods, properties, recent progress, and challenges. Mater Adv 2(6). Royal Society of Chemistry:1821–1871. https://doi.org/10.1039/d0ma00807a

Bebi V, Reddy Prasad P, Sudheer K, Sandhya P, Nageswara Reddy C, Sreedhar NY (2021) A modified sensitive palladium-copper oxide and multiwalled carbon nanotubes electrochemical sensor for detection of ametridione pesticide. J Appl Nat Sci 13(3):798–806. https://doi.org/10.31018/jans.v13i3.2531

Bo J, Su X, Ma H, Song Y (2007) DNA hybridization detection with blue luminescent quantum dots and dye-labeled single-stranded DNA. J Am Chem Soc 129(11):3042–3043

Boruah BS, Biswas R, Deb P (2019) A green colorimetric approach towards detection of arsenic (III): a pervasive environmental pollutant. Opt Laser Technol 111:825–829. https://doi.org/10.1016/j.optlastec.2018.09.023

Bose S, Maity S, Sarkar A (2022) Nano-materials as biosensor for heavy metal detection. In: Food, medical, and environmental applications of nanomaterials. Elsevier, pp 493–526

Cao Y, Wang L, Wang C, Hu X, Liu Y, Wang G (2019) Sensitive detection of glyphosate based on a Cu-BTC MOF/g-C3N4 nanosheet photoelectrochemical sensor. Electrochim Acta 317:341–347. https://doi.org/10.1016/J.ELECTACTA.2019.06.004

Chai F, Wang C, Wang T, Li L, Su Z (2010) Colorimetric detection of Pb2+ using glutathione functionalized gold nanoparticles. ACS Appl Mater Interfaces 2(5):1466–1470. https://doi.org/10.1021/am100107k

Chauhan M, Vatsa P, Vatsa N, Jain A (2011) Aptamer-based fluorescent biosensors. Curr Med Chem 18(27):4175–4184

Cheng N et al (2018) Aptasensor based on fluorophore-quencher nano-pair and smartphone spectrum reader for on-site quantification of multi-pesticides. Biosens Bioelectron 117:75–83. https://doi.org/10.1016/j.bios.2018.06.002

Cheon JY, Park WH (2016) Green synthesis of silver nanoparticles stabilized with mussel-inspired protein and colorimetric sensing of lead(II) and copper(II) ions. Int J Mol Sci 17(12). https://doi.org/10.3390/ijms17122006

Chong ETJ, Ng JW, Lee P-C (2022) Classification and medical applications of biomaterials–a mini review. BIO Integr. https://doi.org/10.15212/bioi-2022-0009

Daniel MC, Astruc D (2009) Gold nanoparticles: assembly, supramolecular chemistry, quantum-size-related properties, and applications toward biology, catalysis, and nanotechnology. Chem Rev 109(11):5828–5878

Dasary NR, Lu J, Liu B, Ganesana M, Vodyanoy V, Ray PC (2009) Cadmium telluride quantum dots cause oxidative stress leading to extrinsic and intrinsic apoptosis in human hepatocellular varcinoma HepG2 cells. J Am Chem Soc 131(17):6439–6445

Do MH et al (2015) Molecularly imprinted polymer-based electrochemical sensor for the sensitive detection of glyphosate herbicide. Int J Environ Anal Chem 95(15):1489–1501. https://doi.org/10.1080/03067319.2015.1114109

Domínguez-González R, González Varela L, Bermejo-Barrera P (2014) Functionalized gold nanoparticles for the detection of arsenic in water. Talanta 118:262–269. https://doi.org/10.1016/j.talanta.2013.10.029

Dou X, Chu X, Kong W, Luo J, Yang M (2015) A gold-based nanobeacon probe for fluorescence sensing of organophosphorus pesticides. Anal Chim Acta 891:291–297. https://doi.org/10.1016/j.aca.2015.08.012

Du X, Du W, Sun J, Jiang D (2022) Self-powered photoelectrochemical sensor for chlorpyrifos detection in fruit and vegetables based on metal–ligand charge transfer effect by Ti3C2 based Schottky junction. Food Chem 385:132731. https://doi.org/10.1016/J.FOODCHEM.2022.132731

Duarte K, Justino CIL, Freitas AC, Gomes AMP, Duarte AC, Rocha-Santos TAP (2015) Disposable sensors for environmental monitoring of lead, cadmium and mercury. Trends Anal Chem 64. Elsevier B.V:183–190. https://doi.org/10.1016/j.trac.2014.07.006

Gangapuram BR, Bandi R, Dadigala R, Kotu GM, Guttena V (2017) Facile green synthesis of gold nanoparticles with Carboxymethyl gum Karaya, selective and sensitive colorimetric detection of copper (II) ions. J Clust Sci 28(5):2873–2890. https://doi.org/10.1007/s10876-017-1264-3

Gaviria-Arroyave MI, Cano JB, Peñuela GA (2020) Nanomaterial-based fluorescent biosensors for monitoring environmental pollutants: a critical review. Talanta Open 2(August). https://doi.org/10.1016/j.talo.2020.100006

Ghica ME, Brett CMA (2008) Glucose oxidase inhibition in poly(neutral red) mediated enzyme biosensors for heavy metal determination. Microchim Acta 163(3–4):185–193. https://doi.org/10.1007/s00604-008-0018-1

Gholivand MB, Akbari A, Norouzi L (2018) Development of a novel hollow fiber- pencil graphite modified electrochemical sensor for the ultra-trace analysis of glyphosate. Sensors Actuators B Chem 272:415–424. https://doi.org/10.1016/J.SNB.2018.05.170

Gilchrist A, Nobbs J (2016) Colorimetry, theory. In: Encyclopedia of spectroscopy and spectrometry. Academic Press, pp 328–333

Goel P, Arora M (2018) Fabrication of chemical sensor for organochlorine pesticide detection using colloidal gold nanoparticles. MRS Commun 8(3):1000–1007. https://doi.org/10.1557/mrc.2018.125

Gong NC, Le Li Y, Jiang X, Zheng XF, Wang YY, Huan SY (2016) Fluorescence resonance energy transfer-based biosensor composed of nitrogen-doped carbon dots and gold nanoparticles for the highly sensitive detection of organophosphorus pesticides. Anal Sci 32(9):951–956. https://doi.org/10.2116/analsci.32.951

Gumpu MB, Sethuraman S, Krishnan UM, Rayappan JBB (2015) A review on detection of heavy metal ions in water - an electrochemical approach. Sens Actuators B Chem 213. Elsevier B.V:515–533. https://doi.org/10.1016/j.snb.2015.02.122

Gunupuru R, Maity D, Bhadu GR, Chakraborty A, Srivastava DN, Paul P (2014) Colorimetric detection of Cu2+ and Pb2+ ions using calix[4]arene functionalized gold nanoparticles. J Chem Sci 126(3):627–635. https://doi.org/10.1007/s12039-014-0600-5

Guo Y, Zhang Y, Shao H, Wang Z, Wang X, Jiang X (2014) Label-free colorimetric detection of cadmium ions in rice samples using gold nanoparticles. Anal Chem 86(17):8530–8534. https://doi.org/10.1021/ac502461r

Gutierrez AM, Dziubla TD, Hilt JZ (2017) Recent advances on iron oxide magnetic nanoparticles as sorbents of organic pollutants in water and wastewater treatment. Rev Environ Health 32(1–2):111–117. https://doi.org/10.1515/reveh-2016-0063

Han J, Sharipov M, Hwang S, Lee Y, Huy BT, Lee YI (2022) Water-stable perovskite-loaded nanogels containing antioxidant property for highly sensitive and selective detection of roxithromycin in animal-derived food products. Sci Rep 12(1). https://doi.org/10.1038/s41598-022-07030-9

Ibrahim RK, Hayyan M, AlSaadi MA, Hayyan A, Ibrahim S (2016) Environmental application of nanotechnology: air, soil, and water. Environ Sci Pollut Res 23(14). Springer Verlag:13754–13788. https://doi.org/10.1007/s11356-016-6457-z

Jain K, Patel AS, Pardhi VP, Flora SJS (2021) Nanotechnology in wastewater management: a new paradigm towards wastewater treatment. Molecules 26(6). MDPI AG. https://doi.org/10.3390/molecules26061797

Kalluri JR et al (2009) Use of gold nanoparticles in a simple colorimetric and ultrasensitive dynamic light scattering assay: selective detection of arsenic in groundwater. Angew Chem 121(51):9848–9851. https://doi.org/10.1002/ange.200903958

Khawashi HP, Chavan VA, Bhagat DS, Deshmukh SU (2023) Recent advances in the detection of lead ions using nanoparticle-based sensors. Biointerface Res Appl Chem 13(5):1–15. https://doi.org/10.33263/BRIAC135.466

Kim S, Kim Y, Kim Y, Kim BH, Choi JW, Lee TS, Choi IS (2001) Microfluidic device for DNA sample preparation and analysis. Anal Chem 73(21):5240–5245

Kohzadi T, Roushani M (2016) Highly sensitive colorimetric determination of malathion using gold nanoparticles. Water Sci Technol Water Supply 16(5):1214–1220. https://doi.org/10.2166/ws.2016.029

Koul B, Taak P (2018) Nanobioremediation. In: Biotechnological strategies for effective remediation of polluted soils. Springer, Singapore, pp 197–220

Kumar S et al (2023) A review on properties and environmental applications of graphene and its derivative-based composites. Catalysts 13(1):111. https://doi.org/10.3390/catal13010111

Lam SM, Sin JC, Abdullah AZ, Mohamed AR (2014) Photocatalytic TiO2/carbon nanotube nanocomposites for environmental applications: an overview and recent developments. Fullerenes Nanotub Carbon Nanostruct 22(5):471–509. https://doi.org/10.1080/1536383X.2012.690458

Lee JH, Hong SW et al (2009) Highly sensitive and selective colorimetric sensors for uranyl. NIH Public Access 130(43):14217–14226

Lee LZ, Zaini MAA, Tang SH (2019) Porous nanomaterials for heavy metal removal. In: Handbook of ecomaterials, vol 1. Springer International Publishing, pp 469–494

Li M, Gou H, Al-ogaidi I, Wu N (2013) Nanostructured sensors for detection of heavy metals: a review. ACS Sustain Chem Eng. Sustainable nanotechnology 1:A–K

Li L, Yuan Z, Peng X, Li L, He J, Zhang Y (2014) Highly selective colorimetric detection of copper ions using cysteamine functionalized gold nanoparticles. J Chin Chem Soc 61(12):1371–1376. https://doi.org/10.1002/jccs.201400188

Liu J, Lu Y (2003) A colourimetric lead biosensors using DNAzyme-directly assembly of gold nanoparticles. J Am Chem Soc 125(22):6642–6643

Liu J, Lu Y (2007a) Non-base pairing DNA provides a new dimension for controlling aptamer-linked nanoparticles and sensors. J Am Chem Soc 129:8634–8643

Liu A, Lu Y (2007b) A colorimetric method for the detection of lead ion based on the use of hemin/G-quadruplex as a peroxidase mimic. Angew Chem Int Ed 46(5):783–786

Liu Z, Li G, Xia T, Su X (2015) Ultrasensitive fluorescent nanosensor for arsenate assay and removal using oligonucleotide-functionalized CuInS2 quantum dot@magnetic Fe3O4 nanoparticles composite. Sensors Actuators B Chem 220:1205–1211. https://doi.org/10.1016/j.snb.2015.06.111

Liu B, Zhuang J, Wei G (2020a) Recent advances in the design of colorimetric sensors for environmental monitoring. Environ Sci Nano 7(8):2195–2213. https://doi.org/10.1039/d0en00449a

Liu B, Zhuang J, Wei G (2020b) Recent advances in the design of colorimetric sensors for environmental monitoring. Environ Sci Nano 7(8). The Royal Society of Chemistry:2195–2213. https://doi.org/10.1039/d0en00449a

Liu H et al (2020c) Electrochemical luminescence sensor based on double suppression for highly sensitive detection of glyphosate. Sensors Actuators B Chem 304:127364. https://doi.org/10.1016/J.SNB.2019.127364

Lofrano G, Libralato G, Brown J (2017) Nanotechnologies for environmental remediation: applications and implications. Springer International Publishing, Cham

Lv Q, Xia N, Gao L, Han B (2022) Detection of mercury ions using graphene oxide sensors assisted by Ag@SiO2. Appl Nanosci:1–10. https://doi.org/10.1007/s13204-022-02555-x

Maji B, Achary LSK, Barik B, Jyotsna Sahoo S, Mohanty A, Dash P (2022) MnCo2O4 decorated (2D/2D) rGO/g-C3N4-based non-enzymatic sensor for highly selective and sensitive detection of Chlorpyrifos in water and food samples. J Electroanal Chem 909:116115. https://doi.org/10.1016/J.JELECHEM.2022.116115

Mane PC et al (2020) Highly sensitive label-free bio-interfacial colorimetric sensor based on silk fibroin-gold nanocomposite for facile detection of chlorpyrifos pesticide. Sci Rep 10(1):1–14. https://doi.org/10.1038/s41598-020-61130-y

May BMM, Oluwafemi OS (2016) Sugar-reduced gelatin-capped silver nanoparticles with high selectivity for colorimetric sensing of Hg 2+ and Fe 2+ ions in the midst of other metal ions in aqueous solutions. Int J Electrochem Sci 11(9):8096–8108. https://doi.org/10.20964/2016.09.29

Moldovan R, Iacob BC, Farcău C, Bodoki E, Oprean R (2021) Strategies for SERS detection of organochlorine pesticides. Nanomaterials 11(2):1–21. https://doi.org/10.3390/nano11020304

Mortimer RJ (2016) Spectroelectrochemistry, applications. In: Encyclopedia of spectroscopy and spectrometry. Academic Press, pp 160–171

Nafisah S et al (2020) Improved sensitivity and selectivity of direct localized surface plasmon resonance sensor using gold nanobipyramids for glyphosate detection. IEEE Sensors J 20(5):2378–2389. https://doi.org/10.1109/JSEN.2019.2953928

Newman D, Grigorova D, Rangel V, Stach E, O'Brien S (2007) Quantum dots in biosensing and bioimaging. Anal Chem 79(14):5483–5491

Le Thao Nguyen N, Park CY, Park JP, Kailasa SK, Park TJ (2018) Synergistic molecular assembly of an aptamer and surfactant on gold nanoparticles for the colorimetric detection of trace levels of As3+ ions in real samples. New J Chem 42(14):11530–11538. https://doi.org/10.1039/c8nj01097h

Nsibande SA, Forbes PB (2020) Development of a turn-on graphene quantum dot-based fluorescent probe for sensing of pyrene in water. RSC Adv 10(21):12119–12128. https://doi.org/10.1039/c9ra10153e

Pooja D, Singh L, Thakur A, Kumar P (2019) Green synthesis of glowing carbon dots from Carica papaya waste pulp and their application as a label-freechemo probe for chromium detection in water. Sensors Actuators B Chem 283:363–372. https://doi.org/10.1016/j.snb.2018.12.027

Parida D et al (2021) Template-free synthesis of hybrid silica nanoparticle with functionalized mesostructure for efficient methylene blue removal. Mater Des 201:109494. https://doi.org/10.1016/j.matdes.2021.109494

Pimsin N et al (2021) Ultratrace detection of nickel(II) ions in water samples using dimethylglyoxime-doped GQDs as the induced metal complex nanoparticles by a resonance light scattering sensor. ACS Omega 6(23):14796–14805. https://doi.org/10.1021/acsomega.1c00190

Priyadarshini E, Pradhan N (2017) Gold nanoparticles as efficient sensors in colorimetric detection of toxic metal ions: a review. Sensors Actuators B Chem 238:888–902. https://doi.org/10.1016/j.snb.2016.06.081

Rassaei M, Olthuis W, van den Berg A (2008) Multi-walled carbon nanotubes-modified glassy carbon electrode as a highly sensitive sensor for L-cysteine. Talanta 74(4):779–784

Regiart M et al (2020) An electrochemically synthesized nanoporous copper microsensor for highly sensitive and selective determination of glyphosate. ChemElectroChem 7(7):1558–1566. https://doi.org/10.1002/celc.202000064

Rossi A et al (2021) Silver nanoparticle-based sensor for the selective detection of nickel ions. Nanomaterials 11(7):1–16. https://doi.org/10.3390/nano11071733

Salcedo ARM, Sevilla FB (2013) Citrate-capped gold nanoparticles as colorimetric reagent for copper(II) ions. Philipp Sci Lett 6(1):90–96

Samuel VR, Rao KJ (2023) A rapid colorimetric dual sensor for the detection of mercury and lead ions in water using cysteine capped silver nanoparticles. Chem Phys Impact 6(January):100161. https://doi.org/10.1016/j.chphi.2023.100161

Seenivasan R, Chang WJ, Gunasekaran S (2015) Highly sensitive detection and removal of lead ions in water using cysteine-functionalized graphene oxide/polypyrrole nanocomposite film electrode. ACS Appl Mater Interfaces 7(29):15935–15943. https://doi.org/10.1021/acsami.5b03904

Sen S, Roy A, Sanyal A, Devi PS (2022) A nonenzymatic reduced graphene oxide-based nanosensor for parathion. Beilstein J Nanotechnol 13(1):730–744. https://doi.org/10.3762/bjnano.13.65

Shi L, Li Y, Rong X, Wang Y, Ding S (2017) Facile fabrication of a novel 3D graphene framework/Bi nanoparticle film for ultrasensitive electrochemical assays of heavy metal ions. Anal Chim Acta 968:21–29. https://doi.org/10.1016/j.aca.2017.03.013

Shim YB, Park SM, Park TS (2009) Electrochemical immunoassay for the simultaneous detection of multiple cancer biomarkers. BiosensS Bioelectron 24(7):2137–2143

Shuai H et al (2020) Fluorescent sensors for detection of mercury: from small molecules to nanoprobes. Dyes Pigm 187(November):109125. 2021. https://doi.org/10.1016/j.dyepig.2020.109125

Silva M, Cesarino I (2019) Evaluation of a nanocomposite based on reduced graphene oxide and gold nanoparticles as an electrochemical platform for detection of sulfamethazine. J Compos Sci 3(2):59. https://doi.org/10.3390/jcs3020059

Sinha R, Ganesana M, Andreescu S, Stanciu L (2010) AChE biosensor based on zinc oxide sol-gel for the detection of pesticides. Anal Chim Acta 661(2):195–199. https://doi.org/10.1016/j.aca.2009.12.020

Sreekanth SP et al (2021) Multi-walled carbon nanotube-based nanobiosensor for the detection of cadmium in water. Environ Res 197:111148. https://doi.org/10.1016/j.envres.2021.111148

Sung YM, Wu SP (2014) Colorimetric detection of Cd(II) ions based on di-(1H-pyrrol-2-yl) methanethione functionalized gold nanoparticles. Sensors Actuators B Chem 201(Ii):86–91. https://doi.org/10.1016/j.snb.2014.04.069

Supraja P, Tripathy S, Krishna Vanjari SR, Singh V, Singh SG (2019) Label free, electrochemical detection of atrazine using electrospun Mn2O3 nanofibers: towards ultrasensitive small molecule detection. Sensors Actuators B Chem 285:317–325. https://doi.org/10.1016/j.snb.2019.01.060

Supraja P, Singh V, Vanjari SRK, Singh SG (2020) Electrospun CNT embedded ZnO nanofiber based biosensor for electrochemical detection of Atrazine: a step closure to single molecule detection. Microsyst Nanoeng 6(1):1–10. https://doi.org/10.1038/s41378-019-0115-9

Talan A, Mishra A, Eremin SA, Narang J, Kumar A, Gandhi S (2018) Ultrasensitive electrochemical immuno-sensing platform based on gold nanoparticles triggering chlorpyrifos detection in fruits and vegetables. Biosens Bioelectron 105:14–21. https://doi.org/10.1016/j.bios.2018.01.013

Tiwari PM, Vig K, Dennis VA, Singh SR (2011) Functionalized gold nanoparticles and their biomedical applications. Nanomaterials 1(1):31–63. https://doi.org/10.3390/nano1010031

Tsai HY, Huang PY, Lee CH, Lu MG (2011) Graphene oxide-based aptasensor for highly sensitive and selective detection of thrombin. ACS Nano 5(12):9789–9795

Uzawa H, Ohga K, Shinozaki Y, Ohsawa I, Nagatsuka T, Seto Y, Nishida Y (2008) A novel sugar-probe biosensor for the deadly plant proteinous toxin, ricin. Biosens Bioelectron 24(4):923–927

Valera E, Ramón-Azcón J, Sanchez FJ, Marco MP, Rodríguez Á (2008) Conductimetric immunosensor for atrazine detection based on antibodies labelled with gold nanoparticles. Sensors Actuators B Chem 134(1):95–103. https://doi.org/10.1016/j.snb.2008.04.023

Vonnie JM, Ting BJ, Rovina K, Aqilah NMN, Yin KW, Huda N (2022) Natural and engineered nanomaterials for the identification of heavy metal ions—a review. Nanomaterials 12(15). Multidisciplinary Digital Publishing Institute (MDPI). https://doi.org/10.3390/nano12152665

Wanekaya AK (2011) Applications of nanoscale carbon-based materials in heavy metal sensing and detection. Analyst 136(21):4383–4391. https://doi.org/10.1039/c1an15574a

Wang C, Yu C (2013) Detection of chemical pollutants in water using gold nanoparticles as sensors: a review. Rev Anal Chem 32(1):1–14. https://doi.org/10.1515/revac-2012-0023

Wang L et al (2016) Facile, green and clean one-step synthesis of carbon dots from wool: application as a sensor for glyphosate detection based on the inner filter effect. Talanta 160:268–275. https://doi.org/10.1016/J.TALANTA.2016.07.020

Wang J et al (2017) Fabrication of fluorescent carbon dots-linked isophorone diisocyanate and β-cyclodextrin for detection of chromium ions. Spectrochim Acta A Mol Biomol Spectrosc 179:163–170. https://doi.org/10.1016/j.saa.2017.02.031

Wei H, Li B, Li J, Dong S, Wang E (2008) DNAzyme-based colorimetric sensing of lead (Pb2+) using unmodified gold nanoparticle probes. Nanotechnology 19(9). https://doi.org/10.1088/0957-4484/19/9/095501

Wen SH, Zhong XL, Di Wu Y, Liang RP, Zhang L, Qiu JD (2019) Colorimetric assay conversion to highly sensitive electrochemical assay for bimodal detection of arsenate based on cobalt oxyhydroxide nanozyme via arsenate absorption. Anal Chem 91(10):6487–6497. https://doi.org/10.1021/acs.analchem.8b05121

Wu Y, Liu L, Zhan S, Wang F, Zhou P (2012) Ultrasensitive aptamer biosensor for arsenic(iii) detection in aqueous solution based on surfactant-induced aggregation of gold nanoparticles. Analyst 137(18):4171–4178. https://doi.org/10.1039/c2an35711a

Xiao Y, Yu X (2010) Folding-upon-binding and signal-on electrochemical DNA sensor with high affinity and specificity. Anal Chem 82(10):4056–4062

Xu D, Chen H, Lin Q, Li Z, Yang T, Yuan Z (2017) Selective and sensitive colorimetric determination of cobalt ions using Ag-Au bimetallic nanoparticles. RSC Adv 7(27):16295–16301. https://doi.org/10.1039/c7ra00900c

Yadav KK, Singh JK, Gupta N, Kumar V (2017a) A review of nanobioremediation technologies for environmental cleanup: a novel biological approach. J Mater Environ Sci 8(2):740–757

Yadav KK, Singh JK, Gupta N, Kumar V (2017b) A review of nanobioremediation technologies for environmental cleanup: a novel biological approach. J Mater Environ Sci 8(2):740–757. [Online]. Available: http://www.jmaterenvironsci.com/

Yan X, Li H, Su X (2018) Review of optical sensors for pesticides. Trends Anal Chem 103. Elsevier:1–20. https://doi.org/10.1016/j.trac.2018.03.004

Yang Y, Wang Z, Yang M, Guo M, Wu Z, Shen G (2006) Inhibitive determination of mercury ion using a renewable urea biosensor based on self-assembled gold nanoparticles. Sensors Actuators B Chem 114(1):1–8. https://doi.org/10.1016/j.snb.2005.04.005

Yola ML, Atar N (2017) Electrochemical detection of atrazine by platinum nanoparticles/carbon nitride nanotubes with molecularly imprinted polymer. Ind Eng Chem Res 56(27):7631–7639. https://doi.org/10.1021/acs.iecr.7b01379

You J, Hu H, Zhou J, Zhang L, Zhang Y, Kondo T (2013) Novel cellulose polyampholyte-gold nanoparticle-based colorimetric competition assay for the detection of cysteine and mercury(II). Langmuir 29(16):5085–5092. https://doi.org/10.1021/la3050913

Yu J, Song N, Zhang YK, Zhong SX, Wang AJ, Chen J (2015) Green preparation of carbon dots by Jinhua bergamot for sensitive and selective fluorescent detection of Hg2+ and Fe3+. Sensors Actuators B Chem 214:29–35. https://doi.org/10.1016/j.snb.2015.03.006

Zhang J, Yi C (2009) A carbon nanotube-based electrochemical immunosensor for detection of C-reactive protein. Sensors Actuators B-Chem 142(1):214–219

Zheng J, Zhang H, Qu J, Zhu Q, Chen X (2013) Visual detection of glyphosate in environmental water samples using cysteamine-stabilized gold nanoparticles as colorimetric probe. Anal Methods 5(4):917–924. https://doi.org/10.1039/C2AY26391B

Zhuang J et al (2008) A novel application of iron oxide nanoparticles for detection of hydrogen peroxide in acid rain. Mater Lett 62(24):3972–3974. https://doi.org/10.1016/j.matlet.2008.05.025

Zúñiga K, Rebollar G, Avelar M, Campos-Terán J, Torres E (2022a) Nanomaterial-based sensors for the detection of glyphosate. Water (Basel) 14(15). Multidisciplinary Digital Publishing Institute:2436. https://doi.org/10.3390/w14152436

Zúñiga K, Rebollar G, Avelar M, Campos-Terán J, Torres E (2022b) Nanomaterial-based sensors for the detection of glyphosate. Water 14(15):2436. https://doi.org/10.3390/w14152436

Chapter 3
Whole-Cell-Based Biosensors for Detection of Environmental Pollutants

Sarika Loni, Ajay Shinde, Sushant Gawali, Atharva Jagtap, Manju Tanwar, and Abhishek Guldhe

Abstract Release of toxic chemicals or pollutants like heavy metals, pesticides, aromatic hydrocarbons, and pharmaceutical products contributes heavily to environmental pollution. Hence, the rapid detection of these pollutants is necessary to reduce possible environmental pollution. Biosensors can be excellent and highly specific tools for detecting pollutants in the environment, which possess significant impacts on ecosystems, human health, and development. Different types of biosensors have been developed for the purposes of pollutant monitoring. This chapter focuses primarily on whole-cell-based biosensors for detecting environmental pollutants. Cell-based biosensors are developed by fusing reporter genes such as *lacZ*, *lux*, and *gfp*, with inducible regulatory genes that react to specific environmental pollutants, like heavy metals and aromatic hydrocarbons. A broad range of microbial biosensors has been developed by using novel recombinant DNA technology.

Keywords Environment · Pollutants · Microorganisms · Biosensors · Heavy metals

---

S. Loni
Centre for Proteomics, Amity Institute of Biotechnology, Amity University Maharashtra, Mumbai, Maharashtra, India

Organic Recycling System Ltd., Navi Mumbai, Maharashtra, India

A. Shinde · A. Jagtap · A. Guldhe (✉)
Amity Institute of Biotechnology, Amity University Maharashtra, Mumbai, Maharashtra, India
e-mail: asguldhe@mum.amity.edu

S. Gawali
Centre for Proteomics, Amity Institute of Biotechnology, Amity University Maharashtra, Mumbai, Maharashtra, India

Department of Chemical Engineering, State University of New York College of Environmental Science and Forestry, Syracuse, NY, USA

M. Tanwar
Organic Recycling System Ltd., Navi Mumbai, Maharashtra, India

A. Guldhe et al. (eds.), *Biomaterials for Environmental Applications: Detection and Remediation*, Environmental Challenges and Solutions,
https://doi.org/10.1007/978-3-032-12701-3_3

3.1 Introduction

The rapid growth in populations is causing a drastic rise in industrialization. The man-made chemicals and some industrial products and byproducts are becoming major concerns for surrounding ecosystems. These pollutants are heavily released into the environment which can cause harmful impacts or serious changes on the environmental health. These hazardous pollutants lead to environmental pollutions which can directly or indirectly affect the health problems in humans and animals (Hashemi Goradel et al. 2018). The farmers are using agrochemicals to increase crop production and to fulfill the food requirement of the growing population. To control plant diseases, certain pesticides, fungicides, and bactericides are applied directly to the plants and soil. These agrochemical compounds are directly resealed into the soil, building up toxic residues that can pollute the soil, interact with the food chain and ecology, and have severe impacts on all living things, including humans (Aynalem and Muleta 2021). Human activities related to industrial and agricultural products are the major sources for release of pollutants. To diminish the pollutants from the environment, it is necessary to detect them on initial levels. Scientists have developed sophisticated tools to detect these deleterious pollutants known as biosensors. Various types of biosensors are available which have the signal transducer to convert biological signal into measurable electrochemical energy.

Biosensor is a portable analytical instrument made up of physicochemical transducer and a biologically derived sensing element. The International Union of Pure and Applied Chemistry (IUPAC) refers to a biosensor as "a self-contained integrated device that uses a biochemical receptor which is maintained in direct contact with a transduction element to provide specific quantitative or semi-quantitative analytical information" (Palchetti and Mascini 2008). Biosensors can detect and quantify environmental pollutants by converting the biochemical signal into a measurable electrical signal. Hence, these are powerful tools which can detect and quantify the harmful chemical substances or the pollutants in the environment (Hashemi Goradel et al. 2018). Here in this chapter, we are focusing on whole-cell-based biosensors for detecting environmental pollutants. Cell-based biosensors are powerful and excellent analytical tools which are developed to detect pollutants from various fields such as environment, biomedical, and food industries. The whole-cell-based biosensors are one of the most prominent tools to detect the target pollutants in which the cells are combined with transducer to get rapid and accurate detection of pollutants from various fields. These biosensors are considered as the first environmental pollutant detecting system which provides sufficient information for routine analysis and screening the samples from the polluted area (Saini et al. 2019). In 1970, the first microbial cell-based biosensor was developed which is combined with organic compound to produce electroactive metabolites and the biological changes in reparation of cell activity is detected by transducer, in this way the first cell-based biosensor phenomenon was studied. Since then, numerous types of microbial cell-based biosensors have been developed for pollutant detection because of quick developments in the domains of synthetic biology and microbial physiology.

Whole-cell biosensor has components such as bio-receptor, transducer, and signal processor. Bio-receptor is a sensitive component derived from biological origin. The types of bio-receptors used for the detection of target pollutants have a large influence on the function of biosensor. Bio-receptor type can change the biosensing parameters like response time, specificity, and sensitivity. In whole-cell-based biosensor, microorganisms act or are used as bio-receptor. The transducer is known as an analyzing tool. The basic principle of the transducer is to convert the biochemical signal into electric signal and that can be further measured or detected. Based on this signal, pollutants can be detected and quantified from the environment. Various types of transducers are used, such as electrochemical, optical-based, and thermal-based. In whole-cell-based biosensor, the living cells are linked to external transducers in order to generate a signal that can be measured and read. The transduction mechanism selection is based on the type of cells used in biosensor for sensing pollutants. Electrochemical and optical-based transducers are used in microbial biosensors. Advanced methodologies, including light addressable potentiometric sensor (LAPS) and electrochemical cell impedance spectroscopy (ECIS) transduction methods are employed in higher eukaryotic cell biosensor. Transducers can convert the biological signal in electric form which can be further displayed by signal processor. The key elements of whole-cell-based biosensor in environment pollutant detection are illustrated in Fig. 3.1.

In order to generate a sensing microorganism, whole-cell-based biosensors are integrated with signal-generating protein molecules that recognize pollutants. Combining sensing microorganism with transducer can convert the identified biological signal into quantifiable output to detect the pollutants. In genetically modified microorganisms, the modified genes for sensor protein can control the synthesis of reporter protein in response to a targeted pollutant. Hence the entire microbial cell functions as a biological sensor, producing a signal that can be measured. Using the specific molecular recognition between proteins and an intended ligand, whole-cell-based biosensor enables the dose-dependent specific identification of the available pollutants. One of the main functions of whole-cell-based biosensor is that it

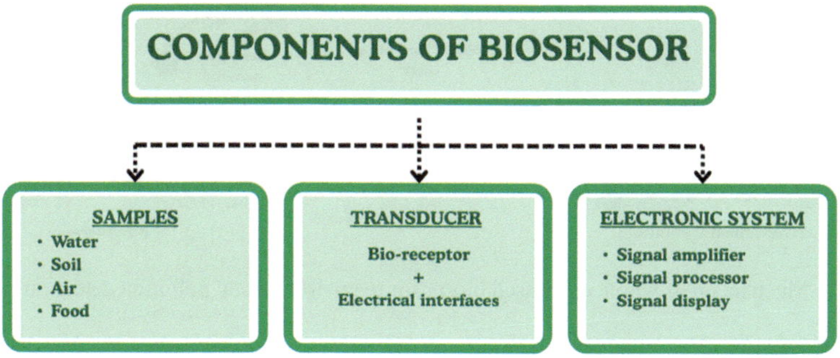

Fig. 3.1 Key elements of biosensor

provides the physiological related data regarding the bioavailability of a particular pollutant, since the target pollutant needs to first contact with the living bacteria in order to generate a response. Whole-cell-based biosensors are cost-effective to construct and store, these can stand with variation in temperature and pH (Struss et al. 2010).

The entire cell-based biosensor has a very simple mechanism in detection of environmental pollutants which is illustrated in Fig. 3.2. The living cells (genetically modified or non-modified) in the biosensors have been integrated with transducers by immobilization method. Cell-based biosensors use such immobilization techniques that cannot affect the cells' viability such as hydrogel, sol-gel entrapment, and encapsulation (Axelrod et al. 2016). After immobilization, the cells and transducer interact with each other. The immobilized cells combine with targeted pollutants to give cell-bound pollutants and that can produce the electronic response that can be further measured by the detector. However, the contaminants or analytes may occasionally be converted into products that release heat, gas (oxygen), electrons, or hydrogen ions. The product-associated changes are then converted into electrical signals by the transducer that can be further measured, amplified, and displayed by the electronic system (Dhewa 2015).

The whole-cell-based biosensor has shown good results in rapid detection of pollutants, but these sensors are not more sensitive to environmental changes. Hence the biosensors are modified or altered by using genetic engineering methods to get high sensitivity toward the pollutants and they can be used to identify a wide range of reactions that occur within the living cells. These modified biosensors can further provide additional information related to pharmacology, cell physiology, and toxicity of sample (Gui et al. 2017).

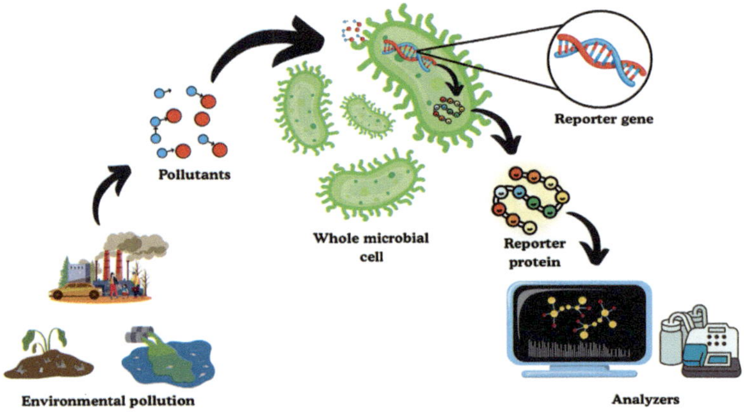

Fig. 3.2 Mechanism of whole cell-based biosensor in environmental pollution detection

3.2 Microorganisms as Whole-Cell Biosensors

Anthropogenic activities which include agricultural and industrial practices have caused the release of various toxic contaminants into the environment (Zeng et al. 2021). These contaminants are mutagenic and carcinogenic, hence increasing uses of chemicals in these two sectors have potentially increased the risk to human health and loss of biodiversity (Bilal and Iqbal 2019). The traditional methods for detection of environmental pollutants include chromatographic and spectroscopic techniques (Brányiková et al. 2020). However, these methods only detect the availability of the pollutants and are expensive with long detecting time periods (Liu et al. 2010). Therefore, a biological perspective is required for rapid and sensitive detection and monitoring the pollutants. Biosensors are devices which can detect signals within a cell or a tissue (Neethirajan et al. 2018). Molecular biosensors detect bioactive compounds like DNA, enzymes, antibodies, and antigens. Whole-cell-based biosensors (WCB) use living cells to provide molecular recognition. They make use of natural or genetically engineered cells like algae, bacteria, fungi, protozoa, or viruses, even tissue pieces as recognition elements. The working of a WCB includes sensing of an analyte species, amplification to an optical or electrical signal, and detection by the immobilized cell which provides the molecular recognition (Gui et al. 2017). The cells are modified to carry a specific reporter gene which is linked to a promotor. When the system is encountered with a specific pollutant, the promotor gene is activated causing the reporter gene to produce a measurable signal. The intensity of the signal correlates with the concentration of the pollutant. The binding of analyte to the regulatory protein and the choice of reporter gene determine the performance of a WCB (Raut et al. 2012). Using live cells in biosensors provides the ability to study and assess the bioavailability, mutagenicity, and cytotoxicity of environment toxins. Hence, WCB can be a vital approach for selective, sensitive, and effectual detection of environmental pollutants (Maduraiveeran and Jin 2017). Cells are susceptible to genetic modifications, allowing WCB to detect a variety of compounds and work under a wide range of conditions like different temperatures and pH (Tian et al. 2017). Due to the rapidity, simplicity, portability, cost efficiency, and better LOD (limit of detection), WCB are not only used for environmental pollutant detection but also equipped in other industries like food, pharmaceuticals, and drug testing (Bilal and Iqbal 2019; Wang et al. 2014). Over the years, WCB has gained attention considering its advantages and applications; however, studies to improve the sensitivity, specificity, and stability of WCBs are still ongoing.

3.2.1 Bacterial Cell-Based Biosensors

Microbial whole-cell-based biosensors (MCBs) are considered to be a modified form of enzymatic biosensors, where microbes are immobilized on the transducers using various immobilization techniques like physical entrapment, adsorption, covalent binding, or cross-linking (Gavrilaş et al. 2022). The cell viability and stability, mechanical resistance, handling, and long-term storage of the biosensor depend on the type of method used to immobilize the cells (Moraskie et al. 2021). MCBs are majorly used for detection of environmental pollutants like heavy metals, aromatic compounds, and other toxic chemicals. *Escherichia coli* is reported to be used in biosensors for detection of heavy metals like cadmium (Cd), mercury (Hg), and arsenic (As) (Frick et al. 2019; Kim et al. 2019). Genetically engineered *E. coli* biosensor is used for environmental monitoring to detect paraoxon and parathion, which are hazardous to aquatic biota and toxic to humans when they encounter the skin. The detection limit for paraoxon was reported to be 9 nM and for parathion to be 10 nM (Tang et al. 2014). A study reported that *Acinetobacter baylyi* incorporated biosensors have the capability to detect heavy metals like chromium (Cr) and lead (Pd) in soil, detection limit of 520 and 2072 mg/kg respectively (Song et al. 2014). Fluorescent biosensor with *Bacillus megaterium* VR1 is used to detect metal contamination of soil, i.e., Cd, Zn, and Cu with 1.42×10^{-4}, 2.42×10^{-4}, and 3.16×10^{-4} LOD, respectively (Rathnayake et al. 2021a). *Pseudomonas putida* is used for environmental monitoring for various contaminants which include Zn, Cd, Hg, and phenanthrene with LOD 5–55 μmol/L, 0.01 μM, <40 μg/kg, and <24 mg/kg respectively (Wei et al. 2014; Liu et al. 2012). Arsenic (As) contamination monitoring can be carried out by genetically engineered *Shewanella oneidensis* with LOD 40 μM (Webster et al. 2014). A recent study revealed that biosensors with *E. coli* strain DH10B could detect the presence of 3-phenoxybenzoic acid with a detection limit of 3 ng/mL (Riangrungroj et al. 2019). Similarly, another study focused on the use of *Pseudomonas aeruginosa* strain PAO1 in detection of sodium dodecyl sulphate (SDS) contamination. The results revealed the detection limit to be around 0.1 ppm (Dey et al. 2020). There are many such examples reported over time about the advantages of using MCBs for environmental pollutant detection. However, such biosensors are also associated with certain disadvantages, which include extended response time, genetic instability of genetically modified microbes, difficulty in maintenance of viability of the cells, societal and technical issues regarding the use of genetically engineered cells and influence of environmental factors on the functionality of the biosensors. Despite the challenges, research is ongoing to overcome the hurdles and improve the application of microbial biosensors (Bilal and Iqbal 2019).

3.2.2 *Algal Cell-Based Biosensors*

Cyanobacteria are considered to be the first group of organisms to implement oxygen-dependent photosynthesis. They are used in biosensors to detect heavy metals, herbicides, biocides, and pesticides. Detection of nickel (Ni) and cobalt (Co) contamination can be done by using *Synechocystis* sp. PCC 6803 biosensors which has been reported to have LOD value 0.2 μM for Ni and 0.3 μM for Co (Peca et al.

2008). Herbicide contamination of water caused by herbicides like atrazine, diuron, and simazine can be detected by using *Anabaena variabilis*, *Chlamydomonas reinhardtii*, and *Dictyosphaerium chlorelloides* biosensors, with detection limit of 0.07 μM, 0.2 μM, and 40.8 μg/L, respectively (Tsopela et al. 2016; Tucci et al. 2019). Green algae are sensitive to a range of pollutants, hence are usually used in biosensors for detecting toxic components in water, BOD, heavy metals, and biocides. An optic fiber-coupled *Chlorella vulgaris* biosensor has been equipped for detection of pesticides and chlorophenols. The same species were used in conductometric biosensors and were found to be sensitive to heavy metals like Co, Ni, Zn, and Pb. Similarly, an optic fiber-coupled *Dictyosphaerium chlorelloides* biosensor has been used for detection of copper ions in reservoirs and water supplies (Peña-Vázquez et al. 2010). A study reported that marine green algae *Ostreococcus tauri* as luminescent biosensor was sensitive for detection of anti-fouling biocides in the coastal waters (Sanchez-Ferandin et al. 2013). *Platimonas subcordiformis*-developed biosensor was tested for sensitivity against Pb, Cu, phenol, and nonylphenol. The results showed detection limits 1.90, 0.65, 2.85, and 5.22 μmol/L for the same. An optical biosensor with a strain of cyanobacteria *Synechoccus* sp. and two strains of *Mesotaenium* sp. were developed for detection of Cd, Cr, and Zn in the aquatic systems (Rathnayake et al. 2021b). With ongoing research, addressing and overcoming the challenges in developing whole-cell biosensors, algal biosensors can be a powerful tool for environmental monitoring.

3.2.3 Fungal Cell-Based Biosensors

Using fungal whole-cell biosensors has advantages over bacterial WCB as it is easy to culture, easy for gene manipulation, and is open to various transducer technologies. *Saccharomyces cerevisiae* is the most widely studied fungi for whole-cell biosensors. A WCB was developed using the organism, which was sensitive to Cu ions, with a detection limit of about 0.5 μM (Roda et al. 2011). In another study, *S. cerevisiae* was genetically modified for producing luminescence and was employed in a biosensor to detect Zn, Cu, and pesticides. The detection was based on the decrease in bioluminescence when the cell encountered and interacted with such chemicals. Similarly, biosensors with filamentous fungi *Mycena citricolor* and *Armillaria mellea* were developed for environmental monitoring by producing luminescent light (Barceló 2009). Co-immobilization using strains *Bacilllus licheniformis* and *Trichosporon cutaneum* was done to develop biosensor for measurement of biochemical oxygen demand (BOD). However, the biosensor showed exhibited low sensitivity.

3.3 Types of Whole-Cell-Based Biosensors

There are two types of whole-cell-based biosensors that are classified into nonspecific whole-cell-based biosensor and specific whole-cell-based biosensor on the basis of their biological ability and sensing element (Kaur et al. 2015).

3.3.1 Nonspecific Whole-Cell-Based Biosensors

Nonspecific whole-cell-based biosensor provides information regarding toxicity and bioavailability of pollutants. Stress-responsive genetic regulation, including heat shock and damages to DNA responses, are the primary building blocks for the construction of nonspecific whole-cell biosensors. In case of heat shock stress, heat shock stress-related genes activate when proteins are damaged, and these genes can subsequently be used as a sensing component in non-specific cell-based biosensors for the detection of pollutants or toxins that cause protein degradation or damage (Eltzov et al. 2009). In case of DNA damage, nonspecific sensing elements could involve SOS response genes whose expression is amplified by DNA damage, and this change in biological response of cell activity can be further used to detect the pollutants (Huang et al. 2023).

3.3.2 Specific Whole-Cell-Based Biosensors

Certain whole-cell-based biosensor microbes use metabolic or detoxifying genes as sensing elements to identify contaminants (Li et al. 2021; Voon et al. 2022). Environmental pollutants like pesticides, heavy metals, alkanes, and organic pollutants are detected by specific regulatory metabolic genes of microorganisms. Operons like *nah*, *alkBAC*, and the transcriptional regulatory gene *DntR* were used in bacterial cell-based biosensor to detect soil and water pollutants like pesticides, linear alkanes, and naphthalene by nitro toluene (Smirnova et al. 2004). The bacterial cells contain certain regulatory genes that can metabolize the organic pollutants like toluene and xylene. TOL plasmid contains *xylR* and *xylS* genes which are specifically constructed for the detection of benzene-related compounds (Li et al. 2008). The specific whole-cell-based biosensor uses resistance genes and these genes are placed in plasmid of microorganisms that can serve as a sensing element for the detection of specific heavy metals in soil and water. Microorganisms use some metals as a nutrient, the greater number of toxic metals inside cells can activate the resistance gene and the function of resistance genes such as redox transformation, active transport, and intracellular or extracellular precipitation is maintained by intracellular metal level (Bontidean et al. 2004).

Nonspecific whole-cell-based biosensors exhibit a medium response time with poor selectivity but moderate sensitivity and good bioavailability. In contrast, specific whole-cell-based biosensors also have a medium response time but offer excellent selectivity while maintaining moderate sensitivity and bioavailability. The key difference between the two lies in their selectivity, with specific whole-cell-based biosensors providing a higher degree of target recognition, making them more suitable for precise biological detections. However, both types share similar response times and bioavailability, ensuring their applicability in various biosensing applications (Huang et al. 2023).

3.3.3 Constitutive and Inducible Whole-Cell-Based Biosensor

Based on the mode of expression of reporter protein, whole-cell-based biosensors are classified into two types: inducible and constitutive (Gu et al. 2004). The early microbial whole-cell-based biosensors were constitutive expression systems, which involve the reporter's constitutive expression at high basal levels. The detection of pollutants or toxins is directly correlated with signal strength generated by whole-cell-based biosensor, but this signal decreases if cells die or show lower metabolic activity. Since constitutive microbial whole-cell-based biosensors can react to any chemical that damages cells, their applicability is limited because they are unable to pinpoint the precise kind of toxin that is present in samples. On the other hand, constitutive MWCBs have been effectively used in environmental monitoring for pollutant identification (Moraskie et al. 2021). Regulatory protein elements and DNA sequence elements interact to control the expression of genes within the microbial genome. Regulatory proteins usually bind to promoter regions in order to control the activity of RNA polymerase. By combining the right promoter elements and regulatory proteins inducible whole-cell-based biosensor can be created which controls the expression of a reporter gene in the presence of a specific pollutant.

Eventually, the need for a target-specific whole-cell-based biosensor and our rapidly expanding understanding of the principles behind microbial gene control prompted the development of inducible microbial cell-based biosensor. To develop inducible microbial cell-based biosensors, regulatory elements from metal and metalloid resistant bacteria are used. In inducible expression system, a protein-analyte complex formed by the basal expression of a regulatory protein that binds specifically to a target analyte triggers the reporter's expression through an inducible promoter. The reporter expresses itself at low basal levels when the target analyte is not present, but it exhibits dose-dependent expression when target analyte is present.

3.4 Development of Whole-Cell-Based Biosensors to Identify Pollutants

The creation of entire microbial cell biosensor uses living cells which are immobilized on transducer. Transducer acts as a detecting tool which identifies the environmental pollutants. When the immobilized cells directly contact pollutants, they will produce biological response which is further converted into electrochemical signal by transducer and detected. The microbial cell immobilized with transducer has a natural reporter gene which produces biological responses against specific pollutants. To detect targeted pollutants, genetically engineered genes will be inserted into microbial cells by using recombinant technology. Transducer is the most crucial component of biosensor; without transducer, biosensors are unable to detect the pollutants. Development of whole-cell-based biosensor includes the selection of cell type, selection of reporter gene, immobilization of cells with transducer, etc. Various sensing techniques like optical and electrochemical are used to identify the pollutants.

3.4.1 Selection of Microbial Cells

Various cells can be used for the fabrication of entire cell-based biosensor, and the selection of host cell is one of the important steps. The kind of cell employed for detecting the pollutants affects a biosensor's responsive time, sensitivity, and specificity. Numerous studies have shown the application of several host cell-based biosensor types for pollutant detection, depending on their specific characteristics. *E. coli*-based host cells are used to detect high concentration pollutants while the *Alcanivorax borkumensis* SK1 cells are used to detect the low concentration pollutants such as octane (Sevilla et al. 2015). Various *E. coli* strains were employed for the construction of whole-cell-based biosensor to detect specific pollutants like nickel from water samples. *Deinococcus deserti* was used for the fabrication of whole-cell-based biosensor which is employed to detect heavy metals like arsenite (Brutesco et al. 2017). The selections of host cells for the development of whole-cell-based biosensor to detect environmental pollutants have an impact on biosensor activity.

3.4.2 Reporter Gene

The selection of suitable reporter gene is a very crucial step while constructing entire cell-based biosensor because the activity of this biosensor to detect environment pollutants depends on reporter gene. The basic principle of reporter gene is that it interacts directly with certain pollutants and generates a signal which shows the presence of specific pollutants (Roggo and van der Meer 2017). As a result, reporter genes are genetically engineered by using recombinant technology for microbial cells that are commonly employed to demonstrate the synthesis of specific biomolecules which can affect gene expression (Park et al. 2013).

The majority of the whole-cell-based biosensors reporter genes utilize the enzymes as a final indicator to detect the pollutants from environment by colorimetric assay (Truffer et al. 2014; Prindle et al. 2012). β-galactosidase (β-gal), which is expressed by the *lacZ* gene, is a frequently utilized enzyme for this purpose. It allows for the quick and easy detection of targeted pollutants from environment by using either colorimetric or fluorescence method. The well-known enzyme luciferase, which stimulates the light that release response and functions as the light reporter gene, is encoded by the bioluminescence gene *lux*, which was replicated in *Vibrio fischeri* and *Photorhabdus luminescens*. These genes are employed in the development of whole-cell-based biosensor to detect the pollutants from environment (Daniel et al. 2008). When oxygen molecules are present, bacterial luciferase has been shown to be helpful in oxidizing long-chain fatty aldehydes, producing a blue-green light (Liu et al. 2010). Hence, this enzyme was widely used to construct the entire cell-based biosensor because of its elevated sensitivity and easy to handle.

Since fluorescent proteins have self-luminescence, they do not need any extra substance to emit light. This gene is used as reporters in microbial whole-cell-based biosensors to detect the pollutants from environment. *Gfp* gene encodes for green fluorescent protein (GFP) which is used to detect environment pollutants by producing light which is easily detected by using modern potentiometer method without harming host cells. Apart from its function as a reporter gene in sensing element, it also acts as pH indicator and fusion tag in the investigation of environment pollutants (Lei et al. 2006). The reporter gene *roGFP2* was highly expressed in *E. coli* which was immobilized on the κ-carrageenan matrix to develop a highly stable and more sensitive biosensor that could rapidly detect substances that cause oxidative stress (Ooi et al. 2015). Furthermore, the maximal detection activity of GFP is lowered as it takes longer to emit a steady fluorescence (Yagi 2007). GFP-based whole-cell biosensors are therefore usually inappropriate for the quick detection of pollutants.

Some researchers developed genetically engineered unique type reporter gene *crtA*, which synthesizes the carotenoid (Fujimoto et al. 2006). By using this unique type of reporter gene, whole-cell-based biosensors were constructed when applied to sample in the presence of pollutants, the color of medial was changed automatically without adding an extra compound. Hence, this reporter gene is considered a good choice for whole-cell-based biosensor construction for rapid detection of environment pollutants. Table 3.1 lists various reporter genes used for developing whole-cell-based biosensor.

Table 3.1 Types of reporter genes employed for developing whole-cell-based biosensor

Reporter gene	Detection method	Target pollutants	Advantages	Disadvantages	References
lux	Bioluminescence technique	Tetracycline, naphthalene	Easy to measure and gives quick response	It has a requirement of oxygen molecules	Hakkila et al. (2002)
luc	Bioluminescence	Heavy metals	More sensitive, thermally more stable and gives quick response	It shows low permeability and has requirements of oxygen and ATP molecules	Gutiérrez et al. (2015)
gfp	Fluorescence	Nalidixic acid	No need of extra compound, highly stable	Less sensitive	Sagi et al. (2003)
lacZ	Bioluminescence, fluorescence, colorimeter	Organic compounds	Highly stable	Require substrate	Mascher et al. (2004)
crtA	Colorimeter	Organophosphorus compounds	Identifiable by naked eyes	Requires substrate	Chong and Ching (2016)

3.4.3 Immobilization of Microorganisms or Whole Cells

The localization or physical confinement of entire cells to a specific surface region while preserving the necessary biological activity is referred to as "immobilization of whole cells." Immobilization gives the microbe a special degree of resistance to external influences and shields the cells from vibration forces such as pH, organic substances, temperature, salts, etc. Immature cells can be kept alive, active, and functional for a very long period (Patel et al. 2020).

The effective functioning of the entire whole-cell-based biosensor depends on the cells immobilization. Immobilization of whole cell increases the cells stability and helps to keep cellular activity for longer time (Zhu 2007). However, the choice of immobilization method is based on the kind of microbial cell and its properties. It's necessary to consider some basic points while selecting the immobilization method such as immobilized microbial cells must be stable, effective, it should not interrupt the cellular activity, the carrier material used to immobilize cells should be biocompatible in nature and it's easy to handle.

Different kinds of immobilization methods were used to construct the whole-cell-based biosensor to detect the environment pollutants such as adsorption, covalent binding, encapsulating, cross linking and entrapment. There are two types of immobilizations active and passive immobilization. Active immobilization uses carrier material like chemical substances, gel entrapments, and flocculating agents.

Most microbes are naturally able to adhere to surfaces, develop there, and become passively immobile on various kinds of carriers (Xu and Ying 2011). Researchers developed various immobilization techniques while developing whole-cell-based biosensor by using *Aliivibrio fischeri*. It was immobilized on cellulose nitrate membrane and integrated with microencapsulation to identify the heavy metals (Futra et al. 2014). Immobilization of *E. coli* cells was used with various alginate-modified cellulose membranes and k carrageenan matrix (Futra et al. 2015; Ooi et al. 2015). *Moraxella* sp. cells were immobilized by sol-gel entrapment method which showed the increase in cell stability (Yu et al. 2005). Table 3.2 lists some immobilization matrices used for immobilizing cells for construction of whole-cell-based biosensor to detect environment pollution.

3.4.4 Transducer

The design's central component is made of genetically modified microorganisms, which are followed by an electrical sensing system (Pham et al. 2022). There are two transducers that are used in whole-cell biosensors. The primary transducer is a synthetic fluorescent protein, which is an essential component of the cell that responds to signals from the environment. The intensity of the primary transducer will be measured and recorded by the secondary transducer, which is an electrochemical or optical detecting device (Flemming et al. 2006). A detectable signal is produced by the transducer upon interpretation of the biological reaction produced by the cells after interacting with the pollutants. It enables the correlation of the signal and pollutants concentration. Optical techniques replaced electrochemical approaches with the development of genetic engineering. Due the genetic engineering, it is possible to construct an optical whole-cell-based biosensor employed to detect environment pollution. The reporter gene is usually coupled with the

Table 3.2 Immobilization of microbial cells on various matrices in whole-cell-based biosensor

Immobilization carrier material	Microbial cells immobilized	Target pollutants	Detection method	References
Agarose	*E. coli*	Cadmium	Optical method	Charrier et al. (2010)
Agarose	*E. coli*	Copper, mercury, and cadmium	Colorimetric method	Kim et al. (2021)
Agarose	*E. coli*	Sodium benzoate	Spectrofluorimeter method	Patel et al. (2020)
Agar	*E. coli*	Arsenite	Optical method	Elcin and Öktem (2020)
Polyurethane	*E. coli*	Mercury heavy metal	Optical method	Biran et al. (2003)

inducible gene. An inducible gene becomes active or inactive in response to a specific pollutant.

The optical intensity linked to the concentration of the target molecule serves as the foundation for the fluorescent technique. Longer fluorescence is emitted by molecules that are excited by shorter wavelengths. Fluorescent whole-cell-based biosensor is developed by applying these fundamental molecular techniques (Ibraheem and Campbell 2010). The fluorescent method can be used in both in vivo and in vitro types. Since microorganisms can generate fluorescence, like green fluorescent protein, that's why in vivo type does not require the addition of an external substance. In the in vitro type, a change in the metabolic activity of the microbe results in changes to the environment around it, because of this electron stimulation occurs in external elements, and emits fluorescent light (Su et al. 2011). In this way, the fluorescent method emits the green, fluorescent light in the presence of particular pollutants and detects its concentration. Various types of environmental pollutants are detected by using fluorescent transducer technique.

Microbial cells contain luminescence after interacting with pollutants. This luminescence may vary on the basis of the type of pollutants. This approach is used to develop bioluminescence whole-cell-based biosensor. Bioluminescent-based technique mostly uses the *lux* reporter gene which codes for luciferase. Bacterial luciferase enzyme catalysis the oxidation of long-chain fatty aldehydes in the presence of oxygen molecules and produces a blue-green light (Liu et al. 2010). The intensity of bioluminescence is dependent on the activity of enzymes rather than protein quantity. Consequently, bioluminescence is a quicker and more accurate method of detection (Sagi et al. 2003). This technique does not need any external compounds because all cofactors and reactants were produced inside the cells. Various heavy metals in the environment were detected by using this technique. Table 3.3 lists the transducer used for developing whole-cell-based biosensor.

Table 3.3 Summary of transducer technique as a sensing element along with microbial cells and targeted pollutants

Microbial cells	Targeted pollutants	Transducer technique	Detection range	References
E. coli	Toluene	Fluorescent	25 μM	Li et al. (2008)
Pseudomonas putida	Naphthalene	Fluorescent	Not determined	Kohlmeier et al. (2008)
Caulobacter crescentus	Uranium	Fluorescent	0.5 μM	Hillson et al. (2007)
Bacillus licheniformis	BOD	Fluorescent	0.2 mM/L	Lin et al. (2006)
Vibrio fischeri	2,4-Dichlorophenol	Bioluminescent	40–200 mg/L	Stolper et al. (2008)
E. coli	Chloramphenicol	Bioluminescent	1 μg/mL	Shapiro and Baneyx (2007)
Acinetobacter sp.	Heavy metals	Bioluminescent	Not determined	Abd-El-Haleem et al. (2006)

3.5 Applications of Microbial Whole-Cell Biosensors for Environmental Pollutant Detection

Microbial whole-cell biosensors are used as an analytical tool for monitoring environmental pollutants. These tools are constructed by the fusion of reporter genes to the regulatory genes responsive to pollutants like aromatic hydrocarbons and heavy metals, organic and inorganic pollutants. Biosensors are developed by using recombinant DNA technology and are applied in various fields such as environmental monitoring and agriculture. By altering regulatory genes, biosensors can increase their sensitivity and target ranges. To enable high-throughput monitoring, biosensor cells have been mounted on semiconductors, optical fibers, and other platforms. Biosensors provide rapid and particular data on contaminated sites, measuring pollutants in complex matrices with minimal sample preparation. Microbial biosensors offer advantages over other sensing elements due to their ability to detect a wide range of chemicals and genetic modification adaptability. Future trends in biosensor development focus on miniaturization, automation, and ubiquitous devices for high-throughput monitoring of environmental variables. Whole-cell biosensors have various applications in detecting environmental pollutants. They provide real-time monitoring of pollutants in the environment, allowing quick detection and response to contamination events. These biosensors offer a cost-effective approach to detecting pollutants, making them suitable for extensive environmental monitoring. These biosensors can be engineered to be highly specific to a particular pollutant, enhancing the accuracy of detection. They can also detect a wide range of pollutants, including heavy metals, pesticides, and organic compounds, making them versatile tools for environmental monitoring. These biosensors can be used for on-site identification of pollutants, allowing rapid evaluation of environmental contamination levels.

The whole-cell-based biosensor is widely applied in environmental monitoring. The application of these biosensors is briefly explained below for the detection of toxins, heavy metals, nitrogen compounds, and pesticides.

3.5.1 Detection of Heavy Metals

Heavy metals are the major cause of environmental pollution hence it's necessary to construct a rapid and sensitive method to monitor these heavy metals in the environment. The heavy metals such as cadmium, chromium, lead, mercury, and arsenic have been criticized for being highly poisonous and for bioaccumulating within the food chain. Chemical analysis is not a reliable method for testing crucial parameters like toxicity, bioavailability, and genotoxicity but we can assess these parameters by using living cells only. A whole-cell biosensor combines the functions of a transducer and bio-receptor into a single reporter, utilizing the entire cell. By combining different reporter genes with metal-responsive regulatory genes (such as *merR* or

Table 3.4 Detection of heavy metals by using microbial cells biosensor

Microorganism used	Target heavy metals	References
E. coli	Cd, Hg, and As	Frick et al. (2019), Kim et al. (2019)
Acinetobacter baylyi	Cr and Pd	Song et al. (2014)
Bacillus megaterium	Cd, Zn, and Cu	Rathnayake et al. (2021a)
Pseudomonas putida	Zn, Cd, and Hg	Liu et al. (2012), Wei et al. (2014)
Shewanella oneidensis	As	Webster et al. (2014)
Synechocystis sp.	Ni and Co	Peca et al. (2008)

arsR), a number of microbial biosensors have been developed to detect heavy metals. The whole-cell-based cadmium biosensor can be constructed to detect the cadmium concentration in water, seawater, and soil samples. Under anaerobic conditions, the system demonstrated the ability to identify cadmium at concentrations ranging from 25 nM in water to 5 μM in contaminated soil samples. Various studies reported the use of whole-cell-based biosensors for detecting heavy metals. Various microorganisms were used to construct the whole-cell-based biosensor for the detection of heavy metals from water and soil samples which are listed in Table 3.4.

3.5.2 Detection of Pesticides and Herbicides

Whole-cell-based biosensors play a crucial role in the monitoring of pesticides. Whole-cell biosensors offer rapid, sensitive, and stable detection methods for pesticide. They can detect toxic organic compounds like polychlorinated biphenyls (PCBs) in the environment. Polychlorinated biphenyls (PCBs) are non-biodegradable chemical insecticides and herbicides used for pest and weed control; these compounds accumulate in the soil and are absorbed by crops, causing serious threats to human health due to their cancer-causing nature (Dhewa 2015).

Organophosphates are widely used as pesticides like insecticides. These chemicals are utilized to combat various insects that pose a threat to plant growth. However, they can also undergo bio magnification in vegetables and grains, resulting in adverse impacts on human growth, development, and health. The use of such chemicals can lead to environmental changes by damaging soil fertility, thereby harming beneficial microbes and insects essential for biodiversity. Biosensors are an effective tool for detecting the presence of pesticides in polluted water and soil.

To address this issue, biosensors utilizing piezoelectric transducers have been developed. Various transducers and whole microbial cells are used to construct the entire cell-based biosensor for identification of pesticides. An optic fiber combined with *Chlorella vulgaris* biosensor has been used for detection of pesticides, herbicide, and chlorophenols (Védrine et al. 2003). Cyanobacteria are employed for the construction of biosensor for the detection of herbicide and pesticide.

3.5.3 Detection of BOD

Biological oxygen demand (BOD) is known as one of the crucial parameters for the detection of amount of biodegradable organic pollutants in the water sample. Numerous cell-based BOD biosensors have been developed for the rapid identification of organic pollutants. Microbial fuel cells have been widely used in the construction of BOD biosensor because of its higher stability and fast response (Dávila et al. 2011). Recombinant *Escherichia coli* cells with the *Vibrio fisheri* gene *lux* AE-based biosensor was created by Nakamura and Karube to measure BOD (Dhewa 2015).

3.5.4 Agriculture Monitoring

The environment includes a complex blend of physical, chemical, and biological elements, intricately balanced and interconnected, where alterations in one aspect can profoundly impact others. Ensuring the efficient and continuous growth of crops is crucial for human well-being, emphasizing the need to closely monitor environmental and agricultural influences such as nutrients, trace metals, and organic pollutants. A diverse range of molecularly wired cell-based biosensors (MWCBs) has been created and utilized for observing environmental and agricultural systems, with many relying on fluorescent, bioluminescent, or electrochemical reporter proteins. The application biosensor in environmental and agricultural monitoring primarily focuses on crafting sensors tailored to specific target of interest (Moraskie et al. 2021). Various nitrogen compounds such as nitrates and other chemical fertilizers were widely applied for agriculture crops to enhance the yield. Overuse of these substances can harm human health, contaminate groundwater and surface water, which can subsequently be harmful to aquatic environments. Various microbial cells are used to develop biosensor to detect the concentration of these nitrogen compounds.

3.6 Conclusion

Environmental pollution arising from heavy metals, pesticides, aromatic hydrocarbons, and pharmaceutical residues poses substantial risks to ecosystems, human health, and socio-economic development. The imperative for advanced, rapid, and precise detection methods is paramount to mitigate these adverse effects and safeguard environmental sustainability. Biosensors have demonstrated immense potential as sophisticated analytical tools due to their high specificity, sensitivity, and adaptability for pollutant detection. Among the various biosensor platforms, cell-based biosensors have gained prominence for their efficacy in real-time environmental monitoring. By leveraging genetic engineering techniques, reporter genes such as *lacZ*, *lux*, and *gfp* have been fused with inducible regulatory elements to construct highly selective biosensors responsive to specific environmental contaminants. Innovations in recombinant DNA technology have further enhanced the versatility and effectiveness of microbial biosensors, enabling their application across a broad spectrum of pollutants. In conclusion, cell-based biosensors exemplify a cutting-edge approach to environmental diagnostics, offering unparalleled precision in the detection and monitoring of pollutants. Their continued development and refinement hold significant promise for advancing environmental stewardship, enabling timely mitigation measures, and fostering a sustainable future.

References

Abd-El-Haleem D, Zaki S, Abulhamd A, Elbery H, Abu-Elreesh G (2006) Acinetobacter bioreporter assessing heavy metals toxicity. J Basic Microbiol 46(5):339–347. https://doi.org/10.1002/jobm.200510122

Axelrod T, Eltzov E, Marks RS (2016) Bioluminescent bioreporter pad biosensor for monitoring water toxicity. Talanta 149:290–297. https://doi.org/10.1016/j.talanta.2015.11.067

Aynalem B, Muleta D (2021) Microbial biosensors as pesticide detector: an overview. J Sens 2021(1):5538857. https://doi.org/10.1155/2021/5538857

Barceló D, Hansen P-D (eds) (2009) Biosensors for environmental monitoring of aquatic systems, vol 5J. Springer, Berlin/Heidelberg. https://doi.org/10.1007/978-3-540-36253-1

Bilal M, Iqbal HMN (2019) Microbial-derived biosensors for monitoring environmental contaminants: recent advances and future outlook. Process Saf Environ Prot 124:8–17. https://doi.org/10.1016/j.psep.2019.01.032

Biran I, Rissin DM, Ron EZ, Walt DR (2003) Optical imaging fiber-based live bacterial cell array biosensor. Anal Biochem 315(1):106–113. https://doi.org/10.1016/S0003-2697(02)00700-5

Bontidean I, Mortari A, Leth S, Brown NL, Karlson U, Larsen MM, Vangronsveld J, Corbisier P, Csöregi E (2004) Biosensors for detection of mercury in contaminated soils. Environ Pollut 131(2):255–262. https://doi.org/10.1016/j.envpol.2004.02.019

Brányiková I, Lucáková S, Kuncová G, Trögl J, Synek V, Rohovec J, Navrátil T (2020) Estimation of Hg(II) in soil samples by bioluminescent bacterial bioreporter *E. coli* ARL1, and the effect of humic acids and metal ions on the biosensor performance. Sensors 20(11):3138. https://doi.org/10.3390/s20113138

Brutesco C, Prévéral S, Escoffier C, Descamps ECT, Prudent E, Cayron J, Dumas L, Ricquebourg M, Adryanczyk-Perrier G, de Groot A, Garcia D, Rodrigue A, Pignol D, Ginet N (2017) Bacterial host and reporter gene optimization for genetically encoded whole cell biosensors. Environ Sci Pollut Res 24(1):52–65. https://doi.org/10.1007/s11356-016-6952-2

Charrier T, Jos M, Affi M, Jouanneau S, Gezekel H, Thou G (2010) Bacterial bioluminescent biosensor characterisation for on-line monitoring of heavy metals pollutions in waste water treatment plant effluents. In: Biosensors. InTech. https://doi.org/10.5772/7210

Chong H, Ching CB (2016) Development of colorimetric-based whole-cell biosensor for organophosphorus compounds by engineering transcription regulator DmpR. ACS Synth Biol 5(11):1290–1298. https://doi.org/10.1021/acssynbio.6b00061

Daniel R, Almog R, Ron A, Belkin S, Diamand YS (2008) Modeling and measurement of a whole-cell bioluminescent biosensor based on a single photon avalanche diode. Biosens Bioelectron 24(4):882–887. https://doi.org/10.1016/j.bios.2008.07.026

Dávila D, Esquivel JP, Sabaté N, Mas J (2011) Silicon-based microfabricated microbial fuel cell toxicity sensor. Biosens Bioelectron 26(5):2426–2430. https://doi.org/10.1016/j.bios.2010.10.025

Dey S, Baba SA, Bhatt A, Dhyani R, Navani NK (2020) Transcription factor based whole-cell biosensor for specific and sensitive detection of sodium dodecyl sulfate. Biosens Bioelectron 170:112659. https://doi.org/10.1016/j.bios.2020.112659

Dhewa T (2015) Biosensors for environmental monitoring: an update. Octa J Environ Res 3(2):212–218. http://www.sciencebeingjournal.com

Elcin E, Öktem HA (2020) Immobilization of fluorescent bacterial bioreporter for arsenic detection. J Environ Health Sci Eng 18(1):137–148. https://doi.org/10.1007/s40201-020-00447-2

Eltzov E, Marks RS, Voost S, Wullings BA, Heringa MB (2009) Flow-through real time bacterial biosensor for toxic compounds in water. Sensors Actuators B Chem 142(1):11–18. https://doi.org/10.1016/j.snb.2009.08.024

Flemming JH, Baca HK, Werner-Washburne M, Brozik SM, López GP (2006) A packed microcolumn approach to a cell-based biosensor. Sensors Actuators B Chem 113(1):376–381. https://doi.org/10.1016/j.snb.2005.03.098

Frick H, Tardif S, Kandeler E, Holm PE, Brandt KK (2019) Assessment of biochar and zero-valent iron for in-situ remediation of chromated copper arsenate contaminated soil. Sci Total Environ 655:414–422. https://doi.org/10.1016/j.scitotenv.2018.11.193

Fujimoto H, Wakabayashi M, Yamashiro H, Maeda I, Isoda K, Kondoh M, Kawase M, Miyasaka H, Yagi K (2006) Whole-cell arsenite biosensor using photosynthetic bacterium *Rhodovulum sulfidophilum*. Appl Microbiol Biotechnol 73(2):332–338. https://doi.org/10.1007/s00253-006-0483-6

Futra D, Heng L, Surif S, Ahmad A, Ling T (2014) Microencapsulated *Aliivibrio fischeri* in alginate microspheres for monitoring heavy metal toxicity in environmental waters. Sensors 14(12):23248–23268. https://doi.org/10.3390/s141223248

Futra D, Heng L, Ahmad A, Surif S, Ling T (2015) An optical biosensor from green fluorescent *Escherichia coli* for the evaluation of single and combined heavy metal toxicities. Sensors 15(6):12668–12681. https://doi.org/10.3390/s150612668

Gavrilaş S, Ursachi CŞ, Perţa-Crişan S, Munteanu F-D (2022) Recent trends in biosensors for environmental quality monitoring. Sensors 22(4):1513. https://doi.org/10.3390/s22041513

Gu MB, Mitchell RJ, Kim BC (2004) Whole-cell-based biosensors for environmental biomonitoring and application. Adv Biochem Eng Biotechnol 87:269–305. https://doi.org/10.1007/b13533

Gui Q, Lawson T, Shan S, Yan L, Liu Y (2017) The application of whole cell-based biosensors for use in environmental analysis and in medical diagnostics. Sensors 17(7):1623. https://doi.org/10.3390/s17071623

Gutiérrez JC, Amaro F, Martín-González A (2015) Heavy metal whole-cell biosensors using eukaryotic microorganisms: an updated critical review. Front Microbiol 6:48. https://doi.org/10.3389/fmicb.2015.00048

Hakkila K, Maksimow M, Karp M, Virta M (2002) Reporter genes lucFF, luxCDABE, gfp, and dsred have different characteristics in whole-cell bacterial sensors. Anal Biochem 301(2):235–242. https://doi.org/10.1006/abio.2001.5517

Hashemi Goradel N, Mirzaei H, Sahebkar A, Poursadeghiyan M, Masoudifar A, Malekshahi ZV, Negahdari B (2018) Biosensors for the detection of environmental and urban pollutions. J Cell Biochem 119(1):207–212. https://doi.org/10.1002/jcb.26030

Hillson NJ, Hu P, Andersen GL, Shapiro L (2007) *Caulobacter crescentus* as a whole-cell uranium biosensor. Appl Environ Microbiol 73(23):7615–7621. https://doi.org/10.1128/AEM.01566-07

Huang C-W, Lin C, Nguyen MK, Hussain A, Bui X-T, Ngo HH (2023) A review of biosensor for environmental monitoring: principle, application, and corresponding achievement of sustainable development goals. Bioengineered 14(1):58–80. https://doi.org/10.1080/21655979.2022.2095089

Ibraheem A, Campbell RE (2010) Designs and applications of fluorescent protein-based biosensors. Curr Opin Chem Biol 14(1):30–36. https://doi.org/10.1016/j.cbpa.2009.09.033

Kaur H, Kumar R, Babu JN, Mittal S (2015) Advances in arsenic biosensor development—a comprehensive review. Biosens Bioelectron 63:533–545. https://doi.org/10.1016/j.bios.2014.08.003

Kim H, Lee W, Yoon Y (2019) Heavy metal(loid) biosensor based on split-enhanced green fluorescent protein: development and characterization. Appl Microbiol Biotechnol 103(15):6345–6352. https://doi.org/10.1007/s00253-019-09908-7
Kim Y, Choi H, Shin WH, Oh J-M, Koo S-M, Kim Y, Lee T, Yu BJ, Park C (2021) Development of colorimetric whole-cell biosensor for detection of heavy metals in environment for public health. Int J Environ Res Public Health 18(23):12721. https://doi.org/10.3390/ijerph182312721
Kohlmeier S, Mancuso M, Deepthike U, Tecon R, van der Meer JR, Harms H, Wells M (2008) Comparison of naphthalene bioavailability determined by whole-cell biosensing and availability determined by extraction with Tenax. Environ Pollut 156(3):803–808. https://doi.org/10.1016/j.envpol.2008.06.001
Lei Y, Chen W, Mulchandani A (2006) Microbial biosensors. Anal Chim Acta 568(1–2):200–210. https://doi.org/10.1016/j.aca.2005.11.065
Li Y-F, Li F-Y, Ho C-L, Liao VH-C (2008) Construction and comparison of fluorescence and bioluminescence bacterial biosensors for the detection of bioavailable toluene and related compounds. Environ Pollut 152(1):123–129. https://doi.org/10.1016/j.envpol.2007.05.002
Li P, Wang Y, Yuan X, Liu X, Liu C, Fu X, Sun D, Dang Y, Holmes DE (2021) Development of a whole-cell biosensor based on an ArsR-P regulatory circuit from Geobacter sulfurreducens. Environ Sci Ecotechnol 6:100092. https://doi.org/10.1016/j.ese.2021.100092
Lin L, Xiao L-L, Huang S, Zhao L, Cui J-S, Wang X-H, Chen X (2006) Novel BOD optical fiber biosensor based on co-immobilized microorganisms in ormosils matrix. Biosens Bioelectron 21(9):1703–1709. https://doi.org/10.1016/j.bios.2005.08.007
Liu X, Germaine KJ, Ryan D, Dowling DN (2010) Whole-cell fluorescent biosensors for bioavailability and biodegradation of polychlorinated biphenyls. Sensors 10(2):1377–1398. https://doi.org/10.3390/s100201377
Liu P, Huang Q, Chen W (2012) Construction and application of a zinc-specific biosensor for assessing the immobilization and bioavailability of zinc in different soils. Environ Pollut 164:66–72. https://doi.org/10.1016/j.envpol.2012.01.023
Maduraiveeran G, Jin W (2017) Nanomaterials based electrochemical sensor and biosensor platforms for environmental applications. Trends Environ Anal Chem 13:10–23. https://doi.org/10.1016/j.teac.2017.02.001
Mascher T, Zimmer SL, Smith T-A, Helmann JD (2004) Antibiotic-inducible promoter regulated by the cell envelope stress-sensing two-component system LiaRS of *Bacillus subtilis*. Antimicrob Agents Chemother 48(8):2888–2896. https://doi.org/10.1128/AAC.48.8.2888-2896.2004
Moraskie M, Roshid MHO, O'Connor G, Dikici E, Zingg J-M, Deo S, Daunert S (2021) Microbial whole-cell biosensors: current applications, challenges, and future perspectives. Biosens Bioelectron 191:113359. https://doi.org/10.1016/j.bios.2021.113359
Neethirajan S, Ragavan V, Weng X, Chand R (2018) Biosensors for sustainable food engineering: challenges and perspectives. Biosensors 8(1):23. https://doi.org/10.3390/bios8010023
Ooi L, Heng L, Mori I (2015) A high-throughput oxidative stress biosensor based on *Escherichia coli* roGFP2 cells immobilized in a k-carrageenan matrix. Sensors 15(2):2354–2368. https://doi.org/10.3390/s150202354
Palchetti I, Mascini M (2008) Nucleic acid biosensors for environmental pollution monitoring. Analyst 133(7):846. https://doi.org/10.1039/b802920m
Park M, Tsai S-L, Chen W (2013) Microbial biosensors: engineered microorganisms as the sensing machinery. Sensors 13(5):5777–5795. https://doi.org/10.3390/s130505777
Patel R, Mukherjee A, Zaveri P, Mehta K, Aswani H, Munshi NS (2020) Optimization of immobilization process and survival study of microbial sensing strains used for aromatic hydrocarbon detection in industrial wastewater. Water Environ J 34(S1):937–948. https://doi.org/10.1111/wej.12604
Peca L, Kós PB, Máté Z, Farsang A, Vass I (2008) Construction of bioluminescent cyanobacterial reporter strains for detection of nickel, cobalt and zinc. FEMS Microbiol Lett 289(2):258–264. https://doi.org/10.1111/j.1574-6968.2008.01393.x

Peña-Vázquez E, Pérez-Conde C, Costas E, Moreno-Bondi MC (2010) Development of a microalgal PAM test method for Cu(II) in waters: comparison of using spectrofluorometry. Ecotoxicology 19(6):1059–1065. https://doi.org/10.1007/s10646-010-0487-y

Pham HL, Ling H, Chang MW (2022) Design and fabrication of field-deployable microbial biosensing devices. Curr Opin Biotechnol 76:102731. https://doi.org/10.1016/j.copbio.2022.102731

Prindle A, Samayoa P, Razinkov I, Danino T, Tsimring LS, Hasty J (2012) A sensing array of radically coupled genetic 'biopixels'. Nature 481(7379):39–44. https://doi.org/10.1038/nature10722

Rathnayake IVN, Megharaj M, Naidu R (2021a) Green fluorescent protein based whole cell bacterial biosensor for the detection of bioavailable heavy metals in soil environment. Environ Technol Innov 23:101785. https://doi.org/10.1016/j.eti.2021.101785

Rathnayake IVN, Munagamage T, Pathirathne A, Megharaj M (2021b) Whole cell microalgal-cyanobacterial array biosensor for monitoring Cd, Cr and Zn in aquatic systems. Water Sci Technol 84(7):1579–1593. https://doi.org/10.2166/wst.2021.339

Raut N, O'Connor G, Pasini P, Daunert S (2012) Engineered cells as biosensing systems in biomedical analysis. Anal Bioanal Chem 402(10):3147–3159. https://doi.org/10.1007/s00216-012-5756-6

Riangrungroj P, Bever CS, Hammock BD, Polizzi KM (2019) A label-free optical whole-cell *Escherichia coli* biosensor for the detection of pyrethroid insecticide exposure. Sci Rep 9(1):12466. https://doi.org/10.1038/s41598-019-48907-6

Roda A, Roda B, Cevenini L, Michelini E, Mezzanotte L, Reschiglian P, Hakkila K, Virta M (2011) Analytical strategies for improving the robustness and reproducibility of bioluminescent microbial bioreporters. Anal Bioanal Chem 401(1):201–211. https://doi.org/10.1007/s00216-011-5091-3

Roggo C, van der Meer JR (2017) Miniaturized and integrated whole cell living bacterial sensors in field applicable autonomous devices. Curr Opin Biotechnol 45:24–33. https://doi.org/10.1016/j.copbio.2016.11.023

Sagi E, Hever N, Rosen R, Bartolome AJ, Rajan Premkumar J, Ulber R, Lev O, Scheper T, Belkin S (2003) Fluorescence and bioluminescence reporter functions in genetically modified bacterial sensor strains. Sensors Actuators B Chem 90(1–3):2–8. https://doi.org/10.1016/S0925-4005(03)00014-5

Saini R, Hegde K, Brar SK, Verma M (2019) Advances in whole cell-based biosensors in environmental monitoring. In: Tools, techniques and protocols for monitoring environmental contaminants. Elsevier, pp 263–284. https://doi.org/10.1016/B978-0-12-814679-8.00013-3

Sanchez-Ferandin S, Leroy F, Bouget F-Y, Joux F (2013) A new, sensitive marine microalgal recombinant biosensor using luminescence monitoring for toxicity testing of antifouling biocides. Appl Environ Microbiol 79(2):631–638. https://doi.org/10.1128/AEM.02688-12

Sevilla E, Yuste L, Rojo F (2015) Marine hydrocarbonoclastic bacteria as whole-cell biosensors for n-alkanes. Microb Biotechnol 8(4):693–706. https://doi.org/10.1111/1751-7915.12286

Shapiro E, Baneyx F (2007) Stress-activated bioluminescent *Escherichia coli* sensors for antimicrobial agents detection. J Biotechnol 132(4):487–493. https://doi.org/10.1016/j.jbiotec.2007.08.021

Smirnova IA, Dian C, Leonard GA, McSweeney S, Birse D, Brzezinski P (2004) Development of a bacterial biosensor for nitrotoluenes: the crystal structure of the transcriptional regulator DntR. J Mol Biol 340(3):405–418. https://doi.org/10.1016/j.jmb.2004.04.071

Song Y, Jiang B, Tian S, Tang H, Liu Z, Li C, Jia J, Huang WE, Zhang X, Li G (2014) A whole-cell bioreporter approach for the genotoxicity assessment of bioavailability of toxic compounds in contaminated soil in China. Environ Pollut 195:178–184. https://doi.org/10.1016/j.envpol.2014.08.024

Stolper P, Fabel S, Weller MG, Knopp D, Niessner R (2008) Whole-cell luminescence-based flow-through biodetector for toxicity testing. Anal Bioanal Chem 390(4):1181–1187. https://doi.org/10.1007/s00216-007-1770-5

Struss A, Pasini P, Ensor CM, Raut N, Daunert S (2010) Paper strip whole cell biosensors: a portable test for the semiquantitative detection of bacterial quorum signaling molecules. Anal Chem 82(11):4457–4463. https://doi.org/10.1021/ac100231a

Su L, Jia W, Hou C, Lei Y (2011) Microbial biosensors: a review. Biosens Bioelectron 26(5):1788–1799. https://doi.org/10.1016/j.bios.2010.09.005

Tang X, Zhang T, Liang B, Han D, Zeng L, Zheng C, Li T, Wei M, Liu A (2014) Sensitive electrochemical microbial biosensor for p-nitrophenylorganophosphates based on electrode modified with cell surface-displayed organophosphorus hydrolase and ordered mesopore carbons. Biosens Bioelectron 60:137–142. https://doi.org/10.1016/j.bios.2014.04.001

Tian Y, Lu Y, Xu X, Wang C, Zhou T, Li X (2017) Construction and comparison of yeast whole-cell biosensors regulated by two RAD54 promoters capable of detecting genotoxic compounds. Toxicol Mech Methods 27(2):115–120. https://doi.org/10.1080/15376516.2016.1266540

Truffer F, Buffi N, Merulla D, Beggah S, van Lintel H, Renaud P, van der Meer JR, Geiser M (2014) Compact portable biosensor for arsenic detection in aqueous samples with *Escherichia coli* bioreporter cells. Rev Sci Instrum 85(1):015120. https://doi.org/10.1063/1.4863333

Tsopela A, Laborde A, Salvagnac L, Ventalon V, Bedel-Pereira E, Séguy I, Temple-Boyer P, Juneau P, Izquierdo R, Launay J (2016) Development of a lab-on-chip electrochemical biosensor for water quality analysis based on microalgal photosynthesis. Biosens Bioelectron 79:568–573. https://doi.org/10.1016/j.bios.2015.12.050

Tucci M, Grattieri M, Schievano A, Cristiani P, Minteer SD (2019) Microbial amperometric biosensor for online herbicide detection: photocurrent inhibition of *Anabaena variabilis*. Electrochim Acta 302:102–108. https://doi.org/10.1016/j.electacta.2019.02.007

Védrine C, Leclerc J-C, Durrieu C, Tran-Minh C (2003) Optical whole-cell biosensor using *Chlorella vulgaris* designed for monitoring herbicides. Biosens Bioelectron 18(4):457–463. https://doi.org/10.1016/S0956-5663(02)00157-4

Voon CH, Yusop NM, Khor SM (2022) The state-of-the-art in bioluminescent whole-cell biosensor technology for detecting various organic compounds in oil and grease content in wastewater: from the lab to the field. Talanta 241:123271. https://doi.org/10.1016/j.talanta.2022.123271

Wang X, Lu X, Chen J (2014) Development of biosensor technologies for analysis of environmental contaminants. Trends Environ Anal Chem 2:25–32. https://doi.org/10.1016/j.teac.2014.04.001

Webster DP, TerAvest MA, Doud DFR, Chakravorty A, Holmes EC, Radens CM, Sureka S, Gralnick JA, Angenent LT (2014) An arsenic-specific biosensor with genetically engineered *Shewanella oneidensis* in a bioelectrochemical system. Biosens Bioelectron 62:320–324. https://doi.org/10.1016/j.bios.2014.07.003

Wei H, Ze-Ling S, Le-Le C, Wen-hui Z, Chuan-Chao D (2014) Specific detection of bioavailable phenanthrene and mercury by bacterium reporters in the red soil. Int J Environ Sci Technol 11(3):685–694. https://doi.org/10.1007/s13762-013-0216-1

Xu X, Ying Y (2011) Microbial biosensors for environmental monitoring and food analysis. Food Rev Int 27(3):300–329. https://doi.org/10.1080/87559129.2011.563393

Yagi K (2007) Applications of whole-cell bacterial sensors in biotechnology and environmental science. Appl Microbiol Biotechnol 73(6):1251–1258. https://doi.org/10.1007/s00253-006-0718-6

Yu D, Volponi J, Chhabra S, Brinker CJ, Mulchandani A, Singh AK (2005) Aqueous sol–gel encapsulation of genetically engineered *Moraxella* spp. cells for the detection of organophosphates. Biosens Bioelectron 20(7):1433–1437. https://doi.org/10.1016/j.bios.2004.04.022

Zeng N, Wu Y, Chen W, Huang Q, Cai P (2021) Whole-cell microbial bioreporter for soil contaminants detection. Front Bioeng Biotechnol 9:622994. https://doi.org/10.3389/fbioe.2021.622994

Zhu Y (2007) Immobilized cell fermentation for production of chemicals and fuels. In: Bioprocessing for value-added products from renewable resources. Elsevier, pp 373–396. https://doi.org/10.1016/B978-044452114-9/50015-3

Chapter 4
Lipids: Biomarkers for Detection of Environmental Pollutants

Sachin Palekar, Sandhya Menon, Nandini Girish, and Shashank Ramakrishnan

Abstract One of the biggest challenges that the world faces today is pollution. The impact of pollution is currently seen around us in the form of damage to the biota. These damages are reflected in the biochemical profile of the organisms depending on the time and quantity of exposure to such pollutants. Studies have reported that both long-term and short-term exposure to pollutants change the internal biochemistry of the cells being exposed, with research being done on humans as well as other model systems like zebra fish. Interestingly, these studies have reported that the effect of pollution causes changes in the lipid profile of these organisms owing to the interaction of the lipids with the pollutants. Many of these studies have reported using analytical techniques like LC-MS, IR and Raman spectroscopy for detecting these changes in the lipid profile in the organisms. These identified molecules can serve as biomarkers for environmental pollution. The use of such biomarkers as indicators to pollution is especially beneficial, as it not only determines the presence of these pollutants but also detects the effect of these pollutants at the molecular, cellular and organism levels.

The present chapter focuses on understanding how lipids serve as biomarkers for detection of environmental pollutants and how we can efficiently harness these as tools towards quick and efficient indicators of detecting pollution levels.

Keywords Pollution · Environmental pollutants · Biomarkers · Indicator · Lipid profile

S. Palekar (✉) · S. Menon · N. Girish · S. Ramakrishnan
Department of Bioanalytical Sciences, S.P. Mandali's Ramnarain Ruia Autonomous College, Mumbai, India

A. Guldhe et al. (eds.), *Biomaterials for Environmental Applications: Detection and Remediation*, Environmental Challenges and Solutions,
https://doi.org/10.1007/978-3-032-12701-3_4

4.1 Introduction

Environmental pollution is a serious hazard that modern society faces today. These anthropogenic pollutants could enter land, air or water, creating hazards that could affect the biota, leading to short-term or long-term effects. The vehicular emissions, power plants, burning garbage, chemical companies and volcanic eruptions releasing sulphur dioxide (SO_2), carbon monoxide (CO), nitrogen oxides (NO_2), heavy metals, biological contaminants, ozone, tobacco smoke, etc. are all released into the air that is inhaled. Studies across the globe have indicated that these pollutants hamper the normal functioning of the body, causing diseases like cancer, cardiovascular, reproductive, prenatal central nervous system, and respiratory health issues. Volatile organic compounds (VOCs) have been largely responsible for accelerating various adverse health effects including allergies, cancers and other respiratory disorders by reacting with other pollutants in air to form secondary air pollutants (Shuai et al. 2018). Similarly, excessive use of fertilisers, extensive industrialisation and population expansion have led to water quality deterioration and water deficiency in many parts of the world (Wang et al. 2010). Pollutants like nitrates, phosphorus, pesticides, soil sediments, salts and pathogens contaminate the water bodies. Mateo-Sagasta et al. (2017) reported that 80% of municipal wastewater, solvents, toxic sludge and other wastes are released into water bodies, leading to the introduction of heavy metals thereby polluting these sources of drinking water.

As per the definition by the National Institute of Environmental Health Sciences, biological markers are nothing but an objective measurement of the changes occurring at a cellular level in the organism at a given point in time. Molecular markers have been described in scientific literature as molecules "whose carbon skeleton can be unambiguously linked to that of a known biological precursor compound" (Tyson et al. 2005). For a given molecule to qualify as a molecular marker, it must possess source specificity and conservative behaviour. Several molecules such as lignin, plant pigments like chlorophylls and carotenoids, lipids and carbohydrates have been successfully used as biomolecular markers for detection of organic matter sources in aquatic environments.

Biomarkers serve as prompt premature signals for the alterations in the condition of an organism. For example, early detection of heart disease risk is indicated by higher levels of cholesterol in blood. Most biomarkers represent the measured data obtained from laboratory tests of blood, urine or tissues. However, sometimes they are composed of simple physical measurements such as blood pressure. They play a crucial role in connecting biology, environmental exposures and diseases or health hazards in living systems.

Normal physiological conditions in aerobes are characterised by a natural balance between endogenous oxidants and the various antioxidant defences of both enzymatic and non-enzymatic nature (Halliwell 1999). During an imbalance event, such oxidants promote degradation of biological macromolecules, including DNA, proteins and lipids. Oxidative degradation contributes significantly to cancer, aging and other degenerative diseases. Scientists can also use these biomarkers to reveal unexplored biochemical pathways and for better understanding of fundamental biochemical processes, resulting in diverse health or environmental applications.

Although biomarkers have several associated advantages as diagnostic markers for environmental quality, the key limitation which remains to be addressed is the identification of diverse responses in different living organisms. Such a plethora of responses makes linking a particular exposure to a specific chemical molecule and a subsequent biological response difficult to achieve. This is therefore suggested to be overcome through the use of a multi-biomarker approach, where an entire range of species, including sentinel species, is studied for better insights into environmental concerns. Harmonisation of biomarker research with genomics can also yield a more holistic understanding of the impacts of environmental toxicants in diverse species.

4.2 Understanding Interaction Between Biotic and Abiotic Components

The interaction between the biotic and the abiotic components of the environment is extremely important to be elucidated well in order to understand how global phenomena and anthropogenic pollutants affect living systems, especially considering the interdependency of the components of the ecosystem on each other. The biotic and the abiotic components thus serve as feedback/control mechanisms for each other. When the responses are elicited towards a given environmental stressor or trigger, within a particular organism as well as in the ecosystem, studies at the molecular level become necessary to understand how the two components fare with respect to toxicity exposures and adaptation (Koelmel et al. 2020). Such studies yield valuable information with important applicability not only in better characterisation of such interactions between anthropogenic modifiers and the biosphere but also in human and environmental protection and care.

Among the various biomolecules that are studied are carbohydrates, proteins and lipids, for their potential as molecular biomarkers (Javed and Usmani 2015). Carbohydrates and proteins are very dynamic components and are always changing in response to internal and external triggers. In contrast, however, are lipid molecules, which have been vastly recognised for their promising role in environmental sciences (Koelmel et al. 2020). Lipids are more of storage molecules and therefore may serve as dynamic yet stable and conserved indicators of changes in the surrounding environment. Being high in abundance and ubiquity across biological species, lipid profiles exhibit dynamism in a way which allows for the conservation of important biological functions such as membrane fluidity when subject to abiotic stresses such as a change in the chemistry of the environment. Environmental lipidomics therefore forms a unique branch of study relying on the extraction and analysis of lipids from biotic components such as plants, animals and humans and their comprehensive study to understand the environment.

Studies involving plants or animal cells exposed to various stress environments and their comparison with controlled groups can be one of the mechanisms underlying such studies. For instance, when healthy and diseased fish, *Oreochromis mossambicus*, commonly called Mozambique tilapia, were analysed for their lipid profiles to understand whether they reflect environmentally induced pansteatitis (fat inflammation) severity, it was observed that the responses of the two groups were largely different (Koelmel et al. 2019). Thirteen different lipid classes known to play a role in inflammation, apoptosis and oxidative damage were observed to be upregulated in pansteatitis-laden adipose tissue. These lipids included ether lipids, sphingolipids, acylcarnitines and short-chained triglyceride oxidation products.

In addition to aiding research on feedback mechanisms operative at the biological level with respect to phenomena such as accelerated climate change, lipid signatures of historical significance can be used to extrapolate recorded responses due to historical chemical changes to those which may possibly be recorded in the future.

4.3 Lipids as Biomarkers

Lipids are water-insoluble organic macromolecules found within the cells of organisms (Derrien et al. 2017). They have a significant role to play at the cellular and organism levels. A primary role of lipids in cells is in the formation of the permeability barrier of cells and subcellular organelles in the form of a lipid bilayer. Lipids form an insulating barrier in cell membranes making up to 7% of cellular dry weight in microorganisms (Langworthy et al. 1983). They are vital components involved in significant functions in cells such as protein stabilisation and nutrient transport across the permeable cell membranes (Van Meer et al. 2008).

At the organism level, the lipid composition within a membrane can give rise to signalling hubs thereby shaping the membrane properties and allowing cells to finely adjust tension, shape, and rigidity, which will ultimately affect the functioning of the cell and/or organisms. These lipids could also be sources of storage of carbon for the organism and, in certain cases, may also act as molecules that establish the identity of the cell and/or organisms. The identity and quantity of these lipids in the cell have been established for decades now, and it is on this basis that many cellular functions of the cell are determined (Levental et al. 2020).

Biotic lipid markers of plant, animal as well as microbial origin can serve as an effective means to achieve improved and timely diagnosis and control of parameters indicating environmental pollution and deterioration. They can be used to generate the quality profile of soil in agriculture to boost productivity as well as sustainability. Several such lipid compounds are capable of being utilised as biomarkers owing to their diagnostic identification of sources, metabolisms and environmental conditions (Finkel et al. 2023).

Lipid biomarkers, in addition, have since been recognised for their tremendous applicability and potential as molecular markers for marine, estuarine systems and inland environments. This is ascribable to the fact that such lipid biomarkers exhibit extreme structural diversity, ubiquitous distribution and diagenetic and chemical stability (Derrien et al. 2017). The most widely studied types of lipid biomarkers include n-alkanes, fatty acids and sterols/stanols as illustrated in Figs. 4.1 and 4.2.

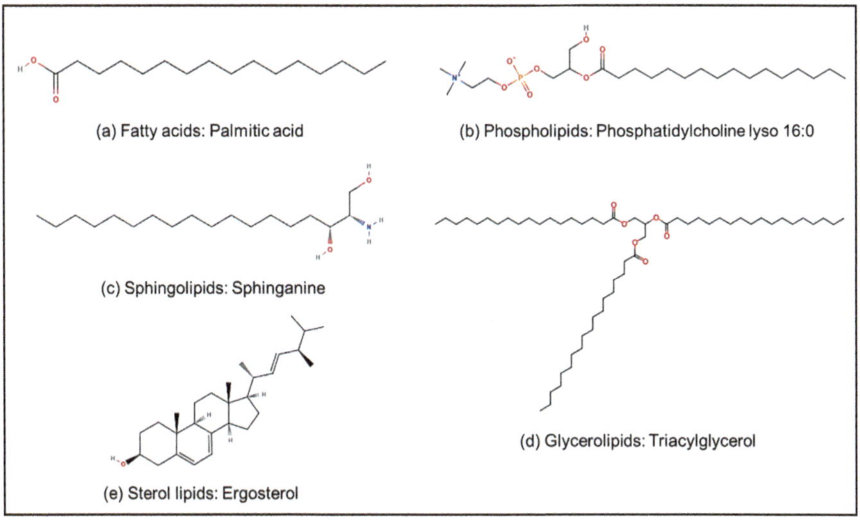

Fig. 4.1 Representative structure of (**a**) fatty acids, (**b**) phospholipids, (**c**) sphingolipids, (**d**) glycerolipids, and (**e**) sterol lipids

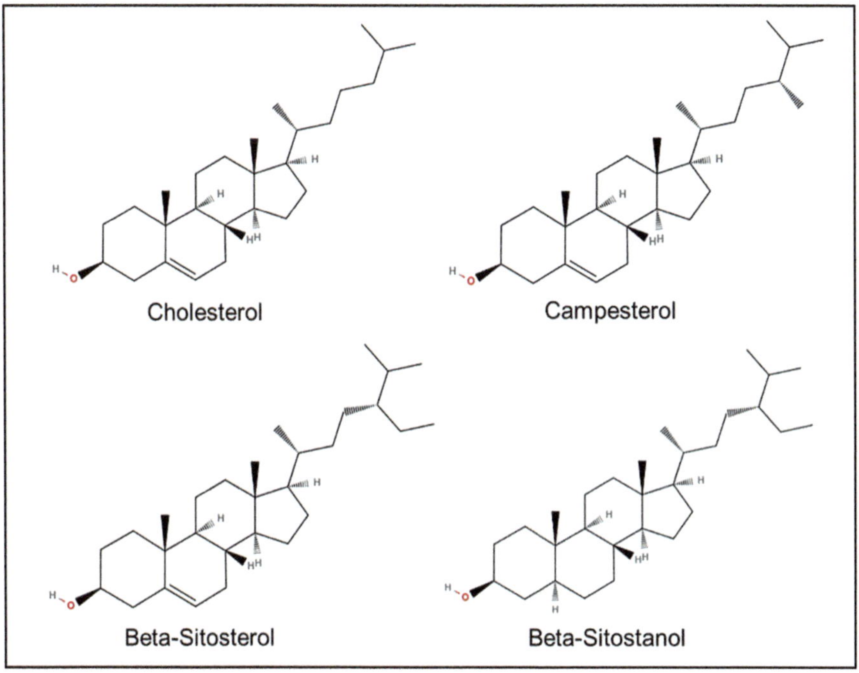

Fig. 4.2 Structures of cholesterol and some plant sterols

4.4 Lipid Biomarkers from Various Biotic Sources

Microorganisms have increasingly become recognised for their biomonitoring potential in the past few decades. Several essential biomolecules in microbial cell membranes, such as phospholipid fatty acids (PLFAs), chitin and ergosterol, are useful as biomarkers for accurate estimations of diverse parameters associated with the microbial community.

Phospholipids in microbial cell membranes undergo rapid degradation upon cell death and can provide an accurate means of analysis of the viable biomass of a given microbial community (White et al. 1979). Most significantly, several PLFAs serve as signature biomarkers of different microbial groups. These are terminally branched saturated fatty acids (for Gram-positive bacteria), monounsaturated fatty acids (for Gram-negative bacteria), mid-branched saturated fatty acids (for *Actinomycetes*) and polyunsaturated fatty acids (PUFAs) (for fungi) (Frostegård and Bååth 1996).

Since PLFAs are present in microbial cell membranes, changes in PLFA content are attributable to intracellular as well as extracellular environmental changes and deviations. PLFAs are thus beneficial for the diagnosis and analysis of changes in the physiology of a living system mediated by functionality. The information which is obtained by PLFA analysis is phenotypic in nature and related to the activity level of the microbial community under question, as opposed to genotypic or relative species-level information derived from other molecular methods. Studies reveal that such PLFA-based biomonitoring is more strongly associated with monitoring of soil processes and physiological capabilities of soil microbes, rather than the number of spacer copies transcribed and quantified with qPCR (Baldrian et al. 2013).

Phospholipid ether lipids are more commonly employed for the characterisation of *Archaea* due to the hydrolytic resistance of archaeal ether lipids using standardised PLFA protocols. In addition, accurate determination of microbial biomass and structure can be achieved through the use of other chemical markers in microbial membranes, namely chitin and ergosterol (Baldrian et al. 2013). Ergosterol analysis has been observed to have a greater ability to differentiate between fungal groups as well as between living and dead fungal biomass (Wallander et al. 2013). Chitin and ergosterol markers are solely employed for fungal characteristics, whereas PLFA analysis has wider applicability in the characterisation of both bacteria and fungi (Frostegård et al. 2011).

A major limitation which has been widely recognised is the considerable difficulty involved in soil microbial study, referred to as the soil microbial "black box". Improved understanding of soil microbiology can be promisingly achieved through amalgamation of lipid biomarker analysis with other analytical tools. One such technique is isotopic labelling, involving use of an isotope such as 13-CO_2 as a label to achieve 13-C PLFA analysis or PLFA-stable isotope probing (SIP). 13-C PLFA analysis confers significant benefits in biomonitoring by revealing crucial links between metabolically active microbial groups and their specific biogeochemical cycling processes and outlining those microbial communities involved in actively assimilating plant-based carbon (Yao et al. 2014). Moreover, it helps in better understanding of how microbial communities utilise mineralisation products of soil organic content, distinguishing microbial groups involved in the biodegradation of environmental contaminants and determining seasonal variability across such communities (Willers et al. 2015). Although PLFA-SIP does not provide any information on the phylogenetic resolution or species composition, it can provide a valuable insight into the structure and functionality of soil microflora when used in conjunction with DNA/RNA-SIP techniques.

Among lipid molecules which are biosynthesised in the *plant kingdom*, the most abundant are the n-alkanes synthesised by terrestrial plants, aquatic plants and phytoplankton alike, though they vary in their chemical structure, enabling their efficient use as signature biomarkers in the plant kingdom. Phytoplankton and zooplankton are characterised by n-alkane molecules with an odd number of carbon atoms (15, 17 or 19 carbon atoms), whereas submerged or floating aquatic macrophytes are predominantly observed to be composed of n-alkanes of medium chain length, ranging between 20 and 25 carbon atoms (Derrien et al. 2017). Odd-numbered long-chain n-alkanes are strongly associated with terrestrial plants (Tyson et al. 2005), with studies reporting C-27 and C-29 n-alkanes being more common among shrubs and trees and the longer C-31 and C-33 n-alkanes distributed in herbs and grasses (Zech et al. 2013).

Long-chain fatty acids (LCFA) are another class of lipid molecules which are abundantly found in leaves, flowers and pollen in vascular plants, as waxy coatings. Polyunsaturated fatty acids (PUFAs) are also commonly observed in higher terrestrial plants and mangroves and can serve as signature markers, as opposed to monounsaturated fatty acids (MUFAs) and branch-chain fatty acids (BrFAs) which are commonly used as bacterial biomarkers due to their marked predominance in bacterial communities. Sterols and stanols can also be used as molecular biomarkers for identifying both natural organic matter features as well as anthropogenic modifications, especially in terrestrial higher plants. The predominant sterols which have been reported to serve as markers in such plants include campesterol, stigmasterol and sitosterol. Studies also indicate that although sitosterol is also observed in marine organisms, unambiguous source assignment is achieved through stereochemical differentiation; the R epimer of sitosterol is assigned to the terrestrial source, while the S epimer is assigned to the marine source. Brassicasterol is the predominant sterol which is associated with marine phytoplankton, while it was noted that cholesterol is predominant in invertebrates and marine zooplankton (Derrien et al. 2017).

Lipid biomarkers from plants have immense significance with respect to an understanding of environmental quality. For instance, short-chain n-alkanes with low odd-to-even carbon preference are utilised as signatures of fossil fuel inputs in land. The use of specific sterols such as coprostanol was suggested for gauging the quality and environmental safety of wastewater discharge impacted by human faecal contamination, although it was soon realised that its use as a signature marker was precluded by its presence in other mammals, causing low specificity (Leeming et al. 1996). Subsequent researchers have since been focusing on the distribution of steroids as a group, rather than individual molecules, to yield what is known as a steroid fingerprint. Such distribution of steroids from anthropogenic sources has been observed to be dependent on the animal diet, biosynthesis of sterols in the animal system and the anaerobic microflora in such animals, which may be capable of biologically hydrogenating sterols into various isomeric stanols (Leeming et al. 1996).

Lipidic biomarkers in blood can also be beneficially used to gauge how *human health* is impacted by environmental parameters, which is derived through a study of the alterations in the internal conditions of exposed organisms (Wallace et al. 2016). The principal lipid biomarkers which have been recognised in humans for such studies include serum triglycerides (serum TGs) and trichain cationic lipids (TC lipids) such as cholesterol. Such lipids undergo either upregulation or downregulation when the liver is affected by damage or metabolic disorders. Serum lipid biomarkers are now known at a preliminary level to reflect changes in hepatic metabolism profiles due to exposure to pollutants such as volatile organic compounds (VOCs) in indoor environments. These responses towards such external parameters should thus be taken into account by healthcare as well as fundamental researchers to understand the tolerance capacity of the human beings towards various xenobiotics.

4.5 Extraction of Lipid Biomarkers

A wide range of analytical techniques have been developed for examining microbial lipids. Given the vast diversity in microbial lipids with respect to chain length, branching patterns, saturation and functionalisation, no single tool is capable of comprehensively analysing such a multitude of microbial lipids (Härtig 2008). As with any analytical method, the principal focus of microbial lipid analysis from environmental samples remains high throughput with maximum automation of the processes involved, namely extraction, fractionation, derivatisation and quantification.

In order to extract these lipids from any cell, their structural and chemical composition should be taken into consideration so as to employ a suitable extraction agent for their removal. Lipids are characterised by a lipophilic tail often connected to a glycerol backbone through ester and/or ether linkages and a hydrophilic, polar head functional group, typically phosphate-based in nature, though other groups like glycosidic-, amino- and sulphate-based groups are also observed (Hölzl and Dörmann 2007). Lipids are thus defined by their insolubility in water and solubility in non-polar solvents, with non-polar organic solvents such as n-hexane being most suitable for their extraction.

Of the four aforementioned steps in lipid extraction from microorganisms, extraction is usually the most crucial but cumbersome step involving organic solvent addition and filtration, followed by gravitation or centrifugation to separate the extract phase containing the extracted lipids (organic layer) from the raffinate phase. The principal factors governing the efficiency of extraction include the time allowed for extraction, possible contaminations or losses of target lipids during post-extractive collection and the extracting solvent chosen (Cescut et al. 2011). The most commonly used extraction techniques employed for microbial lipid profiling are Folch method (Folch et al. 1957), Bligh and Dyer method (Bligh and Dyer 1959) and Soxhlet extraction (Soxhlet 1879). Recent studies have also realised the potential of methyl-tert-butyl ether (MTBE) (Matyash et al. 2008), automated accelerated-solvent extraction (ASE) and supercritical fluid extraction (SFE) for optimal lipid extraction. The advantages and disadvantages of these methods have been summarised in Table 4.1.

The *Folch method* (Folch et al. 1957) was originally developed for lipid extraction from animal tissues such as brain matter, liver and muscle and has since been used to extract lipids for subsequent microbial lipid profiling. The tissues were homogenised using a solvent mixture of chloroform: methanol (2:1) (v/v) and washed with water or a salt buffer, causing residence of lipids in the organic phase. A single washing procedure with water was observed to ensure a minimal lipid loss of 0.3–0.6%, with salt buffers playing a significant role in curbing the loss of acidic lipids into the raffinate phase. The principal advantages of this method are its high extraction efficiency and low interference from non-lipid compounds (Matyash et al. 2008; Sheng et al. 2011). A schematic representation of lipid extraction from microalgae biomass using Folch method has been depicted in Fig. 4.3.

The limitations of the Folch method, namely the need for high volumes of extracting solvents and intermediate washing steps, have in subsequent years served as focal points for method optimisation. Reduction in solvent volumes has been

Table 4.1 Comparative overview of advantages and limitations of different extraction techniques for microbial lipid extraction

Method of extraction	Advantages	Disadvantages
Folch method	High efficiency of extraction; less or no interference from non-lipid components	Extraction volumes are higher; laborious method and safety issues
Bligh & Dyer method	Good extraction efficiency; low volumes of solvent and samples required	Safety issues; longer extraction time
Matyash (MTBE) method	Stable and safer method of extraction; faster and more precise	Expensive due to higher cost of solvent; interference of polar compounds during extraction
Soxhlet extraction	Best compatibility with maximum of samples; choice of solvents available	Higher extraction volumes; laborious method with restricted automation; prolonged time for extraction
Accelerated solvent extraction (ASE)	Excellent extraction efficiency in less time; automated, green and less laborious	Higher temperatures for extraction; expensive due to use of sophisticated instrumentation
Supercritical fluid extraction (SCFE)	High speed; greener method of extraction; recycling of solvent possible; no residual solvents	Expensive; operation at high temperature; restricted solubility of solvent; operation with high pressure

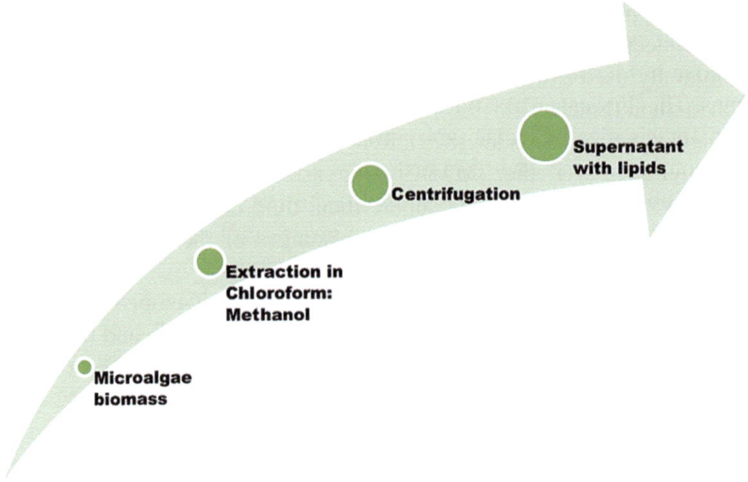

Fig. 4.3 Schematic representation of lipid extraction from microalgae biomass using Folch method

observed to pose no significant impact on recovery of lipids from bacterial cultures (Matyash et al. 2008); use of a modified Folch method without washing processes for soil, manure and compost-based PLFA analysis similarly exhibited no impact on the lipid yield (Gómez-Brandón et al. 2010). More significantly, the modified Folch method yielded the highest recovery of PLFA and specific lipid biomarkers compared to a modified Bligh and Dyer method (White et al. 1979) and microwave-assisted extraction (Lopez-Avila et al. 1994). The longer incubation and larger solvent volumes in the Folch method could have influenced the findings. Although reduction in the extraction time could improve processing time and analysis speed by Folch method, studies indicate that the original exposure time chosen (24 h) yields the highest concentrations of lipids from the subjected samples and is thus the most favourable for extraction.

The *Bligh and Dyer method* (Bligh and Dyer 1959), similar to the unmodified Folch method, was originally optimised for extraction of phospholipids from fish muscles. An initial solvent composition of chloroform: methanol: water (1:2:0.8) (v/v/v) was used, which was observed to yield an excellent extraction efficiency of greater than 95% for the total polar lipids. This method offers significant advantages in the form of smaller sample and solvent volumes compared to the Folch method and has since been modified for microbial lipid extraction from a vast variety of matrices. The most widely used modification with significant applicability is the replacement of water in the extraction mixture with phosphate buffer (White et al. 1979) or citrate buffer (Frostegård et al. 1991) for PLFA analysis. Inclusion of a buffer in the extracting solvent reduces loss of ionisable phospholipids into the aqueous phase and thus improves yield of extraction. Such buffered solvent systems have since been utilised for microbial lipid extraction from macroalgae (Kumari et al. 2011); soil (Gómez-Brandón et al. 2010); activated sludge (Hanif et al. 2012);

estuarine sediments (Guckert et al. 1985); biofilms (Liu et al. 2012) and pure microbial cultures (Macnaughton et al. 1997). Archaeal lipids have also been successfully extracted from microbial matrices using 1% trichloroacetic acid instead of water (Allen et al. 2010). The principal limitations of this method include the hazards associated with chloroform use, prolonged exposure time and risk of target lipid contamination during post-extractive collection (Sheng et al. 2011).

The use of *MTBE* (methyl tert-butyl ether) in lieu of chloroform in the extracting solvent used, with a renewed composition of MTBE: MeOH (1.5:5) (v/v) (Matyash et al. 2008) has provided a suitable means to overcome limitations associated with the use of chloroform, such as its hazardous nature and risk of formation of toxic byproducts such as phosgene upon its decomposition causing chemical modification of the extracted lipids. MTBE offers several significant advantages for extraction—it is stable, much less toxic, does not chemically react with lipids and forms the upper layer in solvent extraction, thus providing ease of collection of the extract phase. In addition, the method offers faster and more precise lipid extraction in comparison to the conventional Bligh and Dyer method, with a yield range between 90% and 98%. The MTBE method yielded an extraction yield of 93% for cyanobacterial lipid extraction, with the same solvent composition mentioned above (Sheng et al. 2011).

Another widely adopted extraction methodology is *Soxhlet extraction*, first developed in 1879 by Frans von Soxhlet. Soxhlet extraction offers a convenient means of extraction of target analytes, including lipids, from a wide variety of sample matrices; the success of extraction is however largely dependent upon the choice of the extracting solvent. For lipid extraction, it is desirable to choose solvents which are non-polar in nature, for instance, dichloromethane, hexane-acetone or isopropyl alcohol. These solvents were observed to provide good extraction efficiencies for lipids from green algae, cyanobacteria (Sheng et al. 2011) and yeast (Cescut et al. 2011). Recent scientific advancements have largely focused on the modification of conventional Soxhlet extraction, with a view to improve efficiency, reduce solvent consumption, prevent degradation of analytes and increase throughput through automation. These advancements include focused microwave-assisted Soxhlet extraction (FMASE) and automated Soxtec® Systems (Luque-García and De Castro 2004). The Soxhlet apparatus is illustrated in Fig. 4.4.

Accelerated solvent extraction (ASE), also called pressurised liquid extraction, was commercialised by Dionex Incorporated, California, as an automated system equipped with ovens as a heating source. The extraction technique employs a combination of high temperature and pressure to ensure that the extracting solvent remains in the liquid state at elevated pressures (>200 bar) and well above its boiling point. Microbial lipid extraction has been achieved through ASE using pure bacterial cultures as well as water and air samples (Macnaughton et al. 1997) and soil (Peterse et al. 2009). ASE inherently possesses several advantages such as reduced time, complete automation, reduced solvent requirements and elimination of the need for filtration. ASE has been demonstrated to improve the speed of extraction by five times and reduce the solvent volumes by twenty times during yeast lipid extraction, compared to Bligh and Dyer method and Soxhlet extraction (Cescut

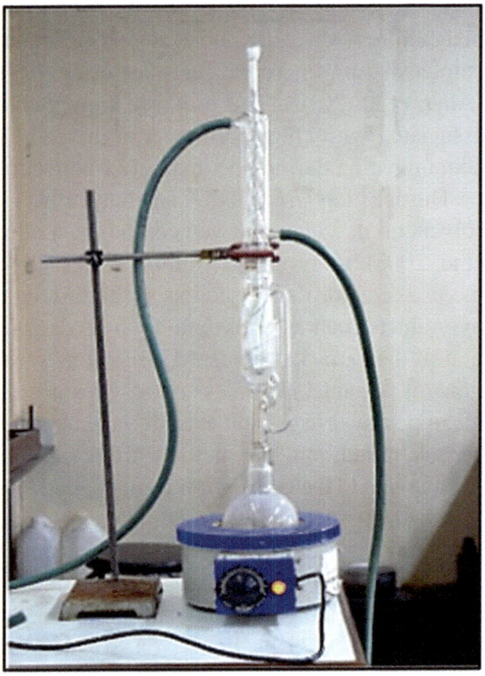

Fig. 4.4 Illustration of a Soxhlet extraction apparatus

et al. 2011); it exhibited greater efficiency in extracting PLFAs and neutral lipids from soil samples compared to the conventional Bligh and Dyer method (White et al. 2009).

Supercritical fluid extraction (SFE) is yet another sophisticated extraction technique which can be employed for lipid extraction from samples. It has been realised as a rapid and safe method for extraction of microbial lipids, making use of one of several supercritical fluids as extracting solvents, such as ammonia, chlorofluorocarbons and carbon dioxide, with carbon dioxide being most effective as an SFE extractant (Cescut et al. 2011). Supercritical fluids possess solubility properties similar to those of liquids but viscosity and diffusivity properties similar to those of gases (Hawthorne et al. 1992). Carbon dioxide is preferred as an extractant due to its non-toxicity, low critical value range and relative eco-friendliness. When SFE was employed for extraction of microbial respiratory quinones, PLFAs and phospholipid ether lipids from sludge which had been digested anaerobically, the need for addition of a modifier such as 10% methanol was realised to improve extractant strength, thus demanding optimisation of experimental parameters (Hanif et al. 2012). Bacterial cultures and archaeal cultures have also been subjected to SFE for lipid extraction.

Figure 4.5a, b depict the sophisticated techniques of supercritical fluid extraction and accelerated solvent extraction schematically, respectively.

For extraction of lipid biomarkers from human serum samples, similar approaches are used. In a multi-omics study conducted to understand how liver and serum lipid

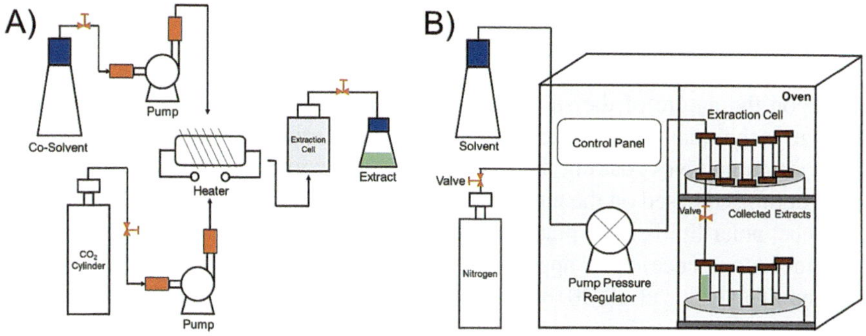

Fig. 4.5 Schematic representation of (**a**) supercritical fluid extraction and (**b**) accelerated solvent extraction

biomarkers reflect the changes in hepatic metabolism caused by indoor volatile organic content (VOC) exposure, the liver tissue and serum samples were subject to homogenisation and vortexing using a solvent mixture of chloroform: methanol (2:1) (v/v) at −20 °C, respectively. Post extraction, mixing of the extracts with deionised water and centrifugation were performed and the lower layer fluid was collected into a fresh tube. This was followed by vacuum drying of the tubes for solvent evaporation, reconstitution using isopropanol and filtration for subsequent analysis by liquid chromatography-mass spectrometry (LC-MS) (Miao et al. 2023). Similar extraction techniques are employed for other animal samples, such as those from fish. For instance, lipids were extracted from the adipose tissue and plasma of *Oreochromis mossambicus* (Mozambique tilapia) and characterised using LC-MS to understand their role as biomarkers in pansteatitis, a disorder recognised by environmentally derived fat inflammation in marine life (Koelmel et al. 2019). The results of the study yielded significant insights into the upregulation of the inflammatory pathways through the lipid biomarkers.

4.6 Analysis of Lipid Biomarkers

Extraction of lipids from a variety of different biological matrices using methods described previously for subsequent analysis serves as a crucial step within the stage of sample preparation. The principal purpose of such an extraction method is not only to break down complex biological matrices into simpler ones devoid of any nonlipid components but also to enrich the target analyte in the sample, providing improved signal-to-noise ratios. It is imperative that a thorough knowledge of the composition of the biological matrix be acquired and the choice of the extraction method be made wisely for the detection and analysis of biomarkers from a sample of biological origin. This is because both these factors also affect the level of contamination in the analytical instrument under consideration, impacting its downtime due to the time involved in maintenance and cleaning (Züllig et al. 2020).

Sample collection and sample preparation strategies vary significantly depending upon the nature of the sample and the biological matrix. Conventional, well-known techniques which are adopted for plant samples include maceration. Maceration involves soaking the sample using an appropriate extractant which has been optimised based on the nature of the target analyte for maximal extraction. For instance, polar lipids from plant samples can be extracted through simple maceration in the presence of an appropriate non-polar solvent such as diethyl ether or hexane (Traversier et al. 2018). For animal samples, homogenisation is a crucial step for ensuring uniform accessibility of lipids to extractants from all different parts of the tissue sample being processed. The most commonly used homogenisation methods are the shear-force-based grinding of tissue in a solvent or the crushing of liquid-nitrogen-frozen tissue by a mortar and pestle (Jurowski et al. 2017).

Sample storage is also a crucial factor which governs the integrity of the sample. It is ideal to use samples which are freshly collected and processed with minimum time lapse to preserve the integrity of lipids, as prolonged storage may result in the metabolism of sample lipids due to enzymatic or chemical activity (Züllig et al. 2020). Studies have shown that concentrations of lysophosphatidylcholine (LPC) and lysophosphatidic acid (LPA) increase significantly in plasma samples which have been stored for prolonged periods at room temperature (Scherer et al. 2009). Samples which cannot be processed immediately should be preserved in ultra-freezing facilities at −80 °C, and even at such low temperatures, the storage should be limited to as short a time period as possible to prevent oxidative or hydrolytic degradation of lipids.

With lipidomes in biological systems being thus characterised by complex structural features, diverse categorisations and variations with respect to physicochemical properties, it becomes important to use appropriate analytical tools for comprehensive analysis of lipids for biomarker discovery. The most commonly used analytical tools today for such applications are powerful chromatographic separation techniques used in tandem with mass spectrometric detection tools (Xu et al. 2020), such as LC-MS and LC-MS/MS. The schematic diagram of LC-MS is depicted in Fig. 4.6. Such techniques offer unparalleled accuracy, precision, sensitivity and resolution, which are of paramount importance in the detection and quantitation of minute changes which may have occurred in the lipid profile of a biological system owing to environmental or anthropogenic factors as an instance. In addition to chromatographic separation and mass spectrometry, it is equally essential to recognise the contributions of biostatistics and bioinformatics in the handling, processing and interpretation of large amounts of data during lipidomics analysis (Hornemann 2022). Chromatography, mass spectrometry and bioinformatics can therefore be termed as components of the "golden triangle" for lipid analysis.

Technological advancements in both chromatography and mass spectrometry have proven instrumental in contributing to the growth of the field of lipidomics and its branches. The advent of ultra high-performance liquid chromatography (UPLC) has enabled direct separation of a wide variety of lipid classes, including

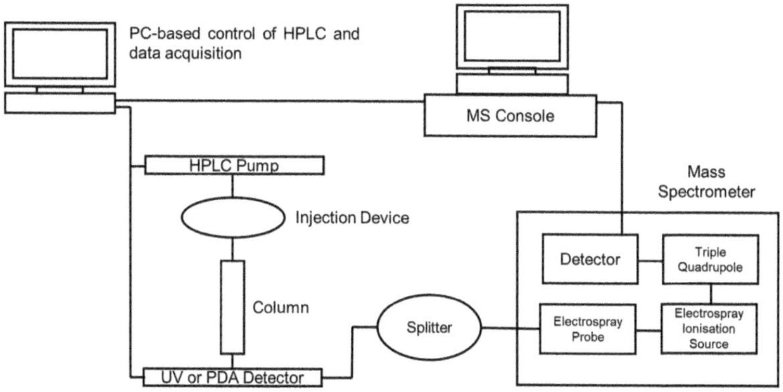

Fig. 4.6 Schematic representation of an LC-MS system

glycerophospholipids, glycerolipids, sphingolipids, sterols and fatty acids (Wenk 2005). In addition, advancements in MS such as ionisation methods have accelerated the rate at which lipids are being analysed from complex matrices. The most widely used ionisation techniques are electrospray ionisation (ESI), atmospheric pressure chemical ionisation (APCI) and atmospheric pressure photoionisation (APPI) (Hornemann 2022).

Some recent applications of LC-MS in the field of environmental lipidomics are illustrated in Table 4.2.

The potential of gas chromatography coupled with mass spectrometry (GC-MS) is also being increasingly established for lipid biomarker analysis. Compound-specific isotopic analysis of lipid biomarkers is achievable using GC-MS post pre-chromatographic derivatisation. Such derivatisation is useful for the introduction of an isotopic label for differential quantitation of the substrate (Züllig et al. 2020). Since autotrophic organisms fix atmospheric inorganic carbon preferentially in its lighter, more abundant isotopic form (12C) opposed to the heavier one (13C), the proportion of the heavier isotope to the lighter isotope can be quantified in the specific compound. Growth will thus be accompanied by a 13C-depleted biomass compared to the substrate, which leads to its interpretation as a signature of one metabolic route of carbon fixation over another (van der Meer et al. 2003). Such stable carbon isotope studies performed on substrates followed by their molecular analysis offer an extremely accurate means for marker identification and their subsequent allocation to a biogenic or an abiogenic origin. A schematic diagram of GC-MS instrumentation is illustrated in Fig. 4.7.

The processing, analysis and interpretation of complex lipidomics data is greatly enhanced by the inputs of advanced biostatistical tools (Hornemann 2022). Several biostatistical tools such as partial least squares discriminant analysis (PLS-DA) and orthogonal projection-potential structure analysis (OPLS-DA) are useful for the understanding of comprehensive datasets. In addition, through comprehensive bioinformatics databases like LIPID MAPS, LipidBank, Lipid Library, SphingoMAP and KEGG, as well as software tools for hyphenated technique data processing such

Table 4.2 Overview of different studies targeting environmental lipidomics through LC-MS analysis

Application	Lipid biomarkers identified	Stationary phase	Mobile phase	Reference
Determination of impact of multiple industrial pollutants on children and adolescents, such as liver dysfunctions and chronic kidney disorders	Lysophosphatidylcholines Phosphatidylcholines Sphingomyelins Phosphatidylinositols	C18 column (2.1 × 100 mm, 1.8 μm, 55 °C), flow rate of 350 μL min^{-1}	Solvent A (water/10 mM ammonium acetate/0.1% formic acid) and solvent B (methanol: isopropanol: water = 60: 40: 1/10 mM ammonium acetate/0.1% formic acid), gradient elution	Chen et al. (2021)
Determination of impact of air pollution on arachidonic acid- and linoleic acid-derived hydroxy metabolites	12-Hydroxyeicosatetraenoic acid (12-HETE) 13-Hydroxyoctadecadienoic acid (13-HODE)	C18 column (2.1 × 50 mm, 1.7 μm, 60 °C), flow rate of 0.25 mL min^{-1}	Solvent A (100% water with 10 mM ammonium acetate and 0.1% formic acid) and solvent B (7:3 acetonitrile: 2-propanol (v/v) with 10 mM ammonium acetate and 0.1% formic acid), gradient elution	Zhu et al. (2021)
Study of mediation and aggravation of risk of incident coronary heart disease by diabetes and ambient air pollution respectively	Cholesteryl esters Ceramides Lysophosphatidylcholines Phosphatidylcholines Sphingomyelins	C8 column (100 × 2.1 mm), flow rate of 0.5 mL min^{-1}	Phase A (acetonitrile / ultra-pure water, v/v 60/40, containing 10 mmol/L ammonium formate) and phase B (isopropanol/ ultra-pure water, v/v 90/10, containing 10 mmol/L ammonium formate), gradient elution	Li et al. (2024)

(continued)

Table 4.2 (continued)

Application	Lipid biomarkers identified	Stationary phase	Mobile phase	Reference
Study of possible lipid mediators of lung and cortex inflammation in mice exposed to diesel air pollution	Arachidonic acid lipids HETEs HODEs Prostaglandin D2 Lipoprotein A	C18 column (3 μ, 2.1 × 150 mm), flow rate of 0.2 mL min^{-1}	Phase A (methanol: water: acetonitrile (10:85:5 v/v)) and phase B (methanol: water: acetonitrile (90:5:5 v/v)), both containing 0.1% ammonium acetate, gradient elution	Massimino et al. (2022)

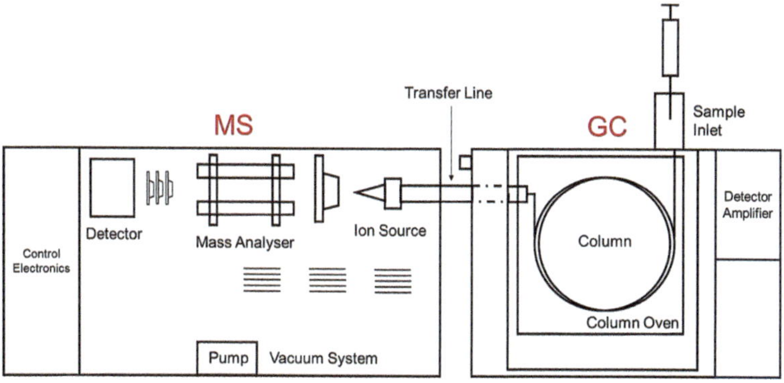

Fig. 4.7 Schematic representation of a GC-MS system

as Mzmine 2 and metAlign, the workflow can be greatly streamlined to achieve convenient accessibility, handling and processing of data (Hornemann 2022). The growing development of artificial intelligence (AI) in recent years is expected to revolutionise the outlook towards lipidomics in the near future. More researchers and scientists would thus opt for a combinatorial approach involving an amalgamation of inputs from chromatography, mass spectrometry, biostatistics and AI to enable robust, precise and insightful lipidomics studies to be carried out. The workflow which is usually adopted for such intensive lipidomics studies is depicted in Fig. 4.8.

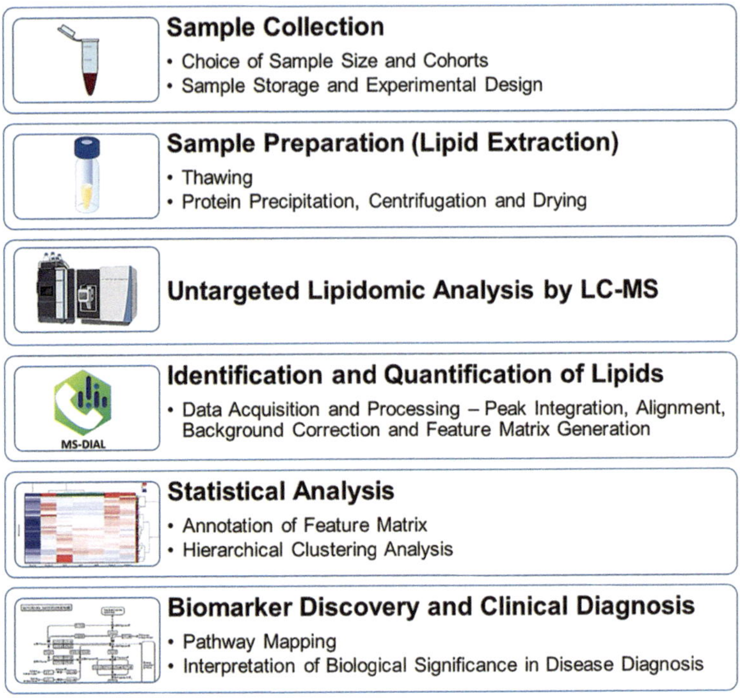

Fig. 4.8 Workflow involved in discovery of human serum biomarkers by using an LC-MS lipidomics analysis and machine learning analysis approach

4.7 Conclusion

Increasing global pollution and environmental deterioration have resulted in significant observable impacts due to pollutants on the internal biochemistry of living organisms, including microorganisms, plants, animals and human beings. For better understanding of the close-knit interaction between the biosphere and the abiotic components such as pollutants of anthropogenic nature, it is essential to identify and characterise biomarkers for the same. Such signature biomarkers at the molecular level are useful as signallers of environment-mediated biochemical changes. Among several biomolecules, lipids are ideal for such studies on account of their dynamism, diversity, ubiquity and chemical stability.

Several groups of lipids are characteristic of specific biotic communities, such as phospholipids in microbial organisms; n-alkanes, long-chain fatty acids and steroids in the plant kingdom; and serum triglycerides and trichain cationic lipids in human beings. The general workflow adopted for any lipidomics study begins with sample collection and pre-processing, followed by sample preparation, usually employing one of several extraction methods. While conventionally described methods such as Folch method, Bligh and Dyer method, Soxhlet extraction and simple maceration retain their utility, recent advancements with greater sophistication, such as pressurised liquid extraction and supercritical fluid extraction, are increasingly being adopted for efficient extraction. Pre-analytical considerations such as the need for homogenisation in animal tissue samples and sample storage conditions should be thoroughly considered and optimised wherever necessary.

Chromatographic techniques coupled with mass spectrometry, such as LC-MS, LC-MS/MS and GC-MS, remain the most popular methods of analysis for lipidomics-based studies, owing to their high sensitivity and resolution. In addition, the contributions of biostatistics and bioinformatics in the processing and prompt retrieval of computationally intense data are now widely recognised. Most lipidomics research today uses a combinatorial, converging approach employing the power of chromatographic separation, the sensitivity of mass spectrometry and the computational benefits of machine learning to yield maximum information out of such studies. Inputs of artificial intelligence are expected to boost the productivity, precision and value of such lipidomics studies to a greater level in the near future.

References

Allen MA, Neilan BA, Burns BP, Jahnke LL, Summons RE (2010) Lipid biomarkers in Hamelin Pool microbial mats and stromatolites. Org Geochem 41(11):1207–1218

Baldrian P, Větrovský T, Cajthaml T, Dobiášová P, Petránková M, Šnajdr J, Eichlerová I (2013) Estimation of fungal biomass in forest litter and soil. Fungal Ecol 6(1):1–11

Bligh EG, Dyer WJ (1959) A rapid method of total lipid extraction and purification. Can J Biochem Physiol 37(8):911–917

Cescut J, Severac E, Molina-Jouve C, Uribelarrea JL (2011) Optimizing pressurized liquid extraction of microbial lipids using the response surface method. J Chromatogr A 1218(3):373–379

Chen CHS, Kuo TC, Kuo HC, Tseng YJ, Kuo CH, Yuan TH, Chan CC (2021) Lipidomics of children and adolescents exposed to multiple industrial pollutants. Environ Res 201:111448

Derrien M, Yang L, Hur J (2017) Lipid biomarkers and spectroscopic indices for identifying organic matter sources in aquatic environments: a review. Water Res 112:58–71

Finkel PL, Carrizo D, Parro V, Sánchez-García L (2023) An overview of lipid biomarkers in terrestrial extreme environments with relevance for Mars exploration. Astrobiology 23(5):563–604

Folch J, Lees M, Sloane Stanley GH (1957) A simple method for the isolation and purification of total lipids from animal tissues. J Biol Chem 226(1):497–509

Frostegård A, Bååth E (1996) The use of phospholipid fatty acid analysis to estimate bacterial and fungal biomass in soil. Biol Fertil Soils 22:59–65

Frostegård Å, Tunlid A, Bååth E (1991) Microbial biomass measured as total lipid phosphate in soils of different organic content. J Microbiol Methods 14(3):151–163

Frostegård Å, Tunlid A, Bååth E (2011) Use and misuse of PLFA measurements in soils. Soil Biol Biochem 43(8):1621–1625

Gómez-Brandón M, Lores M, Domínguez J (2010) A new combination of extraction and derivatization methods that reduces the complexity and preparation time in determining phospholipid fatty acids in solid environmental samples. Bioresour Technol 101(4):1348–1354

Guckert JB, Antworth CP, Nichols PD, White DC (1985) Phospholipid, ester-linked fatty acid profiles as reproducible assays for changes in prokaryotic community structure of estuarine sediments. FEMS Microbiol Ecol 31:147–158

Halliwell B (1999) Antioxidant defence mechanisms: from the beginning to the end (of the beginning). Free Radic Res 31(4):261–272

Hanif M, Atsuta Y, Fujie K, Daimon H (2012) Supercritical fluid extraction of bacterial and archaeal lipid biomarkers from anaerobically digested sludge. Int J Mol Sci 13(3):3022–3037
Härtig C (2008) Rapid identification of fatty acid methyl esters using a multidimensional gas chromatography–mass spectrometry database. J Chromatogr A 1177(1):159–169
Hawthorne SB, Miller DJ, Nivens DE, White DC (1992) Supercritical fluid extraction of polar analytes using in situ chemical derivatization. Anal Chem 64(4):405–412
Hölzl G, Dörmann P (2007) Structure and function of glycoglycerolipids in plants and bacteria. Prog Lipid Res 46(5):225–243
Hornemann T (2022) Lipidomics in biomarker research. In: Prevention and treatment of atherosclerosis: improving state-of-the-art management and search for novel targets. Springer, Cham, pp 493–510
Javed M, Usmani N (2015) Stress response of biomolecules (carbohydrate, protein and lipid profiles) in fish Channa punctatus inhabiting river polluted by thermal power plant effluent. Saudi J Biol Sci 22(2):237–242
Jurowski K, Kochan K, Walczak J, Barańska M, Piekoszewski W, Buszewski B (2017) Analytical techniques in lipidomics: state of the art. Crit Rev Anal Chem 47(5):418–437
Koelmel JP, Ulmer CZ, Fogelson S, Jones CM, Botha H, Bangma JT, Bowden JA (2019) Lipidomics for wildlife disease etiology and biomarker discovery: a case study of pansteatitis outbreak in South Africa. Metabolomics 15:1–11
Koelmel JP, Napolitano MP, Ulmer CZ, Vasiliou V, Garrett TJ, Yost RA, Bowden JA (2020) Environmental lipidomics: understanding the response of organisms and ecosystems to a changing world. Metabolomics 16:1–22
Kumari P, Reddy CRK, Jha B (2011) Comparative evaluation and selection of a method for lipid and fatty acid extraction from macroalgae. Anal Biochem 415(2):134–144
Langworthy TA, Holzer G, Zeikus JG, Tornabene TG (1983) Iso- and anteiso-branched glycerol diethers of the thermophilic anaerobe Thermodesulfotobacterium commune. Syst Appl Microbiol 4(1):1–17
Leeming R, Ball A, Ashbolt N, Nichols P (1996) Using faecal sterols from humans and animals to distinguish faecal pollution in receiving waters. Water Res 30(12):2893–2900
Levental KR, Malmberg E, Symons JL, Fan YY, Chapkin RS, Ernst R, Levental I (2020) Lipidomic and biophysical homeostasis of mammalian membranes counteracts dietary lipid perturbations to maintain cellular fitness. Nat Commun 11(1):1339
Li Y, Wang H, Xiao Y, Yang H, Wang S, Liu L, Qiu G (2024) Lipidomics identified novel cholesterol-independent predictors for risk of incident coronary heart disease: mediation of risk from diabetes and aggravation of risk by ambient air pollution. J Adv Res 65:273–282
Liu J, Zhou XH, Shi HC (2012) Inhibitory effects of pentachlorophenol on wastewater biofilms as determined by phospholipid analysis and microelectrode. Biochem Eng J 66:8–13
Lopez-Avila V, Young R, Beckert WF (1994) Microwave-assisted extraction of organic compounds from standard reference soils and sediments. Anal Chem 66(7):1097–1106
Luque-García JL, De Castro ML (2004) Ultrasound-assisted Soxhlet extraction: an expeditive approach for solid sample treatment: application to the extraction of total fat from oleaginous seeds. J Chromatogr A 1034(1–2):237–242
Macnaughton SJ, Jenkins TL, Wimpee MH, Cormiér MR, White DC (1997) Rapid extraction of lipid biomarkers from pure culture and environmental samples using pressurized accelerated hot solvent extraction. J Microbiol Methods 31(1–2):19–27
Massimino L, Bulbarelli A, Corsetto PA, Milani C, Botto L, Farina F, Rizzo AM (2022) LSEA evaluation of lipid mediators of inflammation in lung and cortex of mice exposed to diesel air pollution. Biomedicine 10(3):712
Mateo-Sagasta J, Marjani S, Turral H (2017) Water pollution from agriculture: a global review. Executive Summary. FAO and IWMI
Matyash V, Liebisch G, Kurzchalia TV, Shevchenko A, Schwudke D (2008) Lipid extraction by methyl-tert-butyl ether for high-throughput lipidomics. J Lipid Res 49(5):1137–1146

Miao G, Wang Y, Wang B, Yu H, Liu J, Pan R, Jin X (2023) Multi-omics analysis reveals hepatic lipid metabolism profiles and serum lipid biomarkers upon indoor relevant VOC exposure. Environ Int 180:108221

Peterse F, Schouten S, Van der Meer J, Van der Meer MTJ, Damsté JSS (2009) Distribution of branched tetraether lipids in geothermally heated soils: implications for the MBT/CBT temperature proxy. Org Geochem 40:201–205

Scherer M, Schmitz G, Liebisch G (2009) High-throughput analysis of sphingosine 1-phosphate, sphinganine 1-phosphate, and lysophosphatidic acid in plasma samples by liquid chromatography–tandem mass spectrometry. Clin Chem 55(6):1218–1222

Sheng J, Vannela R, Rittmann BE (2011) Evaluation of methods to extract and quantify lipids from Synechocystis PCC 6803. Bioresour Technol 102(2):1697–1703

Shuai J, Kim S, Ryu H et al (2018) Health risk assessment of volatile organic compounds exposure near Daegu dyeing industrial complex in South Korea. BMC Public Health 18:528. https://doi.org/10.1186/s12889-018-5454-1

Soxhlet FV (1879) Die gewichtsanalytische bestimmung des milchfettes. Polytech J 232(5):461–465

Traversier M, Gaslondes T, Milesi S, Michel S, Delannay E (2018) Polar lipids in cosmetics: recent trends in extraction, separation, analysis and main applications. Phytochem Rev 17:1179–1210

Tyson R, Killops S, Killops V (2005) Introduction to organic geochemistry, 2nd ed. Geol Mag 143(2):250–250. https://doi.org/10.1017/S0016756806222052

van der Meer MT, Schouten S, Sinninghe Damsté JS, de Leeuw JW, Ward DM (2003) Compound-specific isotopic fractionation patterns suggest different carbon metabolisms among Chloroflexus-like bacteria in hot-spring microbial mats. Applied and environmental microbiology 69(10):6000-6006

Van Meer G, Voelker DR, Feigenson GW (2008) Membrane lipids: where they are and how they behave. Nat Rev Mol Cell Biol 9(2):112–124

Wallace MAG, Kormos TM, Pleil JD (2016) Blood-borne biomarkers and bioindicators for linking exposure to health effects in environmental health science. J Toxic Environ Health, Part B 19(8):380–409

Wallander H, Ekblad A, Godbold DL, Johnson D, Bahr A, Baldrian P, Rudawska M (2013) Evaluation of methods to estimate production, biomass and turnover of ectomycorrhizal mycelium in forest soils–a review. Soil Biol Biochem 57:1034–1047

Wang X, Han J, Xu L, Zhang Q (2010) Spatial and seasonal variations of the contamination within water body of the Grand Canal, China. Environ Pollut 158:1513–1520

Wenk MR (2005) The emerging field of lipidomics. Nat Rev Drug Discov 4(7):594–610

White DC, Davis WM, Nickels JS, King JD, Bobbie RJ (1979) Determination of the sedimentary microbial biomass by extractible lipid phosphate. Oecologia 40:51–62

White PM, Potter TL, Strickland TC (2009) Pressurized liquid extraction of soil microbial phospholipid and neutral lipid fatty acids. J Agric Food Chem 57(16):7171–7177

Willers C, Jansen van Rensburg PJ, Claassens S (2015) Microbial signature lipid biomarker analysis–an approach that is still preferred, even amid various method modifications. J Appl Microbiol 118(6):1251–1263

Xu T, Hu C, Xuan Q, Xu G (2020) Recent advances in analytical strategies for mass spectrometry-based lipidomics. Anal Chim Acta 1137:156–169

Yao H, Chapman SJ, Thornton B, Paterson E (2014) 13C PLFAs: a key to open the soil microbial black box? Plant Soil. https://doi.org/10.1007/s11104-014-2300-9

Zech M, Krause T, Meszner S, Faust D (2013) Incorrect when uncorrected: reconstructing vegetation history using n-alkane biomarkers in loess-paleosol sequences–a case study from the Saxonian loess region, Germany. Quat Int 296:108–116

Zhu K, Browne RW, Blair RH, Bonner MR, Tian M, Niu Z, Mu L (2021) Changes in arachidonic acid (AA)- and linoleic acid (LA)-derived hydroxy metabolites and their interplay with inflammatory biomarkers in response to drastic changes in air pollution exposure. Environ Res 200:111401

Züllig T, Trötzmüller M, Köfeler HC (2020) Lipidomics from sample preparation to data analysis: a primer. Anal Bioanal Chem 412:2191–2209

Chapter 5
Chitosan: An Abundant Biopolymer Used for Water Treatment

Musiiwa V. Sengane and Nomvano Mketo

Abstract Water is a basic need used in our daily lives. Having access to clean water has been a challenge to most developing countries. Water consists of a wide range of contaminants that are detrimental to human health. A need for water treatment is compulsory as it assists in identifying these contaminants also being able to develop remediation strategies. Scientists are here to bridge the gap in saving the planet and keeping everyone alive. There have been a lot of research studies conducted and reported by researchers worldwide. The chitosan biopolymer is crosslinked with various materials to enhance its extraction efficiency. The chitosan has been used for quantitative determination and removal of different contaminants like dyes, heavy metals, pharmaceuticals, pesticides, etc. from wastewater. This has led to so many countries working on some strategies of recycling/purifying and reusing water. Thus, it is achieved by adopting different methods that have been investigated and developed by scientist as a way of bringing a solution to the global problems. For effective removal of these contaminants, a good flocculant needs to be adopted for effective removal of organic pollutants, suspended solids and metal ions that are deposited in water through many activities like agricultural activities, deposition from mining activities and poor sewage drainage management. Chitosan has been favoured because of its higher removal efficiency, it is environmentally friendly, biodegradable, biocompatible and it is non-toxic.

Keywords Chitosan · Contaminants · Purification · Wastewater

M. V. Sengane · N. Mketo (✉)
Department of Chemistry, College of Science and Engineering and Technology, Florida Science Campus, University of South Africa, Roodepoort, Johannesburg, South Africa
e-mail: mketon@unisa.ac.za

A. Guldhe et al. (eds.), *Biomaterials for Environmental Applications: Detection and Remediation*, Environmental Challenges and Solutions,
https://doi.org/10.1007/978-3-032-12701-3_5

Abbreviations

DD	Degree of deacetylation
dl/g	Decilitres per gram
dyn/cm	Dyne second per square centimetre
EC	Emerging contaminants
EPA	Environmental Protection Agency
EU	European Union
H	Hydrogen
H^+	Hydrogen cation
HAc	Acetic acid
MW	Molecular weight
N	Nitrogen
USEPA	United States Environmental Protection Agency
WHO	World Health Organisation
wt	Weight
β	Beta

5.1 Introduction

Access to clean, pure water has been a problem for many nations worldwide. 71% of the world's surface is occupied by water. Given that approximately 97% of this water is saltwater from the ocean while 3% is freshwater. Therefore, provided that already the availability of freshwater is very low this also affects its supply to humans versus the population growth all over the world. This then affects the supply of clean freshwater to humans. Lack of clean freshwater supply to humans all over the world has triggered the health and economy of many countries. Hence, this raises a concern and a need for researchers to come up with the solution to overcome this challenge and improve sustainability. Factors that are impacting a limited supply of freshwater include drastic increase in population, industrialisation and growing demand (Agronomy 2019). Figure 5.1 shows one of the most appealing acts of access to clean water straight from the tap.

Fig. 5.1 Drinking freshwater from the tap.

Fig. 5.2 Illustration of possible activities that contaminate water streams.

For the government to be able to sustain water supply to individuals, they must divert into recycling water. This treatment includes treatment of wastewater, industrial wastewater and agro-industrial processes, and municipal waste processing facilities (Mashile et al. 2022). Major freshwater bodies include rivers, lakes, reservoirs, canals, and rainwater. Contamination of these freshwater bodies has posed health problems to humans. Figure 5.2 shows the possible human activities that contaminate freshwater bodies. A need to develop methods that will impact water purification and enhanced quality has to be reviewed. The main contaminants of these water bodies include organic and inorganic contaminants. Wastewater has been classified as highly polluted water due to the presence of metals and heavy metal ions, bacteria, viruses, garbage, and detergents. Organic pollutants include pesticides from agricultural activities, dyes, and pharmaceuticals from healthcare centres. However, heavy metals, phosphate, and nitrates are examples of inorganic pollutants that are significant contaminants in water. These pollutants are non-biodegradable, persistent in nature with possible chances to end in food chain (Adeola et al. 2022).

In Fig. 5.3, a process for wastewater treatment has been displayed. The first basic step during water treatment first includes coagulation. In order to produce complex precipitates of metal hydroxides at the appropriate pH during the first phase (coagulation), an inorganic coagulant, such as a metal salt, is first introduced to change the physical state of dissolved and suspended particles (Chai et al. 2021). Secondly, flocculation wherein the flocculants are added to increase the rate of sedimentation and treatment efficiency by mechanically aggregating micro-flocs into dense, visible flocs that settle quickly. The filtration step is wherein the solid and liquid phases are being filtered under pressure (Mashile et al. 2020).

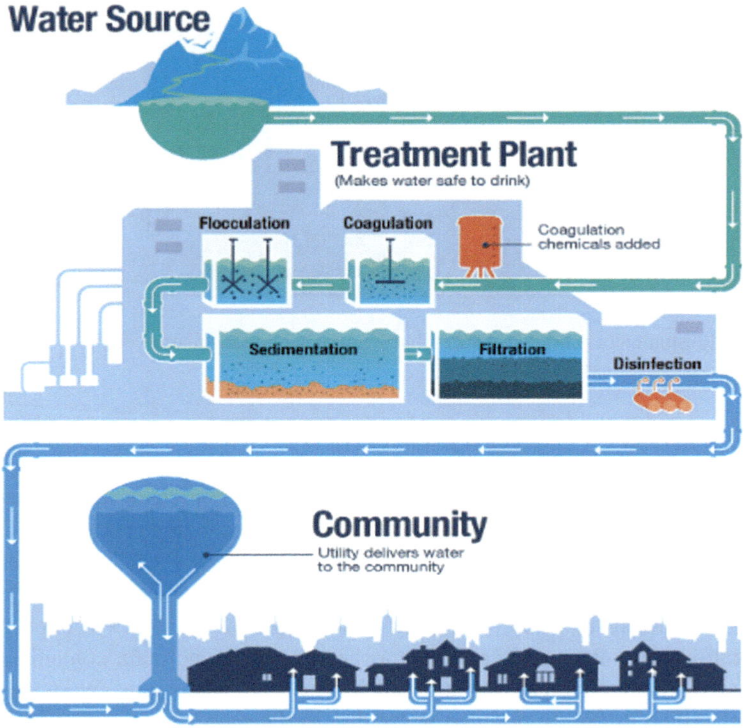

Fig. 5.3 Water treatment plant

5.2 Chitosan: A Biopolymer

To provide people with safe water for drinking, water treatment is a fundamental task that needs to be taken into consideration and put into common practice. Chitosan has been regarded as the second most abundant biopolymer used for water treatment. The chitosan biopolymer and widely developed derivatives are applied on the flocculation step of water treatment. Since chitin and chitosan are abundant, non-poisonous, and have good biocompatibility, they have a wide range of applications (Jin et al. 2021). Figure 5.4 shows a brief summary of different applications that chitosan has found its way in (Suyambulingam et al. 2023).

Chitin and chitosan, as both byproducts of the global seafood industry and as vital parts of many creatures' exoskeletons, are widely distributed in the biosphere (Balakrishnan et al. 2022). The Greek word "chiton," meaning "mail coat," is where the word "chitin" originated. In 1799, a researcher called Hatchett was the first to extract chitin from the shells of prawns, crayfish, and molluscs (crabs and lobsters). In fact, it is a naturally occurring polymer made up of acetyl-D-glucosamine monomers. Figure 5.5 shows the structures of chitin and chitosan (Iber et al. 2022).

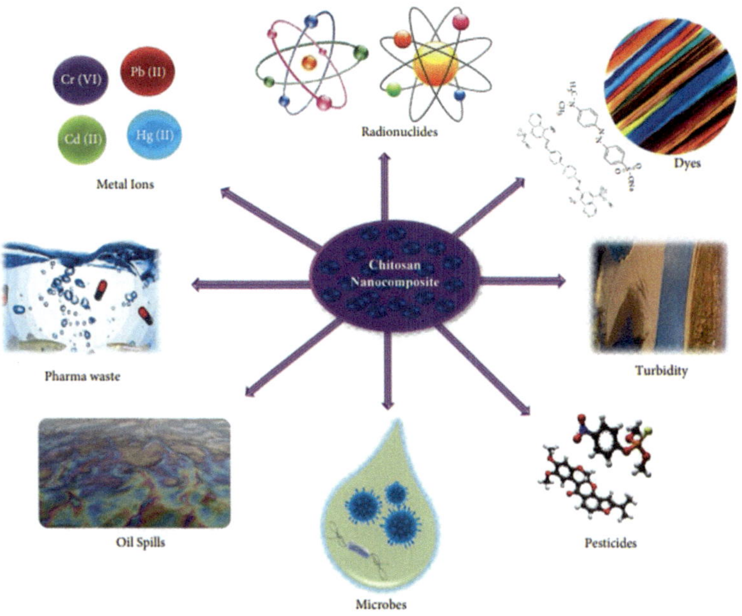

Fig. 5.4 Chitosan-based composites for various applications (Suyambulingam et al. 2023)

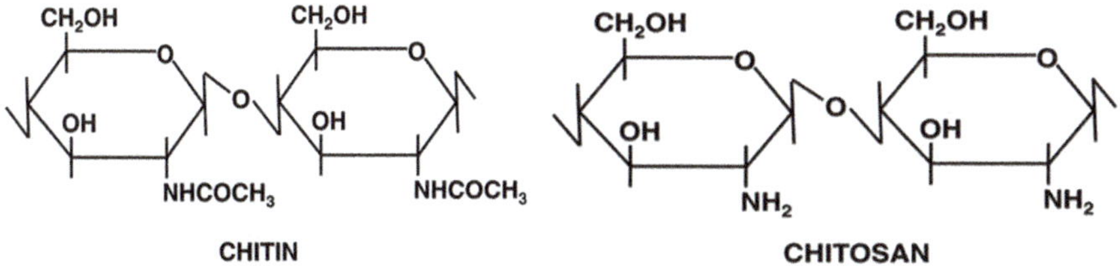

Fig. 5.5 Chitin and chitosan structures

The biopolymer needs to be deacetylated before chitosan can be produced. The deacetylation process has been represented in Fig. 5.6 (Iber et al. 2022). The extraction can be done following the biological procedure for extracting the chitin using microbes. Objectives have not been fulfilled by the production of goods that capitalise on the bioactivities of crustaceans, the main economic source of chitin (Omer et al. 2022). Research for alternative natural sources of chitin and chitosan has also been fuelled by the drawbacks of the current commercial source, including seasonality and rivalry for other purposes. The molecular weight (MW) and degree of deacetylation (DD) of chitosan are the structural parameters that dictate its characteristics, such as its water solubility, conformation, and chain stiffness in solution. The cationic polymer chitosan, which is obtained from the deacetylation of chitin, is extensively found in the exoskeletons of crustaceans, insects, arthropods, and molluscs. It is important to remember that the chemical extraction platform used to

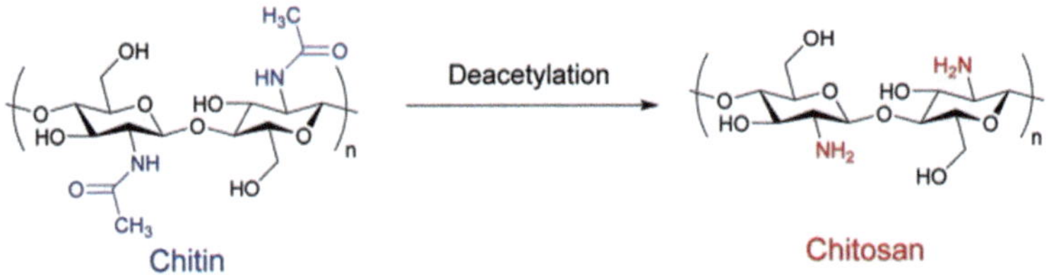

Fig. 5.6 Deacetylation of chitin to chitosan (Suyambulingam et al. 2023)

separate chitin undergoes three stages: deproteinisation, demineralisation, and decolorisation. Chitosan cannot be created without deacetylating chitin (Iber et al. 2022).

Chitosan as a biopolymer used for water treatment possesses the primary benefits of which includes its cost-effectiveness, environmentally friendly, relatively acceptable adsorption properties, less toxic, biodegradable, exceptionally higher thermal and chemical stability, and biocompatibility (Ma 2022). These special qualities allow chitosan to be used for potential catalytic assistance. The hydroxyl and amine groups on the chitosan structure make chitosan significantly insoluble in certain solvents, enabling it to be suitable for the redemption of organic pollutant (Martínez-Huitle and Panizza 2018).

Chitosan, as a byproduct of chitin largely generated through chemical conversion, has a lesser specific weight and acetylation degree than chitin. The product of deacetylated chitin are acetate ions and an $-NH_2$ group produced by hydrolysis of the acetamide group. The glucosamine to N-acetyl glucosamine ratio is typically used to determine chitosan's degree of deacetylation in chitosan (Chadha et al. 2022). Whereas chitosan has a higher proportion of glucosamine than N-acetyl glucosamine, while chitin has a higher proportion of glucosamine. Depending on the experimental circumstances, the degree of deacetylation has an impact on chemical and biological characteristics like adsorption enhancer, biodegradability, antioxidant, bioavailability, and biocompatibility. The chemical properties of the chitosan include tensile strength, viscosity, solubility, porosity, surface area, conductivity, and flexibility (Iber et al. 2022).

Due to its solid structure and strong intra- and intermolecular hydrogen bonds, chitin is immiscible with water and most organic solvents. On the other hand, chitosan dissolves readily in acidic solutions (e.g., hydrochloric acid, citric acid, and acetic acid) with a pH below 6 (Jiménez-Gómez and Cecilia 2020). Additionally, the usage of chitosan as a condensed solution promotes heavy metal adsorption at the free amine functions or homogenous modification reactions. However, in chitosan, the density of accessible primary amine groups which are principally in charge of the electrostatic interaction increases with increasing deacetylation (Tiwari et al. 2022). The increased percentage of amorphous regions in chitosan enhances its sorption capacity and availability, and it also makes it a perfect adaptable linker between various colloids in the coagulation/flocculation process. In Fig. 5.7, different methods of chitin extraction have been highlighted.

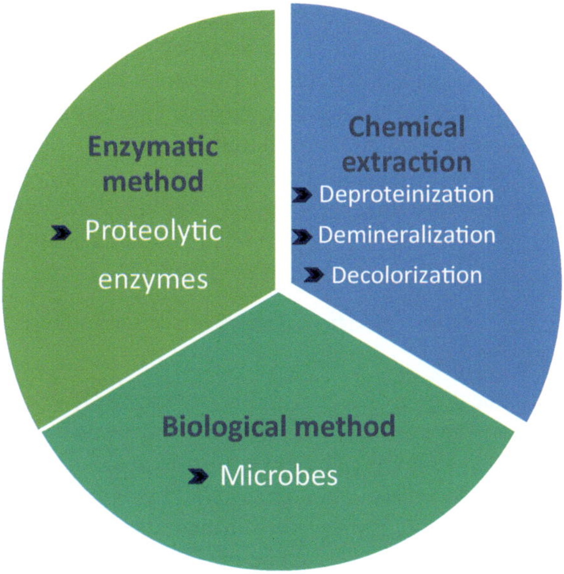

Fig. 5.7 Methods for chitin extraction from different sources (Iber et al. 2022)

5.3 Physicochemical Properties of Chitosan

5.3.1 Solubility

The wide range of solutions like water, alkaline solutions, diluted sulfuric and phosphoric acid solutions, and majority of organic solvents are unable to dissolve chitosan. This feature of chitosan's insolubility in water makes it a very more compatible biopolymer for water treatment. However, it was found to be soluble in dilute hydrochloric and nitric acid. Factors affecting chitosan's solubility are the type and concentration of solvents, its molecular mass and degree of deacetylation (Sangamnere et al. 2023). Literature have reported that diluted nitric acid had impact with obtaining 1–5 wt% chitosan solution. Furthermore, alkali breaks down the H bonds in chitosan at low temperatures, creating an alkali-chitosan complex and the process is known as dissociation (Adeola et al. 2022). Additionally, the urea hydrates contain an alkali-chitosan complex as an intermediary, creating an inclusion compound. However, when heated, the solution would gel since it is stable at low temperatures (Wang and Zhuang 2022).

5.3.2 pH

The polymer chitosan is pH sensitive. Chitosan has a pKa value of about 6.5. The pH level has a significant impact to reversibly change chitosan's solubility. When chitosan dissolves in acidic solutions, the amino group's N atom binds H+ from solutions by swapping electrons, creating a positively charged polyelectrolyte. The repulsion of identical charges forces the molecules unfold on the molecular chain (Gunatilake 2015). Chitosan becomes soluble when its intra- and intermolecular hydrogen bonds are gradually disrupted. This process can be aided by additional procedures like heating and stirring, albeit at the expense of some deterioration (Haripriyan et al. 2022). Additionally, the β (1 → 4) glycosidic link in chitosan has an acid-sensitive hemiacetal structure (No and Meyers 2000). There's possible degradation of the chitosan solutions after being stored for a long time.

5.3.3 *Deacetylation Degree*

Since the degree of deacetylation and the molecular weight of commercial chitosan vary depending on the raw materials and production process, the chitosan deacetylation degree is typically higher than 55%, which makes it soluble in 1% nitric acid solution. As a result, commercial chitosan typically bears different degrees of acetylation groups in its structure in addition to the prominent amino and hydroxyl groups. The chitosan structure depicted in Figure 5.6 should have a fully deacetylation degree, but it was found to be challenging to achieve the ideal structure (Wang and Zhuang 2022).

5.3.4 *Viscosity*

When the concentration, molecular weight, and degree of deacetylation of the chitosan increase, the viscosity of the chitosan solution falls as the temperature increases (Cohen and Poverenov 2022). The solvent has a significant effect on the chitosan. It will cause chain breakage, leading to chitosan solutions viscosity to drop during extended storage in acidic solutions. Furthermore, the early phases exhibit a non-linear and less sensitive time-dependent drop in molecular weight and viscosity (Riaz Rajoka et al. 2019). Chitosan's intrinsic viscosity decreased from 7.4 dl/g to 2.2 dl/g following 48 h of storage in 0.1 M HAc solution. Surface tension increased from 54.6 dyn/cm to 31.5 dyn/cm when the concentration of HAc was increased from 10% to 90%, but there was no appreciable change in viscosity (Wang and Zhuang 2022).

5.4 Different Methods for Water Treatment

There are several methods that have been reported for water treatment. The section highlights the most reported methods amongst the others along with their advantages and limitations. Moreover, the methods outlined in this section made use of chitosan biopolymer along with the chitosan-based functionalised. Some reported methods include adsorption, coagulation and flocculation, membrane filtration, chemical precipitation, ion exchange and electrochemical treatment (Zhao et al. 2016).

5.4.1 Flocculation

For method development and optimisation of reaction conditions, the flocculation mechanisms and kinetic processes are needed to direct the creation of innovative, high-performing flocculants (Sun et al. 2019). The interaction of contaminants and flocculants at a broad range via kinetic and flocculation mechanisms has been shown in Fig. 5.8. During wastewater treatment, the flocculant is used at its suitable concentration under flocculation kinetics using polymeric flocculants. The flocculant will be immersed in wastewater as part of the wastewater treatment processes, and the macromolecular flocculant will come into contact with the suspended colloids via adsorption, electrostatic interactions, hydrogen bonds, van der Waals forces, etc. (Lichtfouse et al. 2019).

The different polymeric flocculants, such as chitosan and its derivatives, have different flocculation mechanisms that are categorised as charge patching, sweeping, bridging, and simple charge neutralisation. Since the structure of the adsorbed polymer changes as a result of the kinetics of flocculation by polymeric flocculants, the adsorbed suspended particles accumulate to form sizable flocs that effectively settle down (Velasco-Garduño et al. 2020). Firstly, the type of flocculant was used in the wastewater along with its ionic, charge, functional, and adsorption qualities were, and MW all affect the major flocculation mechanism and secondly, the characteristics of the dissolved solutes and colloidal dispersion, including their functional groups, surface charge density, and hydrophobicity.

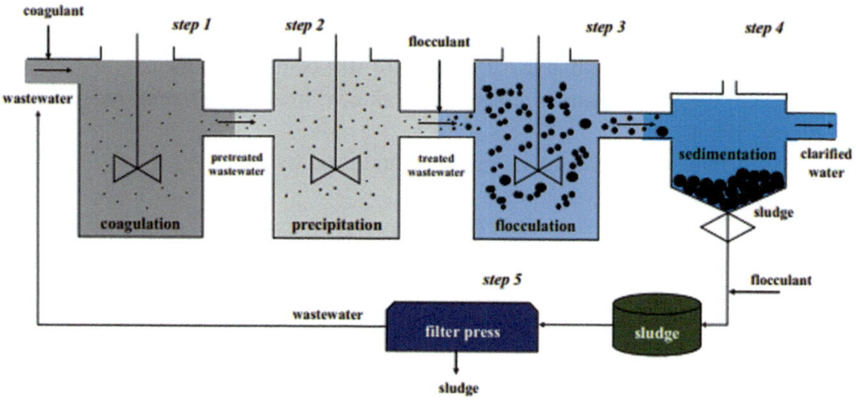

Fig. 5.8 The flocculation process (Lichtfouse et al. 2019)

The study conducted by Kaczmarek et al. in 2022 used chitosan biopolymer for eradicating colloidal particles from water by coagulation and flocculation processes (Ma 2022). The process showed the highest reduction percentage when using chitosan as a flocculant compared to other flocculants used in this study, respectively (Ma 2022). In 2022, another report was published by Hamdon et al. where chitosan was used for water purification in Egypt (Hamdon et al. 2022). The study showed a significant impact that the chitosan dosage has on adsorption of the microbes during water treatment. Furthermore, as the chitosan dosage increases the amount of adsorbed and removed pollutants also increases, hence this showed a direct proportionality relationship between the two (Hamdon et al. 2022).

5.4.1.1 Flocculation-Based Applications of Chitosan Biopolymer

The main characteristics and properties of chitosan as a flocculant include and not limited to:

- The biopolymer is highly hydrophilic with high reactivity.
- Chitosan is environmentally friendly and eco-friendly due to its biodegradability, non-toxicity, and biocompatibility.
- Chitosan structure has an amino polysaccharide which accounts for high nitrogen content.
- Compared to metal salts, chitosan is less susceptible to change in pH.
- The interactions on a chitosan structure are able to form hydrogen bonds and other non-covalent bonds.
- It has the ability to possess the adsorption, ion-exchange and chelation properties.
- Chitosan can exceptionally remove pollutants like colour and odour.
- Microalgae cells are strongly adsorbed due to their negatively charged surface (Karimi-Maleh et al. 2021).

There are drawbacks of using chitosan biopolymer as a bio-flocculant during the process of water and wastewater treatment including:

- It can clog the pores of the filters.
- Different technologies that use chitosan are still being developed.
- Since chitosan is insoluble in water, a weak acid treatment is employed in order to dissolve it.
- Its performance depends on its raw type and its activation.
- Chitosan is pH sensitive.
- Chitosan's improved performance requires chemical modification in order to facilitate its pH sensitivity and improve its potential application (Lichtfouse et al. 2019).

Due to high contamination of our freshwater systems, several techniques have been employed in order to overcome a challenge of having access to clean and safe drinking water.

5.4.1.2 Factors Impacting the Flocculation Mechanism

Numerous elements significantly affect the flocculant's structure and solution characteristics and consequently, the flocculation process and performance. The following are the factors that influence the flocculation mechanisms: effects of pH, temperature, ionic strength, and flocculant dose (Kanmani et al. 2017).

5.5 Techniques Adopted for Water Treatment

5.5.1 Adsorption

As one of the most prevalent methods, adsorption has been the most adopted method for removal of contaminants in different matrices. The method has been favoured due to its cost-effectiveness. The process of adsorption may be reversible, and an appropriate desorption technique can regenerate adsorbents (Peralta et al. 2019). Adsorbents can be renewed through the use of an appropriate desorption technology when the adsorption process is occasionally reversible. Adsorbent materials can be selected based on the surrounding conditions and the type of metal contamination present in the groundwater.

For adsorption process different solid materials can be used like activated carbons that have been reported to have high surface area, metal oxides, and biomass generated from agricultural waste. Moreover, these methods have three major stages which involve: Pollutant diffusion from solution to adsorbent surface. Adsorption of the pollutant onto the surface of the adsorbent and adsorption into the adsorbent's structure. In 2021, Omar et al. reported about adsorptive removal of the heavy metal ions using a chitosan-based adsorbent in wastewater (Ramírez Calderón et al. 2020). The study has reported that the chitosan biopolymer has shown a greater potential to influence to a certain degree metal ions removal from wastewater treatment. From a review published in 2016 by Yang et al., chitosan was used for flocculation of some of the organic pollutants, respectively (Yang et al. 2016).

5.5.2 Biosorption

Biosorption refers to the ability of biological materials to accumulate contaminants from wastewater through metabolically mediated or spontaneous physicochemical pathways of uptake, or as a characteristic of specific kinds of non-living, dormant microbial biomass that bind and concentrate the pollutants from even extremely weak water solutions (Ramírez Calderón et al. 2020). Cell physiology, physicochemical parameters like pH, temperature, contact time, ionic strength, and metal concentration, as well as the chemistry of pollutants of interest, all play a role in this intricate process and cell wall composition of microorganisms. In 2017, Holder et al. have reported about the use of cross-linked chitosan-graphene oxide membranes for the treatment of municipal wastewater (Holder et al. 2017).

5.5.3 Ion Exchange

Ion exchange processes, in which heavy metals and their ions are physically absorbed, qualify as the first step in the ion exchange process. The reversibility of ions serves as the fundamental idea behind this technique (Zhao et al. 2016). The development of the counter-ion complex and functional group is the primary factor influencing the removal efficiency of this approach. Reversible exchange method involving a resin is used to remove ions from an electrolytic solution between the two phases of liquid and solid. Additionally, a resin substance is used to remove metals from water. There are different factors that affect the effectiveness of removing various metals and their ions, which are mostly dependent on the chemical properties of resins, temperature, pH, sorbents, anions, and the duration of contact, as well as the initial concentration of the adsorbent (Sarode et al. 2019). Several advantages establish this process apart from other physico-chemical methods, including excellent ion removal efficiency, reduced sludge volume, metal recovery, good selectivity, and a unique resin structure throughout the entire process. Ion exchange drawbacks include cost cost-effectiveness of resin, used exclusively at a small scale, and resin rejuvenation causes secondary pollutants (Rafique et al. 2022).

5.6 Common Wastewater Pollutants

Water pollution is a universal problem. All pollutants found in wastewater come from different human activities. These pollutants include organic pollutants, inorganic pollutants, and bacterial suspensions. There have been many reports about water treatment. This accounts for different pollutants that contribute to water pollution and contamination. Out of many pollutants reported, the heavy metals were found to be more prevalent, hence many studies were reporting about their remediation strategies (Begum et al. 2021).

5.6.1 Heavy Metals

Periodic table in Fig. 5.9 consists of 90 naturally occurring elements. However, from the list of these 90 elements, there are 21 non-metals, 16 light metals and the remaining 53 including arsenic are heavy metals. Heavy metals are usually defined as metals with a density that is more than 5 g/cm^3. Since they can be needed in our

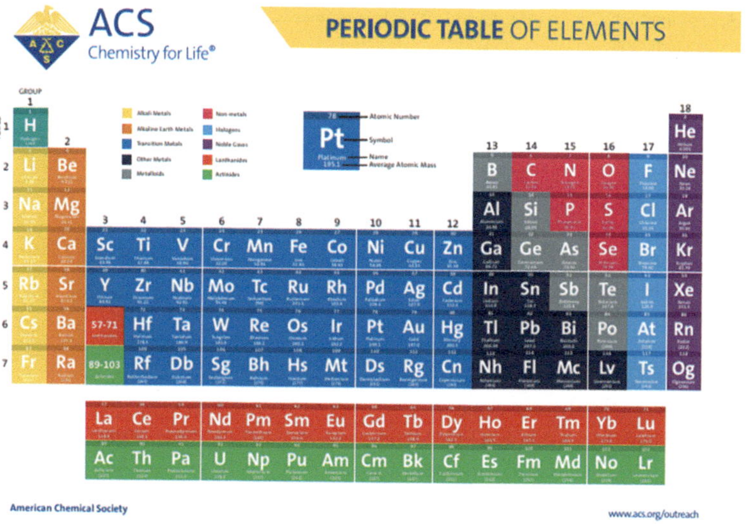

Fig. 5.9 Periodic table of elements (Pt 9 78, no. 262, p. 262)

bodies at trace levels, we have metals that have been regarded as essential such as copper, zinc and nickel (Dinu et al. 2021). These metals have been regarded as being essential because they are required for normal cellular growth. Furthermore, those metals that have no effect on biological growth have been regarded as non-essential and these include mercury, lead and cadmium (Vakili et al. 2014). At any concentration, these kinds of metals are hazardous.

Mostly, transition metals on the periodic table are heavy metals because their d orbital is not fully filled. Heavy metal cations can form complex compounds with or without redox activity due to the incompletely filled d orbital. Heavy metal trace elements are crucial cofactors for metalloproteins and enzymes and play a significant part in biological reactions (Sakib et al. 2021). Depending on the complex compounds that the heavy metal ions form, their toxicity is observed when their concentration is larger in the cells (Yang et al. 2021). Heavy metals contamination comes from various industrial activities such as mining, energy and fuel production and use, tanneries, pipelines corroding, municipal waste processing facilities, and agro-industrial operations (Wang et al. 2022). Nevertheless, natural processes like rock erosion and weathering can also release heavy metals. Heavy metals distribution to the environment includes dispersion and transport through water bodies.

When humans are exposed to heavy metals in amounts that are adequately measurable and depending on the kind of metal ion, they might result in a variety of symptoms such as nausea, diarrhoea, teeth fluorosis, skin allergies, and even cancer (Sumaila et al. 2022). Additional health effects of heavy metals include congenital conditions such as cerebral palsy, cardiovascular disease, mutations, damage to neurons kidney and liver damage, and effects on a child's developing brain (Bhatt et al. 2023). An assortment of rules and laws has been created in light of the negative

Table 5.1 Standard acceptable levels of selected heavy metals in drinking water (Ramírez Calderón et al. 2020)

Heavy metals	Anthropogenic sources	WHO guideline (μg/L)	Maximum permissible level for drinking water (μg/L)	
			EC	USEPA
Copper	Corroded interior copper plumbing systems	2000	2000	1300
Nickel	Stainless steel and nickel alloys production	70	20	–
Chromium	Steel and pulp mills, tanneries	50	50	100
Selenium	Mines and petroleum refineries	40	10	50
Lead	Lead-acid batteries, solder, alloys, and corroded plumbing systems	10	10	15
Arsenic	Glass and electronics production	10	10	10
Mercury	Refineries and factories, runoff from landfills and agriculture, and electrical appliances	6	1	2
Cadmium	Batteries and paints, steel and plastic industries, metal refineries, corroded galvanised pipes	3	5	5

WHO World Health Organisation, *EC* Emerging contaminants, *USEPA* United States Environmental Protection Agency

effects that heavy metal exposure can have on human health in order to limit exposure to amounts that could cause harm. The prevalent man-made sources and advised concentrations of significant heavy metals in drinking water, including the World Health Organisation (WHO) recommended levels and legally permitted upper limits in the EU (Nasrollahzadeh et al. 2021). Table 5.1 briefly shows the acceptable concentration levels of some heavy metals. As is evident, guidelines established by the World Health Organisation serve as the foundation for many laws.

These values, however, are determined for particular metals based on practical removal technologies and currently available detection limits determined using the practical removal technologies and detection thresholds that are now accessible (Keshvardoostchokami et al. 2021). Not only must the release of heavy metals be controlled, but these pollutants which are essentially secondary metal resources must be extracted and recovered in order to avoid negative impacts on the environment and public health (Aizat and Aziz 2019). The EPA claims that heavy metals may not be effectively eliminated from conventional wastewater treatment facilities or may wind up in the sludge. Therefore, if the intent is to extract or recover the heavy metals, then further steps and techniques are required.

5.7 Conclusion

Water is a basic need that we use in our daily lives. Due to poor access to clean water all around the world, water treatment has been an adopted solution to this gigantic catastrophe. Implications and employment of different techniques for water purification can assist with meeting the supply of high demand of clean and safe water. In this chapter, we were looking into chitosan as a biopolymer and its application in water treatment. Chitosan, a natural biocompatible and biodegradable polymer, has gained interest for water treatment in recent years. Adding to its advantages due to its availability naturally, its high charge density, excellent turbidity removal compared to its inorganic coagulants and the reactive amine and hydroxyl groups that are present in its structure.

Many studies of its application in water treatment have been reported wherein it showed excellent performance in flocculation. Chitosan also shows its strong potential to be used for water treatment. However, more studies to explain its characteristics and mechanisms still need to be investigated in order to gain knowledge on its absorption/coagulation/flocculation ability. Furthermore, with different reported methods of water treatment involving the chitosan biopolymer, there's a need to investigate the developed methods and employ improved strategies in developing the most cost-effective, more environmentally friendly, and less toxic method and chitosan-based nanocomposites. The crosslinking of the chitosan with different composites to fit the flocculation of the type of pollutants for a certain purification study has also shown significantly good and acceptable performance. Different characteristics and parameters such as pH, chitosan-based nanocomposite dosage, molecular weight, degree of diacylation, and settling time also need to be part of the figures to be investigated when improving the developed methods applied for water treatment.

References

Adeola AO, Abiodun BA, Adenuga DO, Nomngongo PN (2022) Adsorptive and photocatalytic remediation of hazardous organic chemical pollutants in aqueous medium: a review. J Contam Hydrol 248(January):104019. https://doi.org/10.1016/j.jconhyd.2022.104019

Agronomy D (2019) The clean water: the basic need of human and agriculture related papers. https://doi.org/10.1027/IJCS.2022218235

Aizat MA, Aziz F (2019) Chitosan nanocomposite application in wastewater treatments. Elsevier Inc. https://doi.org/10.1016/B978-0-12-813902-8.00012-5

Balakrishnan A, Appunni S, Chinthala M, Vo DVN (2022) Biopolymer-supported TiO2 as a sustainable photocatalyst for wastewater treatment: a review. Environ Chem Lett 20(5):3071–3098. https://doi.org/10.1007/s10311-022-01443-8

Begum S, Yuhana NY, Saleh NM, Kamarudin NHN, Sulong AB (2021) Review of chitosan composite as a heavy metal adsorbent: material preparation and properties. Carbohydr Polym 259(September 2020):117613. https://doi.org/10.1016/j.carbpol.2021.117613

Bhatt P, Joshi S, Urper Bayram GM, Khati P, Simsek H (2023) Developments and application of chitosan-based adsorbents for wastewater treatments. Environ Res 226(February):115530. https://doi.org/10.1016/j.envres.2023.115530
Chadha U et al (2022) Advances in chitosan biopolymer composite materials: from bioengineering, wastewater treatment to agricultural applications. Mater Res Express 9(5). https://doi.org/10.1088/2053-1591/ac5a9d
Chai WS et al (2021) A review on conventional and novel materials towards heavy metal adsorption in wastewater treatment application. J Clean Prod 296:126589. https://doi.org/10.1016/j.jclepro.2021.126589
Cohen E, Poverenov E (2022) Hydrophilic chitosan derivatives: synthesis and applications. Chem Eur J 28(67). https://doi.org/10.1002/chem.202202156
Dinu MV, Humelnicu D, Lazar MM (2021) Analysis of copper(ii), cobalt(ii) and iron(iii) sorption in binary and ternary systems by chitosan-based composite sponges obtained by ice-segregation approach. Gels 7(3). https://doi.org/10.3390/gels7030103
Gunatilake SK (2015) Methods of removing heavy metals from industrial wastewater. Methods 1(1):12–18
Hamdon RSMA, Salem A, Ahmed HGI, ElZahar MMH (2022) Use of chitosan for enhancing the process of surface water purification in Egypt. Int J Environ Sci Dev 13(2):26–34. https://doi.org/10.18178/ijesd.2022.13.2.1368
Haripriyan U, Gopinath KP, Arun J, Govarthanan M (2022) Bioremediation of organic pollutants: a mini review on current and critical strategies for wastewater treatment. Arch Microbiol 204(5):1–9. https://doi.org/10.1007/s00203-022-02907-9
Holder SL, Lee CH, Popuri SR (2017) Simultaneous wastewater treatment and bioelectricity production in microbial fuel cells using cross-linked chitosan-graphene oxide mixed-matrix membranes. Environ Sci Pollut Res 24(15):13782–13796. https://doi.org/10.1007/s11356-017-8839-2
Iber BT, Kasan NA, Torsabo D, Omuwa JW (2022) A review of various sources of chitin and chitosan in nature. J Renew Mater 10(4):1097–1123. https://doi.org/10.32604/JRM.2022.018142
Jiménez-Gómez CP, Cecilia JA (2020) Chitosan: a natural biopolymer with a wide and varied range of applications. Molecules 25(17). https://doi.org/10.3390/molecules25173981
Jin T, Liu T, Lam E, Moores A (2021) Chitin and chitosan on the nanoscale. Nanoscale Horiz 6(7):505–542. https://doi.org/10.1039/d0nh00696c
Kanmani P, Aravind J, Kamaraj M, Sureshbabu P, Karthikeyan S (2017) Environmental applications of chitosan and cellulosic biopolymers: a comprehensive outlook. Bioresour Technol 242:295–303. https://doi.org/10.1016/j.biortech.2017.03.119
Karimi-Maleh H et al (2021) Recent advances in using of chitosan-based adsorbents for removal of pharmaceutical contaminants: a review. J Clean Prod 291. https://doi.org/10.1016/j.jclepro.2021.125880
Keshvardoostchokami M, Majidi M, Zamani A, Liu B (2021) A review on the use of chitosan and chitosan derivatives as the bio-adsorbents for the water treatment: removal of nitrogen-containing pollutants. Carbohydr Polym 273(September):118625. https://doi.org/10.1016/j.carbpol.2021.118625
Lichtfouse E et al (2019) Chitosan for direct bioflocculation of wastewater. Environ Chem Lett 17(4):1603–1621. https://doi.org/10.1007/s10311-019-00900-1
Ma P (2022) Backwash water treatment
Martínez-Huitle CA, Panizza M (2018) Electrochemical oxidation of organic pollutants for wastewater treatment. Curr Opin Electrochem 11(1):62–71. https://doi.org/10.1016/j.coelec.2018.07.010
Mashile PP, Mashile GP, Dimpe KM, Nomngongo PN (2020) Occurrence, quantification, and adsorptive removal of nodularin in seawater, wastewater and river water. Toxicon 180(April):18–27. https://doi.org/10.1016/j.toxicon.2020.03.009

Mashile GP, Selahle SK, Mpupa A, Nqombolo A, Nomngongo PN (2022) Remediation of emerging pollutants through various wastewater treatment processes. Elsevier Inc. https://doi.org/10.1016/B978-0-12-822850-0.00005-3

Nasrollahzadeh M, Sajjadi M, Iravani S, Varma RS (2021) Starch, cellulose, pectin, gum, alginate, chitin and chitosan derived (nano)materials for sustainable water treatment: a review. Carbohydr Polym 251(May 2020):116986. https://doi.org/10.1016/j.carbpol.2020.116986

No HK, Meyers SP (2000) Application of chitosan for treatment of wastewaters. Rev Environ Contam Toxicol 163(February):1–27. https://doi.org/10.1007/978-1-4757-6429-1_1

Omer AM, Dey R, Eltaweil AS, Abd El-Monaem EM, Ziora ZM (2022) Insights into recent advances of chitosan-based adsorbents for sustainable removal of heavy metals and anions. Arab J Chem 15(2):103543. https://doi.org/10.1016/j.arabjc.2021.103543

Peralta ME et al (2019) Highly efficient removal of heavy metals from waters by magnetic chitosan-based composite. Adsorption 25(7):1337–1347. https://doi.org/10.1007/s10450-019-00096-4

Pt 9 78, no. 262, p 262

Rafique M, Hajra S, Tahir MB, Gillani SSA, Irshad M (2022) A review on sources of heavy metals, their toxicity and removal technique using physico-chemical processes from wastewater. Environ Sci Pollut Res 29(11):16772–16781. https://doi.org/10.1007/s11356-022-18638-9

Ramírez Calderón OA, Abdeldayem OM, Pugazhendhi A, Rene ER (2020) Current updates and perspectives of biosorption technology: an alternative for the removal of heavy metals from wastewater. Curr Pollut Rep 6(1):8–27. https://doi.org/10.1007/s40726-020-00135-7

Riaz Rajoka MS, Zhao L, Mehwish HM, Wu Y, Mahmood S (2019) Chitosan and its derivatives: synthesis, biotechnological applications, and future challenges. Appl Microbiol Biotechnol 103(4):1557–1571. https://doi.org/10.1007/s00253-018-9550-z

Sakib MN, Mallik AK, Rahman MM (2021) Update on chitosan-based electrospun nanofibers for wastewater treatment: a review. Carbohydr Polym Technol Appl 2(February):100064. https://doi.org/10.1016/j.carpta.2021.100064

Sangamnere R, Misra T, Bherwani H, Kapley A, Kumar R (2023) A critical review of conventional and emerging wastewater treatment technologies. Sustain Water Resour Manag 9(2):1–26. https://doi.org/10.1007/s40899-023-00829-y

Sarode S et al (2019) Overview of wastewater treatment methods with special focus on biopolymer chitin-chitosan. Int J Biol Macromol 121:1086–1100. https://doi.org/10.1016/j.ijbiomac.2018.10.089

Sumaila A, Sumaila AO, Abdullahi AS, Usman AO, Ekwoba L (2022) Application of chitosan-silver nanocomposites for heavy metals removal: a review study. J Mater Environ Sci 2022(8):869–883

Sun Y et al (2019) Novel chitosan-based flocculants for chromium and nickle removal in wastewater via integrated chelation and flocculation. J Environ Manag 248(July):109241. https://doi.org/10.1016/j.jenvman.2019.07.012

Suyambulingam I et al (2023) Chitosan biopolymer and its nanocomposites: emerging material as adsorbent in wastewater treatment. Adv Mater Sci Eng 2023. https://doi.org/10.1155/2023/9387016

Tiwari AK et al (2022) Bioadsorbent and adsorbent-based heavy metal removal technologies from wastewater: new insight. Biomass Convers Biorefin:0123456789. https://doi.org/10.1007/s13399-022-02343-1

Vakili M et al (2014) Application of chitosan and its derivatives as adsorbents for dye removal from water and wastewater: a review. Carbohydr Polym 113:115–130. https://doi.org/10.1016/j.carbpol.2014.07.007

Velasco-Garduño O, Martínez ME, Gimeno M, Tecante A, Beristain-Cardoso R, Shirai K (2020) Copper removal from wastewater by a chitosan-based biodegradable composite. Environ Sci Pollut Res 27(23):28527–28535. https://doi.org/10.1007/s11356-019-07560-2

Wang J, Zhuang S (2022) Chitosan-based materials: preparation, modification and application. J Clean Prod 355(March):131825. https://doi.org/10.1016/j.jclepro.2022.131825

Wang F, Gao J, Jia L, Wang S, Ning P (2022) Green synthesis of a novel functionalized chitosan adsorbent for Cu(II) adsorption from aqueous solution. Environ Sci Pollut Res 29(1):989–998. https://doi.org/10.1007/s11356-021-15684-7

Yang R, Li H, Huang M, Yang H, Li A (2016) A review on chitosan-based flocculants and their applications in water treatment. Water Res 95(2015):59–89. https://doi.org/10.1016/j.watres.2016.02.068

Yang J, Chen X, Zhang J, Wang Y, Wen H, Xie J (2021) Role of chitosan-based hydrogels in pollutants adsorption and freshwater harvesting: a critical review. Int J Biol Macromol 189(235):53–64. https://doi.org/10.1016/j.ijbiomac.2021.08.047

Zhao M, Xu Y, Zhang C, Rong H, Zeng G (2016) New trends in removing heavy metals from wastewater. Appl Microbiol Biotechnol 100(15):6509–6518. https://doi.org/10.1007/s00253-016-7646-x

Chapter 6
Alginate-Based Biosorbents for Environmental Remediation

Lama Sirkhot, Pragati Prakash Sabale, and Jaya Lakkakula

Abstract Biosorbents are materials with capability to adsorb pollutants from wastewater solutions offering a promising solution for environmental remediation. Among these materials, sodium alginate (SA) is a naturally occurring biodegradable polysaccharide derived from brown seaweed. SA has gained significant attention due to its hydrophilic ability, cost-effectivity as well as versatility. With the evolution in industries an ever-increasing concern regarding the wastewater treatment of industrial effluents, which contain combination of contaminants like heavy metals, oils, and dyes, raises an urgent need for effective and sustainable water treatment methods. Despite the utility, the conventional methods are limited in their efficacy and scalability highlighting the critical demand for innovative approaches. This chapter focuses on the development and application of alginate-based biosorbents specifically designed to address the limitations of traditional water treatment methods through modifying sodium alginate through techniques like cross-linking with nanoparticles and combining with biopolymers to form bio-composites—these novel biosorbents exhibit enhanced adsorption capacity and stability along with antibacterial properties. The chapter delves into various formulations and their effectiveness in eliminating a wide range of contaminants from aqueous solutions by showcasing structural and functional advantages of alginate-based solutions focusing on both a viable alternative for effective wastewater treatment along with its emphasis on environmental and economical benefits associated with utilizing biodegradable materials addressing the growing challenge of water pollution.

L. Sirkhot · P. P. Sabale
Amity Institute of Biotechnology, Amity University Maharashtra, Mumbai-Pune Expresway, Panvel, Maharashtra, India

J. Lakkakula (✉)
Amity Institute of Biotechnology, Amity University Maharashtra, Mumbai-Pune Expresway, Panvel, Maharashtra, India

Centre for Computational Biology and Translational Research, Amity University Maharashtra, Mumbai-Pune Expressway, Panvel, India
e-mail: jrlakkakula@mum.amity.edu

A. Guldhe et al. (eds.), *Biomaterials for Environmental Applications: Detection and Remediation*, Environmental Challenges and Solutions,
https://doi.org/10.1007/978-3-032-12701-3_6

Keywords Sodium alginate · Bioadsorbents · Wastewater remediation · Heavy metals · Dyes

6.1 Introduction

In the last decade, a pronounced increase in oil spills as well as areas with industrial wastewater discharge has led to severe water contamination, posing a substantial threat to aquatic ecosystems and human health. Traditional methods, such as air flotation, absorption of oil, centrifugation in addition to membrane separation, have been utilized for oil-water separation. Hydrophilic membranes, particularly inspired by antifouling properties in marine organisms, have developed as a new promising technique owing to its simplicity, efficacy, and cost-effectiveness. However, these membranes often excel at addressing singular pollutants, making them less effective in handling the complex composition of real-world wastewater. To address this challenge, researchers have sought inspiration from antifouling properties observed in marine organisms to create innovative hydrophilic membranes that resist oil contamination. Simultaneously, addressing bacterial interference, which compromises separation efficiency, involves exploring antibacterial agents like ZnO nanoparticles, Ag nanoparticles, graphene, and metal-organic frameworks. Quaternary ammonium compounds, known for their broad-spectrum effectiveness, are considered promising. Additionally, prevalent water pollutants such as dyes can be effectively removed through a combined adsorption-filtration methodology (Li et al. 2023a).

Sodium alginate (SA), an intrinsic hydrophilic biopolymer, stands out for its degradable nature, cost-effectiveness, and versatility. Modifying SA to fabricate a calcium alginate (CA) aerogel membrane enhances its proficiency in oil-water separation and facilitates heavy metal ion adsorption. In the context of this study, the integration of SA, tetrabutylammonium hydroxide (TBA), and waterborne polyurethane (WPU) has resulted in the development of a sustainable and environmentally benign CTW membrane. This membrane, characterized by super-hydrophilicity and subaquatic super-oleophobicity, demonstrates efficacy in the partition of oil-water emulsions, microorganism filtration, and dye adsorption from water. A comprehensive investigation into the structural, compositional, wettability, and performance aspects of the CTW membrane has been undertaken (Shen et al. 2023)

Concerns have arisen regarding water contamination due to cultural and industrial progress, with industrial effluents containing diverse pollutants like dyes, heavy metals, plastic materials, pesticides, as well as oils. The prevalent use of plastics has further compounded the environmental predicament, introducing micro-plastics and even smaller nano-plastics, posing intricate challenges due to their diminutive size and intricate composition. Existing methodologies, including filtration, encounter difficulties with particles below 20 μm, while alternative techniques exhibit limitations. Notably, membrane separation technology, employing selectively permeable membranes, stands out for its user-friendliness and low energy consumption.

Graphene oxide (GO), despite inherent drawbacks in its pure form, emerges as a promising material for membranes. Addressing this, composite membrane of a three-layer SA/GO/CS was developed through the modification of GO with sodium alginate (SA) and chitosan (CS), representing negatively and positively charged polymers, respectively. This biomimetic membrane leverages the positive charges as well as negative charges on both sides, mimicking natural magnetic principles, thereby enhancing its capacity to repel or attract nanoparticles and charged dyes (Zahakifar et al. 2021).

The composite membrane demonstrated notable permeability and selectivity in the efficient purification of wastewater containing emulsified oil, dyes, and nanoplastics. The study rigorously evaluated long-term performance, reusability, and structural stability of the eco-friendly membrane, with the overarching goal of addressing various pollutants in wastewater removal. The widespread utilization of organic dyes across diverse sectors has resulted in the contamination of water bodies, potentially engendering adverse consequences for human health and ecosystem. The wastewaters comprising these dyes present risks such as skin sensitization, irritation, and carcinogenic effects. Existing methods for dye removal often exhibit limitations, necessitating the exploration of environmentally benign alternatives (Zahakifar et al. 2021; Garg and Sabouni 2023).

Within this framework, the investigation is directed toward the formulation of an efficacious and environment loving adsorbent tailored for the elimination of hazardous dyes, exemplified by Brilliant Green (BG) and Methylene Blue (MB) from wastewater. This study introduces a pioneering composite material denoted as Alginate/AgNPs modified with mica (Alg-AgNPs/MC), which manifests promising adsorption capabilities. This composite is engineered by amalgamating silver nanoparticles (AgNPs) and mica with sodium alginate, a biocompatible and biodegradable polysaccharide. Leveraging the distinctive chemical and physical attributes of silver nanoparticles, the composite material significantly enhances adsorption capacity. The engineered Alg- AgNPs/MC bio-nanocomposite is designed with ecological considerations and economic viability, aiming to mitigate secondary water pollution. This study assesses its efficacy in removing the Brilliant Green (BG) and Methylene Blue from aqueous solutions. The material undergoes comprehensive characterization using various techniques, including BET, EDX, XRD, TGA, FTIR, SEM, and TEM. Operational parameters, such as pH, dose, concentration, and duration, are systematically adjusted to optimize adsorption efficiency. The investigation utilizes isotherm and kinetic models to validate experimental data, offering a thorough comprehension of the adsorption process. Thermodynamic analysis is employed to elucidate the nature and spontaneity of the adsorption. Furthermore, regeneration and desorption studies are directed to enhance the affordability as well as practicality of the adsorption process. Collectively, the findings highlight the Alg-AgNPs/MC bio-nanocomposite as a potentially promising, sustainable approach for the secure elimination of deleterious dyes from wastewater. This, in turn, contributes to resource sustainability and environmental safeguarding (Garg and Sabouni 2023).

Uranium, a crucial raw material for advancing the nuclear industry, plays a dual role as both essential and potentially environmentally hazardous. It is vital to extract uranium effectively from wastewater to mitigate radioactive contamination and protect the environment. China, facing a shortage of terrestrial uranium resources, is witnessing a rising demand for natural uranium owing to the quick growth of the nuclear business. Various techniques, including membrane treatment, precipitation, extraction, and adsorption, are employed for uranium enrichment and separation. From mentioned, adsorption is notable for its vast removal efficiency, stability, cost-effectiveness, and recyclability. However, conventional adsorption materials made with uranium have limitations like slow adsorption rates, weak selectivity, and stability. To address these challenges, there is a need for the development of adsorbent materials with enhanced efficiency. Ferricyanide ions, known for their ability to form potent complexes with U(VI), have been explored for uranium adsorption. Significant progress has been made by modifying substances like potassium manganese ferrocyanide and hydrated titanium dioxide, resulting in remarkable adsorption capabilities. Recent research indicates that the crosslinked variant of sodium alginate, specifically calcium alginate, exhibits promising capabilities for adsorbing uranyl ions. In this study, the investigation of KMnFC/CNT@SA, as a novel adsorption material for its ability in removing U(VI) from aqueous solutions, was performed. Adsorption kinetics and isotherms were analyzed using diverse models. Fourier-transform infrared spectroscopy (FTIR) and X-ray photoelectron spectroscopy investigations were conducted to gain insights into the structural changes of the adsorbent before and after the process of adsorption.

This study contributes to the ongoing efforts to develop efficient uranium adsorbent materials, crucial for mitigating the risks associated with uranium pollution and safeguarding human health and the environment. Adsorption technology, recognized for its simplicity, efficiency, and cost-effectiveness, relies on the advancement of high-performing adsorbents. Despite the promise of biochar due to its wide availability and low cost, unmodified biochar exhibits suboptimal cadmium removal capabilities. Attempts to enhance biochar adsorption using NaOH or H3PO4 have led to facility corrosion. Alginate, characterized by its egg-box structured gel sphere, demonstrates ion exchange capabilities but lacks mechanical strength (Wang et al. 2023; Li et al. 2023b; Wen et al. 2023; Sánchez-Castro et al. 2021).

A novel approach is presented by combining alginate, gelatin, and biochar in a ternary composite (SA/GE@BC). This synthesized composite addresses the shortcomings of raw biochar, improving cadmium removal efficiency while mitigating corrosion damage and the risk of secondary metal ion pollution. The adsorption process was systematically adjusted through experimentation with parameters like the dosage curve, pH, time, humic acid, starting cadmium level, and coexisting ions. This work contributes both technological and theoretical insights into the potential applications of ternary composites in the removal of cadmium from contaminated water (Zong et al. 2023; Sami Abboud et al. 2023; Sadoun et al. 2023).

6.2 Alginate-Based Biosorbents for Remediation

6.2.1 Alginate-Based Biosorbents for the Removal of Dyes

Zhao et al. investigated a novel hydrogel bead system, denoted as Ag-DE@CAH, engineered by incorporating nano-silver/diatomite and calcium alginate aiding in the removal of MB from water. Throughout a 30-day release experiment, the AgDE@CAH beads consistently released Ag+ ions within the World Health Organization (WHO) standards. Structural analysis via scanning electron microscopy (SEM) unveiled diverse pore structures distributed across the diatomite surface following acid pickling. High-resolution transmission electron microscopy (TEM) of Ag-DE beads disclosed the presence of silver nanoparticles (Ag NPs) distributed within the diatomite's pore structures and surface, with silver particles measuring less than 100 nm. FTIR indicated that the absorption peaks in Ag-DE and Ag-DECAH were accredited to the antisymmetric stretching vibration of Si-O-Si at 1080 cm^{-1}, affirming the effective embedding of Ag-DE into CAH. X-ray diffraction (XRD) patterns of DE, Ag-DE, and Ag-DECAH materials offered insights into their structural properties, including the presence of specific crystal phases and their arrangement. BET analysis determined the total pore volume (Vtotal) at PP0 0.99. The disinfection effectiveness of Ag-DE@CAH beads increased over time, achieving a sterilization rate for *Escherichia coli* that rose from 54.0% to 98.7% and for *Staphylococcus aureus* from 26.5% to 68.5% after 15 to 60 min of incubation, while the released silver ion concentration remained below the WHO limit of 100 ppb, measured at 41.27 ppb after 30 days (Fig. 6.1). Hydrogel beads of Ag-DE@CAH demonstrated the highest MB adsorption ability of 128.21 mg/g. At adsorption equilibrium concentration (C_e), the capacity of adsorption of Ag-DECAH beads outperformed other materials in methylene blue adsorption tests and exhibited reusability. The incorporation of diatomite (DE) as a cost-effective and naturally occurring matrix improved the synthesis of Ag-DECAH beads, enhancing their water purification capacity. In conclusion, the study suggests that these dual-functional hydrogel beads effectively contribute to both adsorption and disinfection processes, showcasing potential applications in water purification (Zhao et al. 2021).

In a research endeavor led by Verma et al., porous hybrid hydrogel of novelty was engineered by employing sodium alginate (SA) and graphite, with the primary aim of serving as a highly effective source of adsorbent for elimination of malachite green (MG) from aqueous solutions. The synthesis process involved graft polymerization of acrylic acid onto SA to improve capacity to adsorb. Subsequently, graphite powder was incorporated into the modified sodium alginate to create the hydrogel composite. A varied range of analytical techniques, including SEM, thermogravimetric analysis, Raman spectroscopy, infrared spectroscopy, x-ray photoelectron spectroscopy, and XRD, were meticulously incorporated to comprehensively examine structural and chemical characteristics of hydrogel composite. To assess the adsorption performance, batch adsorption studies were conducted. Remarkably, the hydrogel composite exhibited robust adsorption efficiency, maintaining a high level

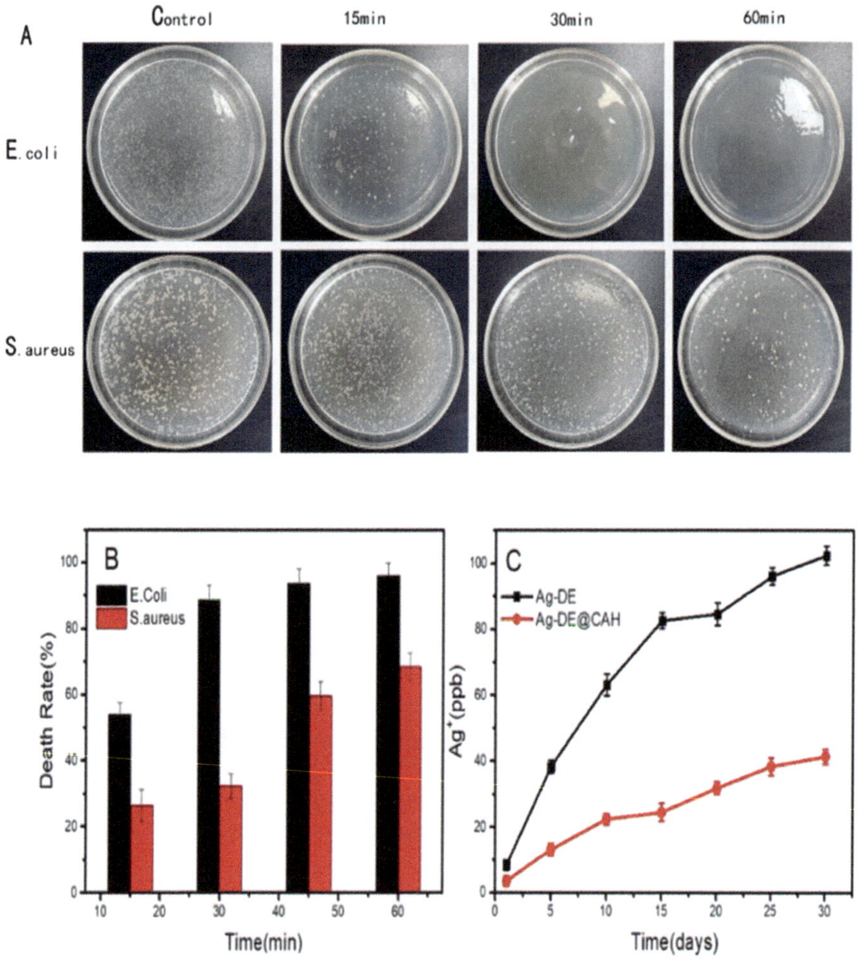

Fig. 6.1 (**a**) Photographs of the colonies of the *E. coli* and *S. aureus* under different incubation time for Ag-DE@CAH beads. (**b**) The death rate of *E. coli* and *S. aureus* different incubation time for Ag-DE@CAH beads. (**c**) Release profiles of Ag+ from Ag-DE and Ag-DE@CAH beads over time. (Reproduced with permission from Zhao et al. (2021))

of adsorption (91%) even with three succeeding cycles of dye adsorption-desorption. The overarching objective of the project was successfully realized by developing a hydrogel composite built on SA and graphite, showcasing promising potential for the effective elimination of MG dye from environments. This innovative approach holds significant promise for addressing water pollution challenges associated with MG dye contamination (Verma et al. 2020).

You et al. conducted a study centered on creating effective adsorbing materials for elimination of dye from polluted aqueous bodies, focusing on sodium alginate as a template hydroxyapatite/calcium silicate composite. Conventional sodium alginate (SA) composite microspheres are prone to instability and damage in solution due to their high water-absorbing capacity and low mechanical strength. To overcome this drawback, hydroxyapatite (HA) and calcium silicate (CS) were combined to enhance the mechanistic robustness of HA compound adsorbent. While maximum HA-CS materials which are porous are typically used as formed powder, posing challenges in separation, reusing after recycling. The exchange of ions among the SA and Ca^{2+} led to the formation of SA templated hydroxyapatite/calcium silicate composite (SCHG), which was subsequently freeze-dried and calcined at 800 °C for 1 h. The morphological characteristics and adsorption capabilities of the composite adsorbent (SCHG) were comprehensively examined. Evaluation of its adsorption performance was carried out through EBBR tests, demonstrating the effective elimination of dyes in the contaminated water. The outcomes indicate that the synthesized composite adsorbent shows promise for applications in water treatment, highlighting its potential effectiveness in addressing dye contamination issues in aqueous environments (You et al. 2019; Zhang et al. 2023)

Khapre et al. conducted a study outlining a synthesis approach for a microporous composite comprising chitosan and alginate cross-linked with glutaraldehyde. The key target of this research is to identify the efficacy of this composite as an adsorbent material for removing three distinct dyes: Patent Blue V, Methyl Orange, and Brilliant Green. The process of adsorption was evaluated using Freundlich adsorption isotherm and pseudo-second-order kinetics. Chitosan along with alginate, recognized biopolymers with favorable adsorption properties, underwent pretreatment before being combined in predetermined ratios. The synthesis of the composite remained achieved through modifying chitosan with analine along with utilizing glutaraldehyde as a cross-linking agent, which enhanced the binding of modified chitosan with alginate. Spectral methods were employed for a comprehensive characterization of the resulting composite. Batch and column adsorption studies on three different dyes were performed. The highest monolayer adsorption capacity for each dye was determined at different pH levels, with the composite's point of zero charge (PZC) identified as 7.5. The overarching objective of this project is to address the extraction of organic dyes from aqueous solutions through synthesizing and characterizing the chitosan-alginate composite cross-linked with glutaraldehyde. The study highlights valuable aspects of the adsorption efficiency of the synthetic composite and its possible application in remediating water systems contaminated with dyes (Khapre et al. 2021).

Dubey et al. conducted research detailing the synthesis of hematite/alginate nanocomposites and their effectiveness in eliminating organic dyes, specifically methylene blue (MB). The nanocomposites underwent comprehensive examination using UV-visible, SEM, FTIR, and XRD methods. To eliminate MB dye, hematite/alginate beads with a diameter of 2-3 mm were employed. A room temperature adsorption process, utilizing 1 g/L hematite/alginate beads nanocomposite, achieved 100% elimination of MB dye (25 ppm) in 40 min under pH 9 conditions, with the adsorption parameters investigated using a first-order process with linear fitting. The study emphasizes the development of hematite/alginate nanocomposites and their potential application in environmental remediation. Hematite nanoparticles possess hydrophilicity, magnetism, non-toxicity, excellent chemical stability, and compatibility with biological systems. In addressing the removal of certain organic dyes like MB, hematite/alginate nanocomposites offer a promising solution. An improved hydrothermal process using urea as the sole precursor was employed for hematite nanoparticle synthesis, and XRD and SEM methods were used for nanocomposite characterization. Room temperature adsorption test was executed to identify the kinetic behavior of MB adsorption on hematite/alginate beads. The nanocomposite exhibited effective adsorption and removal of MB dye, achieving nearly 100% elimination with 1 g/L of the nanocomposite. Significantly, MB adsorption was most effective at pH 9, indicating enhanced adsorption performance under basic conditions. The nanocomposite demonstrates promise for cost-effective water cleanup in dye industries, with the adsorption process exhibiting remarkable efficiency (Dubey et al. 2020).

Bayat et al. performed a detailed study on the effective elimination of water-soluble methylene blue dye. This was achieved through the fabrication of a bio-nanocomposite comprising zeolite, nickel ferrite, and sodium alginate, using a co-precipitation approach. The synthesis was carried out via the process of co-precipitation. Surface characteristics and morphology regarding the modified zeolite/nickel ferrite nanocomposite were evaluated using SEM images. The process of adsorption was systematically investigated by both pseudo-first-order and pseudo-second-order models. The findings revealed that the synthesized bio-nanocomposite demonstrated effective removal capabilities for water-soluble MB dye. The utilization of various models provided a complete understanding of adsorption kinetics and mechanisms, highlighting along with the efficacy of bio-nanocomposite as a potential adsorbent for water treatment applications, particularly in eradicating the MB dye (Bayat et al. 2018).

In the research conducted by Belhouchat et al., a cross-linked composite of activated organo-bentonite (AOBent) and sodium alginate (SA) was developed to effectively extract both anionic and cationic dyes from aqueous solutions. The key objective was to synthesize a cost-effective material for the same from water. The composite, characterized through FTIR, TGA, and XRD analyses, displayed significant potential in removal of anionic and cationic pigments from wastewater. Beyond offering an economical solution, the developed composite aligns with environmentally friendly approaches to dye removal. The study suggests the application of this composite in sustainable and eco-friendly wastewater treatment practices (Belhouchat et al. 2017).

In the study led by Mallakpour et al., the research focused on assessing the effectiveness of kinetic models and both linear and non-linear isotherms to evaluate and compare the adsorption behavior of various materials. The specific aim was to investigate MB adsorption properties of alginate enhanced by activated carbon black (ACB) derived using old compact discs (CDs). This utilization of polymer-based materials has been on the rise, but concerns about their non-biodegradability have raised environmental issues. Repurposing polymeric waste, such as old CDs, was considered as a sustainable solution to address this environmental concern. The researchers developed a cost-effective and eco-friendly adsorbent by extracting

ACB from used CDs. This ACB was then encapsulated in alginate, a biocompatible and inexpensive polymer, to create a composite adsorbent. Upon incorporating ACB into the alginate matrix, the mean diameter decreased to around 2 nm, indicating reduced particle aggregation and uniform dispersion. The ability of film to diffuse was recognized as the governing apparatus regulating the methylene blue adsorption process, while pore diffusion-imposed constraints. The study emphasizes the successful removal of MB using a composite adsorbent composed of alginate and ACB derived from waste CDs. Film diffusion governs the adsorption process, and pore diffusion sets limitations. The research advocates for the use of waste products, especially polymeric wastes, as alternative adsorbents in water remediation endeavors (Mallakpour and Behranvand 2021).

6.2.2 *Alginate-Based Biosorbents for Removal of Cadmium*

Research led by Ayouch et al. concentrated on enhancing cadmium ion retention using alginate beads, investigating their effectiveness in removing cadmium ions from aqueous solutions. Air-drying and freeze-drying were compared to evaluate their impact on the mechanistic properties of beads. The result showed that the freeze-dried beads (AL-FD) had an even and smooth surface and maintained their original size, while air-dried beads (AL-AD) exhibited structural changes. The specific surface area of AL-FD was significantly higher (56.83 m^2/g) compared to AL-AD (0.039 m^2/g), contributing to its enhanced adsorption performance. The kinetic sorption process was described the best by Avrami fractional-order model, and the adsorption capacity of AL-FD surpassed AL-AD, underlining its superior performance. Overall, freeze-dried alginate beads demonstrated improved adsorption capacity and surface properties compared to air-dried beads, emphasizing their potential for cadmium ion elimination from water (Ayouch et al. 2020a).

Sidgel et al. examined the effectiveness of alginate beads impregnated with powdered activated carbon (ABPAC) for eliminating benzene and cadmium. ABPAC exhibited a significant sorption capacity of 89.60 mg g^{-1} for benzene and 149.32 mg g^{-1} for cadmium representing a chemisorption mechanism. The ABPAC beads were synthesized by incorporating powdered activated carbon and sodium alginate into a calcium chloride solution, followed by a cleaning and drying process before use. The study highlights the significance of addressing heavy metal ions and volatile compounds and their damaging effects on environment and human well-being. Notably, ABPAC proved effective in the simultaneous removal of these pollutants, as no competition effects were observed in the binary-contaminant system. These findings position ABPAC as a promising sorbent for water purification applications (Sigdel et al. 2017).

In a study led by Ayouch et al., an economically viable method for removing Cadmium (Cd) from aqueous solutions was achieved using composite beads incorporating bentonite (BE) and phosphate washing sludge (PS). The combined adsorption capabilities of these beads surpassed those of individual BE and PS components.

New adsorbents, for example, Moroccan clay and alginate, were synthesized for the removal of low concentrations of cadmium. Phosphate washing sludge, a by-product of phosphate enrichment, consists of carbonates, clays, quartz, and fluorapatite. Alginate, sourced from sea algae, offers gelling and adsorption qualities. To address the issue of adsorbent separation, a solvent mixing method was employed to encapsulate the composite materials within alginate beads, enabling their application in water treatment procedures. The beads underwent drying and lyophilization processes, and their structural, textural, morphological, thermal, chemical composition, and size distribution characteristics were comprehensively characterized. SEM analysis revealed a distinct sponge-like structure formed through lyophilization (Fig. 6.2). Batch studies were conducted to evaluate the composite beads' ability to

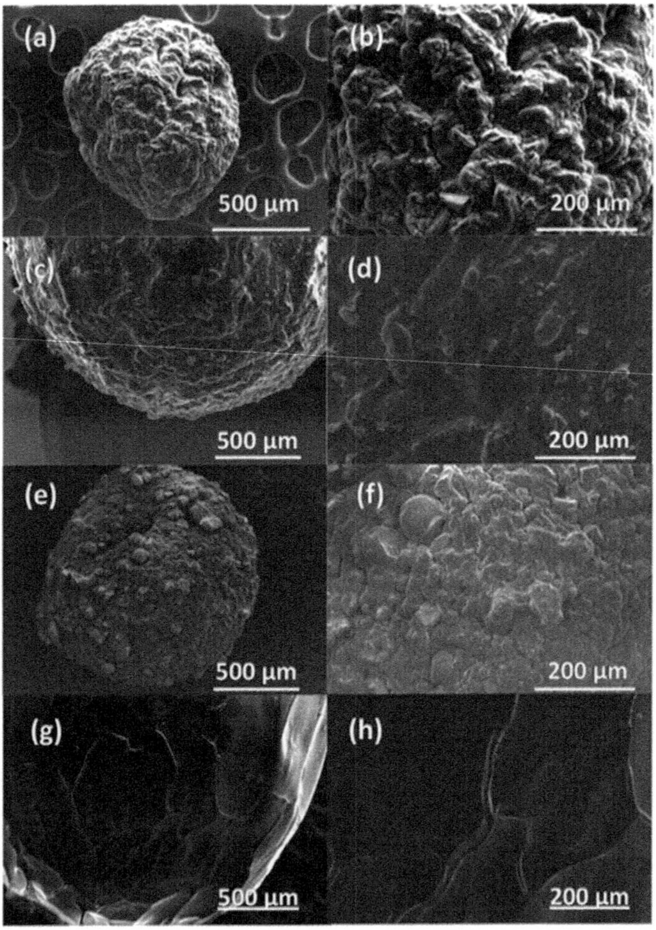

Fig. 6.2 SEM images at low and high magnification of (**a**, **b**) BE@AL-AD, (**c**, **d**) BE@AL-FD, (**e**, **f**) PS@AL-AD, and (**g**, **h**) PS@AL-FD composite beads

adsorb Cd ions, considering factors such as adsorbent dosage, equilibrium time, temperature, starting pH, and Cd content. XRD analysis revealed that the phosphate sludge contained fluorapatite, quartz, dolomite, and calcite, while the primary component of the raw clay, according to XRD analysis, was montmorillonite. Alginate's structure was identified as amorphous. The Langmuir isotherm models showed applicability during the adsorbing process, with equilibrium attained in under 180 min suggesting temperature was a crucial factor influencing cadmium recovery. The synergistic combination of Moroccan clay in addition to phosphate washing sludge, and alginate within composite beads demonstrated considerable prospect as a cheaper alternative material for the elimination of cadmium (Ayouch et al. 2020b).

A study led by Thakur et al. focused on developing sustainable materials for elimination of metals, specifically targeting cadmium (Cd), chromium (Cr), and lead (Pb) in wastewater. Three different beads—Zero Valent Iron (ZVI), Activated Carbon (AC), and ZVI-AC—combinations were tried with ZVI-AC-Na alginate beads emerging with maximum efficacy, achieving a removal efficiency hierarchy of Pb (96%) > Cr (59%) > Cd (39%). The adsorption of these metals is linked to the existence of functional groups, specifically carboxyl (–COO) and hydroxyl (–OH). With metal contamination from agricultural and industrial sources posing a significant environmental concern, alginate beads have gained attention for their metal-removal potential. The study utilized materials such as calcium chloride dihydrate, sodium borohydride, sodium alginate, and ferric chloride hexahydrate with ZVI synthesized via $NaBH_4$ reduction. Pb exhibited the highest adsorption rate, followed by Cr and Cd with ZVI-AC-Na alginate beads highlighting the superior adsorption capabilities of synthesized beads and emphasizing their potential as an eco-friendly and effective material in the elimination of metals from site contamination (Thakur and Kumar 2021).

Shasha Wu et al. directed research focusing on the development of a modified sodium alginate aerogel for purpose of eliminating lead (Pb) and cadmium (Cd) ions. The SA-DMSA adsorbent was formulated using sodium alginate (SA) modified with dimercaptosuccinic acid (DMSA), followed by examining the morphology and surface area using SEM. The resulting indicated a rough and porous surface, providing increased binding sites for metal ions. The average pore width was measured at 9.1726 nm, and was determined to be 0.0034 cm^3/g. The BET surface area of SA-DMSA was calculated to be 2.3890 m^2/g, indicating a favorable structural configuration for enhanced adsorption capabilities. The modified sodium alginate aerogel, SA-DMSA, demonstrated promising results in adsorbing lead and cadmium ions from wastewater. The rough and porous surface of SA-DMSA contributed to an augmented adsorption capacity along with increasing contact area for effective elimination of heavy metals (Wang et al. 2020).

Castro et al. devised an effective remediation strategy for effluents with uranium generated by the nuclear energy sector, aiming to prevent their release into the environment. The research focused on employing microbial-based approaches, specifically utilizing the *Stenotrophomonas sp.* Br8. The experimental procedure involved refining the immobilization of Br8 biomass within alginate beads to safeguard the cells and enhance the feasibility of large-scale implementation. Remarkable results were achieved, with uranium immobilization rates reaching approximately 98% after a single 72-h cycle using Br8-doped alginate beads. The approach combines slow active biomineralization and passive sorption onto the bead and/or cell surfaces by immunized cells, showcasing the effectiveness of microbial-based techniques in uranium removal. The study addresses the challenges of upscaling such approaches, particularly in exposing cells to adverse environmental conditions. The immobilized biomaterials underwent analysis using techniques such as FTIR with Attenuated Total Reflectance, Environmental Scanning Electron Microscopy (ESEM). The application of biocompatible alginate beads for Br8 cell immobilization demonstrated high uranium immobilization rates, offering a practical method for extraction of uranium from mining waters. The findings contribute valuable insights into the biological and physico-chemical aspects of an effective uranium bioremediation method utilizing immobilized bacterial cells, with implications for the development of microbial-based remediation techniques in the nuclear energy sector (Sánchez-Castro et al. 2021).

Liu et al. (2020) focused on removing cadmium (Cd(II)) from wastewater utilizing water hyacinth biochar alginate beads. Water hyacinth-derived biochar has previously exhibited efficacy in managing invasive species and eliminating heavy metals. The objective of the research is to investigate the extraction of Cd(II) from aqueous solutions using water hyacinth biochar encapsulated in alginate. The sorption process was evaluated at varying temperatures, with maximal sorption of Cd(II) observed at 37 °C and pH 6. The Freundlich and Langmuir equations were employed to model sorption equilibria, with the Langmuir isotherms exhibiting a superior fit. Pseudo-first-order kinetics governed the sorption of Cd(II), demonstrating increased effectiveness at temperatures above 500 °C. The encapsulated water hyacinth biochar alginate beads exhibited efficient removal of Cd(II) from wastewaters. The research utilized water hyacinth biochar processed with ferric/ferrous sulfate and encapsulated with alginate. Measurements of surface area, pore size, and pH were performed on biochar as well as modified biochar. Magnetically modified biochar treatments demonstrated distinct surface morphological characteristics, including iron distribution. The study emphasizes the potential applications of water hyacinth biochar for wastewater treatment, particularly in the efficient removal of Cd(II). The sorption dynamics were influenced by Cd concentration, pH, and solution temperature, with higher temperature-generated biochar demonstrating more effective Cd(II) elimination. The encapsulated water hyacinth biochar alginate beads hold promise for practical use in wastewater treatment applications (Liu et al. 2020).

Shim et al. aimed to develop an economical wastewater treatment system to remove heavy metal ions, specifically arsenic (As) and cadmium (Cd). The study introduced a novel composite, alginate beads impregnated with MnO_2 (CABs- MO), and assessed its adsorption capacity for As and Cd. The investigation included an analysis of adsorption isotherms, removal kinetics, and the effectiveness of various regeneration agents for the adsorbent. The creation of CABs-MO involved encapsulating nanocrystalline MnO_2 in calcium alginate microcapsules. The study also assessed the performance of different regeneration agents, such as distilled water and 10 mM HCl. Physicochemical characteristics and surface texture of the adsorbents were evaluated using BET, Langmuir, SEM-EDS, and XRD techniques. In terms of regeneration, the agent 10 mM HCl was more effective than pure water for removing As. The study underscores the potential of synthesizing sustainable and cost-effective techniques for treatment of wastewater. However, the research acknowledges the need for further investigation to understand its efficacy in removing emerging contaminants. The innovative composite beads with diverse physicochemical characteristics demonstrate potential for decontaminating wastewater economically and environmentally (Shim et al. 2019).

Oulguidoum et al. (2019) synthesized sulfonate-functionalized hydroxyapatite (wBDS-HAp) by incorporating benzoenedisulfonate molecules into natural phosphate. The objective was to enhance the ability of adsorption of hydroxyapatite for eliminating cadmium (Cd) from water, addressing environmental and occupational concerns associated with this heavy metal. Successful integration of sulfonate groups, specifically –SO3 acid groups, into the hydroxyapatite matrix made the modified apatites adaptable adsorbents for Cd(II) cations. The hybrid apatites demonstrated promising results in cadmium removal, exhibiting the highest capacity for adsorption of 457 mg/g. The experiments were directed to analyze the effectiveness of wBDS-HAp for cadmium removal, comparing its kinetics with those of pure hydroxyapatite (HAp) and natural phosphate (NP). The experimental data for chemical adsorption aligned with the Ho-second-order kinetic model. Various characterization techniques, including nitrogen adsorption-desorption, XRD, and thermal gravimetry (TG), were employed to analyze the modified apatites. The addition of benzoenedisulfonate molecules and the resulting sulfonate functionalization aimed to create new active sites on the hydroxyapatite surface, enhancing its interaction with hazardous cadmium species in water. The study suggests that the Cd(II) removal kinetics of wBDS-HAp are superior to those of pure HAp and NP. The findings underwrite to the development of effective and versatile adsorbents for heavy metal cleanup, addressing concerns in the scientific community regarding water and soil remediation methods (Oulguidoum et al. 2019).

Wang et al. fabricated phosphate-embedded calcium alginate beads for the effective sorption and immobilization of Pb(II) and Cd(II) in solutions. The optimum pH range for Pb(II) and Cd(II) sorption was determined to be 4.0 and 5.5, respectively. The sorption mechanism involved the penetration of heavy metal ions into the beads, followed by precipitation with phosphate radicals. The sorption demonstrated capacities of 263.16 mg/g and 82.64 mg/g, respectively, following the Langmuir isotherm model. Overall, the research highlights the critical importance of developing efficient sorbents for various environmental applications, addressing the pressing issue of heavy metal contamination (Wang et al. 2016).

6.2.3 Alginate-Based Biosorbents for Removal of Uranium

Saha et al. conducted research with the aim of developing an effective sorbent for the deletion of uranium, resulting in the creation of a novel sorbent named HAP-AC-Alginate. This sorbent involved coating nano-hydroxyapatite onto activated carbon, followed by immobilization within the alginate matrix. The study focused on assessing the sorbent's efficacy in eliminating uranium from portable water, with demonstrated efficiency exceeding 92%. The HAP-AC-Alginate hydrogel bead exhibited a maximum sorption capacity of 18.66 mg $g^{(-1)}$. The study also investigated the sorbent's performance in the presence of interfering ions (Na+, Ca2+, and Fe3+), demonstrating the continued effectiveness of the synthesized beads. In conclusion, the HAP-AC-Alginate beads proved to be a promising extraction sorbent for eliminating uranium from water, showcasing high efficiency and potential applicability in water purification processes (Saha et al. 2020).

Zahakifar et al. delved into fabrication of hybrid nanostructure adsorbent crafted using a casting approach, incorporating sodium alginate (SA), polyvinyl alcohol (PVA), polyethylene oxide (PEO), and ZSM-5 zeolite. The chief objective of the research was to assess the efficiency of this nanohybrid adsorbent in sequestering uranium (VI) ions from aqueous solutions. The nanohybrid adsorbent, designated as SAPVAPEOZSM-5, was synthesized through the casting procedure, and its characteristics were assessed through BET and SEM. The highest adsorption capacity for uranium ions was identified to be 92.76 mg g^{-1} under conditions of 45 °C and pH 5.0. Furthermore, the nanohybrid adsorbent exhibited promising regeneration capabilities, with the adsorption capability changing by <20% after about five adsorption-desorption cycles. The findings deliver valuable understandings into the development of effective nanohybrid adsorbents for the elimination of uranium (Zahakifar et al. 2021).

In this study, Akhtar et al. observed that the efficiency of Ca-alginate immobilized Trichoderma harzianum in the extracting as well as recovering uranium was investigated. Experimental results showcased the effectiveness of packed bed column operations under varying conditions, including flow rates, bed heights, and initial uranium concentrations. Breakthrough curves indicated that a 1.5 g of immobilized material in column can effectively purify 8.5 L of bacterial leach liquor with the early uranium concentration of 58 mg/L before saturation occurred, achieving a total purification of 25 L under optimum conditions. The study illustrated that uranium removal was maximized at a flow rate of 5 ml/min and a bed height of 25 cm, with the immobilized biomass achieving removal rates of up to 97.3%, compared to 75.3% for the free biomass at 100 mg/L uranium concentration. The bio-regeneration of the biosorbent was effectively performed using 0.1 N HCl, resulting in elution efficiency of 98.1–99.3%. The research concludes that using Ca-alginate as a support matrix not only enhances biosorption performance but also contributes to environmental remediation efforts by providing an efficient method for uranium extraction from wastewater streams (Akhtar et al. 2009).

In a study led by Jing Yu and collaborators, the efficacy of calcium alginate gel beads in removing uranium from water solution was investigated. The presence of uranium in the environment positions a substantial risk to human health and ecological systems, prompting the exploration of various techniques, including adsorption, for uranium extraction. Alginate, recognized as a biopolymer with exceptional selectivity and effectiveness as an adsorbent, has been extensively researched in this context. The study identified that maximum adsorption was observed at 237.15 mg U/g. Furthermore, the consequence of varying temperature on U(VI) adsorption was examined. The findings suggest that calcium alginate gel beads exhibit promise as an effective means for removing uranium from aqueous solutions, providing insights into their potential application in addressing environmental uranium pollution (Yu et al. 2017).

Li and colleagues conducted a study focusing on immobilizing uranium (VI) from aqueous solutions, utilizing a novel adsorbent termed GO/FeS. This adsorbent was developed by attaching graphene oxide onto nanoscale iron sulfide (FeS). XPS was employed to explore the mechanistics of immobilizing U(VI) onto GO/FeS. The significance of immobilizing uranium, a hazardous radioactive element prevalent in nuclear technology, is underscored by its extreme toxicity and mobility, necessitating the use of suitable materials for environmental and public health protection. The experimental setup involved the use of substances such as uranyl nitrate, iron sulfate heptahydrate, and sodium sulfide nonahydrate. The modified Hummers technique was employed in the preparation of graphene oxide (GO), while the coprecipitation process using iron sulfate hexahydrate and sodium sulfide nonahydrate facilitated the creation of GO/FeS nanocomposites. The integration of FeS nanoparticles onto the GO surface generated more reactive sites in comparison to pure FeS or GO. The SEM-EDS results for the said composite sample treated with U(VI) solutions at acidic conditions revealed contents of 37.46% carbon, 21.01% oxygen, 14.59% sulfur, 6.47% uranium, and 20.48% iron (Fig. 6.3). The study demonstrated a notable adsorption ability of 347.2 mg/g for U(VI) by GO/FeS. The primary mechanism identified for U(VI) immobilization on GO/FeS was a synergistic interplay in the surface adsorption and reduction/precipitation processes. This discovery positions GO/FeS as a possible material in the remediation of contaminated water with U(VI), showcasing its ability for application in effectively treating uranium-containing effluents (Li et al. 2021).

Wei Lu and colleagues conducted a study focusing on the development of a composite hydrogel, referred to as PCG, designed for adsorbing uranium (VI) from aqueous solutions. Structural analysis using FTIR revealed distinctive peaks indicative of functional groups present in the composite hydrogel. Furthermore, equilibrium outcomes were effectively explained by the Langmuir model, suggesting monolayer adsorption on PCG. An important feature of PCG highlighted in the study is its outstanding regeneration capabilities, coupled with the ease of isolation from aqueous solutions. These characteristics, combined with the high adsorption capability and selective adsorption of U(VI), position PCG as a promising and valuable material for the absorption of uranium from wastewaters. The composite

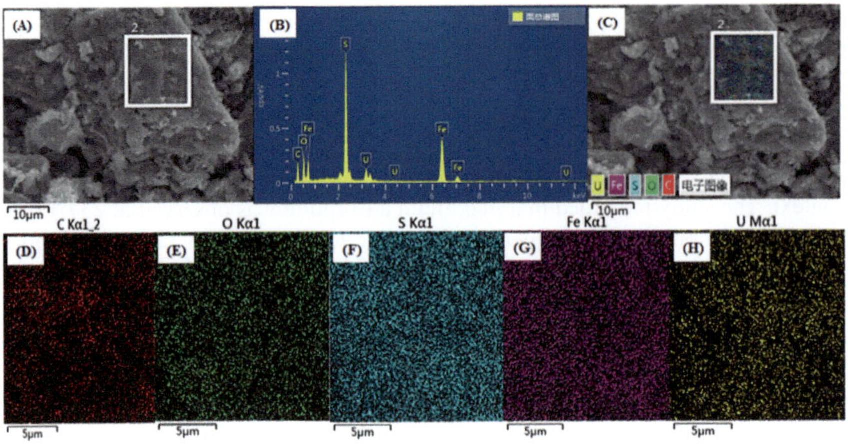

Fig. 6.3 SEM-EDS results of GO/FeS reacted with U(VI) at acidic condition. (Reproduced with permission from Li et al. (2021))

hydrogel offers potential applications in addressing environmental concerns related to uranium contamination (Lu et al. 2020).

Yan-Rong et al. focused on the preparation and evaluation of melamine-formaldehyde alginate (MF-A) aerogel for the adsorption with huge size and pores were instrumental in facilitating a comprehensive understanding of the U(VI) adsorption process. Porous organic polymers (POPs) have expanded attention for U(VI) elimination due to their chemical stability and three-dimensional network structure. The highest adsorption capacity of MF-A aerogel for U(VI) was observed at pH 4.0, reaching 211.22 mg $g^{(-1)}$. The alginate content in MF-A aerogel played a significant role in determining its U(VI) removal efficiency. The pH of the solution emerged as a critical factor affecting the adsorption capacity, with optimal removal efficiency observed in the pH range of 2.0–5.5. Beyond pH 5.5, a decline in U(VI) removal efficiency was noted. The study highlighted the promising performance of MF-A aerogel in U(VI) adsorption, facilitated by its larger size and network structure with pores. Alginate proved to be a crucial component in the U(VI) adsorption process. The findings contribute a novel approach to the development of U(VI) adsorbents, emphasizing the potential of MF-A aerogel for effective elimination of uranium from the aqueous solutions (He et al. 2019).

El-Maghrabi et al. conducted a study focused on the synthesis and characterization of magnetically modified hydroxyapatite nanoparticles (MNHA) designed for the extraction and eradication of uranium (VI) ions from wastewater. The research aimed to optimize MNHA's adsorption capacity and investigate its efficiency, mechanisms, and reusability in treating uranium-contaminated water. MNHA nanoparticles were efficiently synthesized using a microwave-assisted method. The adsorption efficacy of MNHA was evaluated by shaking the adsorbent with solutions containing uranyl ions at different concentrations and ionic strengths. The XRD patterns confirmed the presence of a well-crystallized single-phase hydroxyapatite and α-Fe2O3 phase after magnetic modification. Equilibrium was achieved after 120 min of contact time at 25 °C, with a maximum adsorption capacity of 310 mg/g. The Langmuir adsorption isotherm model effectively explained the experimental data, and the pseudo-second-order model accurately represented the kinetics of the adsorption process. The study also delved into the recovery and reusability of MNHA beads, providing valuable insights into the practical application of the synthesized nanoparticles for uranium removal from wastewater. Overall, the successful development and characterization of MNHA nanoparticles demonstrated high adsorption capacity and efficiency, making them promising candidates for addressing uranium contamination in water treatment processes (El-Maghrabi et al. 2019).

Jie Meng et al. led a study focusing on the production and characterization of a gelatin/PVA composite nanofiber band coated with bayberry tannin (GPNB-BT) for the eradication of uranium from simulated seawater. The synthesized composite exhibited a nano-network topology with a fiber diameter ranging from 200 to 400 nm, contributing to its enhanced uranium adsorption capability. Decreasing fiber diameter during the curing process led to an increase in both tannin content and uranium adsorption capacity. The synthesis of GPNB-BT involved dissolving bayberry tannin (BT) in water, combining it with gelatin/PVA, adding glutaraldehyde, and subsequently forming the resulting material. The highest capacity of GPNB-BT to adsorb uranium was determined to be 170 mg/g under certain conditions, including an initial uranium concentration of 80 mg/L and a pH of 5.5. Notably, GPNB-BT exhibited remarkable adsorption capacity even at extremely low initial uranium concentrations, such as 3 μg/L in simulated seawater. The unique nano-network structure and strong uranium adsorption ability of GPNB-BT make it an ideal material for applications in the enrichment and separation of uranium in radioactive wastes and saltwater (Xiao et al. 2023).

6.2.4 Alginate-Based Biosorbents for Eradication of Chromium

The research by Daradmare et al. is centered on the utilization of metal-organic frameworks (MOFs) combined with alginate beads (MOFs@ABs). The study demonstrates a notable Cr(VI) removal efficiency of about 98% at the initial pH. Two different MOFs, UiO-66 and ZIF-8, were incorporated into alginate beads (UiO-66@ABs and ZIF-8@ABs, respectively), and their Cr(VI) adsorption capabilities were compared. The observations revealed that UiO-66@ABs exhibited superior Cr(VI) adsorption compared to ZIF-8@ABs. The study also explored the desorption process using 2 M HCl, suggesting that UiO-66@ABs could potentially be regenerated after adsorption. Optimization experiments were conducted with UiO-66@ABs, varying the loading volumes of UiO-66, and the results were analyzed using the DPC method to determine Cr(VI) concentrations. The adsorption ability and the extraction percentage were optimized using UiO-66@ABs with specific loading volumes, with the maximum equilibrium adsorption capacity reaching 2.181 mg $g^{(-1)}$ and a removal percentage of 76%. Increased loading volumes of UiO-66 further improved the adsorption ability and removal efficiency, highlighting the potential of MOFs/alginate composite beads, particularly UiO-66@ABs, as effective adsorbents in the elimination of Cr(VI) (Liu et al. 2023; Cheng et al. 2023).

In the research conducted by Khan et al., a biobased amphoteric aerogel was developed using amine-modified clay-enriched chitosan/alginate for the purpose of adsorbing organic dyes and chromium (VI) ions from aqueous solutions. The aerogel exhibited notable features like high thermal stability and an organized, fishing net-like microstructure with interconnected holes, facilitating effective adsorption. The adsorption capabilities of the aerogel were assessed for MB, acid black, and hexavalent chromium ions. The study aimed to address the challenge of finding effective amphoteric adsorbents for the treatment of complex wastewater using naturally occurring, renewable macromolecular polysaccharides. The aerogel was prepared through a process involving post-ionic crosslinking and freeze-drying. The aerogel demonstrated excellent thermal stability and a unique microstructure suitable for efficient adsorption. The adsorption capacities for MB, acid black, and hexavalent chromium ions were reported as 73.1 mg/g, 71.5 mg/g, and 62.4 mg/g, respectively. The combination of great performance, affordability, and recyclability positions the biobased aerogel as a potential alternative in the treatment of industrial wastewater (Guan et al. 2022; Wu et al. 2023).

In the study by Venkatrajan Gopalakannan et al., a sustainable bio-composite, termed alumina/alginate (AlAlg) composite, was synthesized for effective removal of chromium(VI). The synthesis involved the even dispersion of alumina particles within the alginate polymeric material, ensuring an environmentally friendly process. The composite was characterized using FTIR, SEM, and EDAX analyses to elucidate its structural and elemental properties. The SEM micrographs of the biocomposite before and after Cr(VI) sorption illustrate a distinct disparity, indicating that sorption occurred onto the biocomposite (Fig. 6.4). The AlAlg composite exhibited a superior maximum sorption capacity (17.45 mg/g) compared to the CaAlg composite. Thermodynamic parameters were calculated to understand the nature of chromium sorption, with electrostatic adsorption identified as the primary mechanism governing the sorption process. The AlAlg composite proved to be an inexpensive, environmentally benign, non-toxic as well as biocompatible and biodegradable material suitable in the selective removal of chromium(VI). The research underscores the development of sustainable and efficient methods for chromium remediation (Gopalakannan et al. 2016).

This study focuses on the development of hydrothermally aided magnetic nanohydroxyapatite entrapped alginate beads for efficient Cr(VI) absorption from water. The highest recorded sorption capacity within 40 min was 29.14 mg/g, showcasing the efficacy of the adsorption process in removal of chromium. The adsorbents were characterized using an FTIR spectrometer, revealing the presence of crucial functional groups. SEM analysis provided insights into the surface morphology of both the artificial magnetic hybrid beads and the magnetic hybrid beads after sorption with Cr(VI). Positive changes in entropy (ΔS°) values indicated a favorable affinity

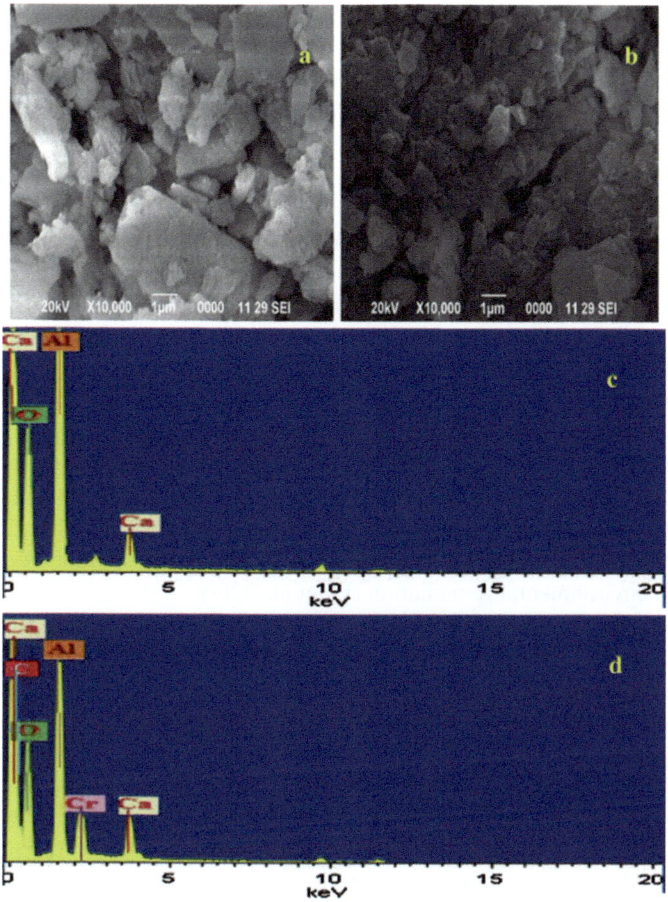

Fig. 6.4 SEM images of (**a**) AlAlg composite and (**b**) Cr(VI) sorbed AlAlg composite, EDAX spectra of (**c**) AlAlg composite and (**d**) Cr(VI) sorbed AlAlg composite. (Reproduced with permission from Gopalakannan et al. (2016))

of the produced adsorbents for binding chromium ions. For the formation of beads, the components utilized were calcium chloride dihydrate and sodium alginate. The adaptability of these beads in field circumstances was examined using field water contaminated with Cr(VI). Overall, the developed magnetic nano-hydroxyapatite encapsulated alginate beads, with hydrothermal assistance, demonstrated effective and promising uptake of Cr(VI) from water, showcasing potential applications in field settings for the removal of chromium contamination (Periyasamy et al. 2018).

The study investigated the potential of *Turbinaria turbinata*, a brown alga prevalent in tropical regions, as a sorbent in the elimination of hexavalent chromium from aqueous media. The sorbent utilized was biomass derived from *Turbinaria turbinata*, and multiple techniques were used to modify its structure and surface, aiming to enhance its adsorption capabilities. Notably, surface modification through acid treatment and structural modification involving activated carbon preparation were performed (Yacou et al. 2018).

The research by Lv et al. focuses on the synthesis of Fe0 NPs embedded in reduced graphene oxide-alginate beads (Fe@GA beads) for the efficient removal of chromium (VI) from aqueous solutions. The study addresses challenges associated with technologies utilizing zero- valent iron nanoparticles (Fe0 NPs) where the manufacturing process involves trapping Fe0 NPs within an alginate gel modified with graphene oxide, creating Fe@GA beads reduction via in-situ. The efficacy of Fe@GA beads in Cr(VI) removal was assessed by considering various parameters such as initial Fe and alginate concentrations, temperature, solution pH, and ionic strength. The beads exhibited remarkable mechanical strength and efficiency in removing Cr(VI). The rate-limiting step in the procedure was identified as the adsorption of Cr(VI) onto the bead surface, highlighting the role of Fe0 NPs in subsequent reduction. The study concludes that Fe@GA beads, enhanced with reduced graphene oxide and pre-reduction treatment, effectively eradicate Cr(VI) from aqueous solutions. The combination of adsorption and reduction processes contributes to the beads' ability to address chromium pollution in water bodies. The results pinpoint the potential of Fe@GA beads as an upcoming solution for pollutant removal in environmental remediation (Lv et al. 2017).

The study directed by Gokila et al. investigates the efficacy of chitosan and alginate nanocomposites for the elimination of chromium (VI) from wastewater. Nanocomposites, composed of two or more constituent materials at the nanoscale, were created by combining chitosan and alginate in appropriate proportions. The nanocomposites were subjected to physicochemical characterization methods, where SEM analysis revealed that the surface morphology of chitosan-alginate nanocomposites was rough and porous. The addition of glutaraldehyde for cross-linking altered the nanocomposites' shape and enhanced their adsorption capacity. The nanocomposites exhibited high sorption capacity, and the adsorption equilibrium data fitting well with the Freundlich isotherm indicated multilayer adsorption. The kinetics of chromium (VI) elimination was perfectly described by the second-order reaction kinetics model, whereas the glutaraldehyde-infused CS-AL nanocomposites displayed a single glass transition (Tg), indicating good miscibility of components. Overall, the study concludes that chitosan and alginate nanocomposites are superior biosorbents for removing chromium (VI) from wastewater. The demonstrated high sorption capacity, multilayer adsorption behavior, and good component miscibility suggest potential applications in heavy metal removal and wastewater treatment (Gokila et al. 2017).

In the research conducted by Mishra et al., the focus was on synthesizing hollow carbon nanofibers (CNFs) grafted with triethanolamine and supported on carbon beads using a template technique for the extraction of hexavalent chromium (Cr(VI)) ions from water. The hollow CNF was cultured on carbon beads through a templating process with in situ integrated Cr nanoparticles serving as catalysts. Subsequent removal of nanoparticles from the CNF tips resulted in the formation of hollow CNFs. The synthesis involved chemical vapor deposition of CNFs on carbon beads with in situ integrated Cr nanoparticles, followed by ultrasonication to remove the nanoparticles and create hollow CNFs. The triethanolamine-grafted hollow carbon nanofibers supported on carbon beads exhibited a significant Cr(VI)-adsorption capacity of approximately 51 mg/g at an initial concentration of 125 ppm and a solution pH of 4.5. Importantly, the material demonstrated insignificant interference toward the coexisting metal ions already present in water, highlighting its selectivity for Cr(VI) adsorption. The templating method utilized in this research holds promise for creating patterns for the eradication of other metals from wastewater, suggesting potential applications and further developments in the future (Mishra and Verma 2016; Šillerová et al. 2015).

In the study conducted by Hana Sillerová et al., Brewer's draff and grape waste, both serving as biosorbents, were encapsulated within calcium alginate beads to assess their effectiveness in removing Cr(VI) from aqueous solutions. The investigation revealed that the type and proportion of waste significantly influenced both the diameter of the beads and the efficiency of Cr(VI) removal. Particularly, beads containing 4% (w/w) of grape waste exhibited the highest performance in removing Cr(VI). The sorption efficiencies of the encapsulated wastes were notably higher compared to using the wastes in their original states. The study aimed to evaluate the impact of encapsulation factors on bead diameter and Cr(VI) sorption kinetics. Brewer's draff and grape waste were incorporated into calcium alginate beads, resulting in twenty-one distinct types of beads with variations in waste type, proportion, and CaCl2 curing period. Statistical analysis using SPSS software was employed for data interpretation. The study observed that the diameter of the beads increased with a higher amount of contained biomaterial, with grape waste-containing beads being larger than brewer's draff-containing beads. Notably, beads with 4% (w/w) of grape waste exhibited optimal Cr(VI) removal efficiency. Additionally, the Cr(VI) sorption kinetics in grape waste beads did not exhibit significant variations concerning curing time. The findings highlighted the substantial influence of waste type and proportion on bead diameter and Cr(VI) removal efficiency. Encapsulated biosorbents, particularly those containing grape waste, demonstrated enhanced sorption efficiency compared to their native waste counterparts (da Costa et al. 2022; Li et al. 2019).

6.3 Conclusion

Alginate-based biosorbents emerge as a transformative solution to address the pressing issues of water pollution due to heavy metals and dye-based contaminants due to their biocompatibility, and cost-effectivity along with enhanced adsorption capacity. The chapter highlights the unique advantages of sodium alginate over the traditional treatment methods, particularly by modifying to create composites and nanoparticles leveraging the advancements in material sciences. The incorporation of nanoparticles with biopolymers demonstrates superior performance in removing a wide array of pollutants demonstrating superiority. The sustainable alginate-derived products mitigate the environmental pollution along with aligning with growing demand for eco-friendly remediation strategies. A refine an optimized biosorbents will hold a significant potential for future applications in water purification, paving the way for more sustainable and effective solutions to global water quality challenges.

References

Akhtar K, Khalid AM, Akhtar MW, Ghauri MA (2009) Removal and recovery of uranium from aqueous solutions by Ca-alginate immobilized Trichoderma harzianum. Bioresour Technol 100(20):4551–4558. https://doi.org/10.1016/j.biortech.2009.03.073

Ayouch I, Barrak I, Kassab Z, El Achaby M, Barhoun A, Draoui K (2020a) Impact of the drying process on the efficiency of alginate beads for cadmium removal from water: kinetic, isotherm and thermodynamic study. Environ Technol Innov 20:101157. https://doi.org/10.1016/j.eti.2020.101157

Ayouch I, Barrak I, Kassab Z, El Achaby M, Barhoun A, Draoui K (2020b) Improved recovery of cadmium from aqueous medium by alginate composite beads filled by bentonite and phosphate washing sludge. Colloids Surf A Physicochem Eng Asp 604:125305. https://doi.org/10.1016/j.colsurfa.2020.125305

Bayat M, Javanbakht V, Esmaili J (2018) Synthesis of zeolite/nickel ferrite/sodium alginate bionanocomposite via a co-precipitation technique for efficient removal of water-soluble methylene blue dye. Int J Biol Macromol 116:607–619. https://doi.org/10.1016/j.ijbiomac.2018.05.012

Belhouchat N, Zaghouane-Boudiaf H, Viseras C (2017) Removal of anionic and cationic dyes from aqueous solution with activated organo-bentonite/sodium alginate encapsulated beads. Appl Clay Sci 135:9–15. https://doi.org/10.1016/j.clay.2016.08.031

Cheng R et al (2023) An improved preparation method of carbon aerogels derived from alginate/chitosan with a superiority of promoting nZVI for chromium(VI) ions removal from aqueous solution. Microporous Mesoporous Mater 352:112512. https://doi.org/10.1016/j.micromeso.2023.112512

da Costa TB, da Silva TL, Costa CSD, da Silva MGC, Vieira MGA (2022) Chromium adsorption using Sargassum filipendula algae waste from alginate extraction: batch and fixed-bed column studies. Chem Eng J Adv 11:100341. https://doi.org/10.1016/j.ceja.2022.100341

Dubey M, Wadhwa S, Kumar R (2020) Synthesis of hematite/alginate beads nanocomposite and its application in organic dye removal. Mater Today Proc 28:70–73. https://doi.org/10.1016/j.matpr.2020.01.302

El-Maghrabi HH, Younes AA, Salem AR, Rabie K, El-shereafy E (2019) Magnetically modified hydroxyapatite nanoparticles for the removal of uranium (VI): preparation, characterization and adsorption optimization. J Hazard Mater 378:120703. https://doi.org/10.1016/j.jhazmat.2019.05.096

Garg R, Sabouni R (2023) Efficient removal of cationic dye using ZIF-8 based sodium alginate composite beads: performance evaluation in batch and column systems. Chemosphere 342:140163. https://doi.org/10.1016/j.chemosphere.2023.140163

Gokila S, Gomathi T, Sudha PN, Anil S (2017) Removal of the heavy metal ion chromiuim(VI) using chitosan and alginate nanocomposites. Int J Biol Macromol 104:1459–1468. https://doi.org/10.1016/j.ijbiomac.2017.05.117

Gopalakannan V, Periyasamy S, Viswanathan N (2016) One pot eco-friendly synthesis of highly dispersed alumina supported alginate biocomposite for efficient chromium(VI) removal. J Water Process Eng 10:113–119. https://doi.org/10.1016/j.jwpe.2016.02.005

Guan X, Zhang B, Li D, He M, Han Q, Chang J (2022) Remediation and resource utilization of chromium(III)-containing tannery effluent based on chitosan-sodium alginate hydrogel. Carbohydr Polym 284:119179. https://doi.org/10.1016/j.carbpol.2022.119179

He Y-R, Li X-L, Li X-L, Tan Z-Y, Zhang D, Chen H-B (2019) Aerogel based on melamine-formaldehyde and alginate: simply removing of uranium from aqueous solutions. J Mol Liq 289:111154. https://doi.org/10.1016/j.molliq.2019.111154

Khapre MA, Pandey S, Jugade RM (2021) Glutaraldehyde-cross-linked chitosan–alginate composite for organic dyes removal from aqueous solutions. Int J Biol Macromol 190:862–875. https://doi.org/10.1016/j.ijbiomac.2021.09.026

Li Z, Xu S, Xiao G, Qian L, Song Y (2019) Removal of hexavalent chromium from groundwater using sodium alginate dispersed nano zero-valent iron. J Environ Manag 244:33–39. https://doi.org/10.1016/j.jenvman.2019.04.130

Li L et al (2021) Anchoring nanoscale iron sulfide onto graphene oxide for the highly efficient immobilization of uranium (VI) from aqueous solutions. J Mol Liq 332:115910. https://doi.org/10.1016/j.molliq.2021.115910

Li Z, Xie W, Zhang Z, Wei S, Chen J, Li Z (2023a) Multifunctional sodium alginate/chitosan-modified graphene oxide reinforced membrane for simultaneous removal of nanoplastics, emulsified oil, and dyes in water. Int J Biol Macromol 245:125524. https://doi.org/10.1016/j.ijbiomac.2023.125524

Li T, Xu Y, Wang F, Xia L (2023b) Effective separation of U(VI) in acidic uranium-containing wastewater by potassium manganese ferrocyanide and carbon nanotube encapsulated calcium alginate beads. J Environ Chem Eng 11(5):111133. https://doi.org/10.1016/j.jece.2023.111133

Liu C et al (2020) Removal of Cadmium (II) using water hyacinth (Eichhornia crassipes) biochar alginate beads in aqueous solutions. Environ Pollut 264:114785. https://doi.org/10.1016/j.envpol.2020.114785

Liu J, Dai Q, Xiao R, Zhou T, Han J, Fu B (2023) Immobilization of ZnIn2S4 on sodium alginate foam for efficient hexavalent chromium removal. Int J Biol Macromol 236:123848. https://doi.org/10.1016/j.ijbiomac.2023.123848

Lu W et al (2020) Preparation of composite hydrogel (PCG) and its adsorption performance for uranium(VI). J Mol Liq 303:112604. https://doi.org/10.1016/j.molliq.2020.112604

Lv X et al (2017) Zero-valent iron nanoparticles embedded into reduced graphene oxide-alginate beads for efficient chromium (VI) removal. J Colloid Interface Sci 506:633–643. https://doi.org/10.1016/j.jcis.2017.07.024

Mallakpour S, Behranvand V (2021) Methylene blue contaminated water sanitization with alginate/compact discs waste-derived activated carbon composite beads: adsorption studies. Int J Biol Macromol 180:28–35. https://doi.org/10.1016/j.ijbiomac.2021.03.044

Mishra S, Verma N (2016) Carbon bead-supported hollow carbon nanofibers synthesized via templating method for the removal of hexavalent chromium. J Ind Eng Chem 36:346–354. https://doi.org/10.1016/j.jiec.2016.02.025

Oulguidoum A, Bouyarmane H, Laghzizil A, Nunzi J-M, Saoiabi A (2019) Development of sulfonate-functionalized hydroxyapatite nanoparticles for cadmium removal from aqueous solutions. Colloid Interface Sci Commun 30:100178. https://doi.org/10.1016/j.colcom.2019.100178

Periyasamy S, Gopalakannan V, Viswanathan N (2018) Hydrothermal assisted magnetic nano-hydroxyapatite encapsulated alginate beads for efficient Cr(VI) uptake from water. J Environ Chem Eng 6(1):1443–1454. https://doi.org/10.1016/j.jece.2018.01.007

Sadoun L, Benmounah A, Ait-Ramdane-Terbouche C, Seffah K, Terbouche A (2023) A novel spherical hybrid material based on the combination of humic acid/alginate/Algerian Zeen oak sawdust for removing chromium (VI) from wastewater. Mater Today Chem 34:101819. https://doi.org/10.1016/j.mtchem.2023.101819

Saha S, Basu H, Rout S, Pimple MV, Singhal RK (2020) Nano-hydroxyapatite coated activated carbon impregnated alginate: a new hybrid sorbent for uranium removal from potable water. J Environ Chem Eng 8(4):103999. https://doi.org/10.1016/j.jece.2020.103999

Sami Abboud A, Ghaffarinejad A, Mollahosseini A (2023) Removal of lead and cadmium ions from wastewater using magnetic alginate impregnated sulfur. Environ Nanotechnol Monit Manag 20:100874. https://doi.org/10.1016/j.enmm.2023.100874

Sánchez-Castro I, Martínez-Rodríguez P, Abad MM, Descostes M, Merroun ML (2021) Uranium removal from complex mining waters by alginate beads doped with cells of Stenotrophomonas sp. Br8: novel perspectives for metal bioremediation. J Environ Manag 296:113411. https://doi.org/10.1016/j.jenvman.2021.113411

Shen Y, Li B, Zhang Z (2023) Super-efficient removal and adsorption mechanism of anionic dyes from water by magnetic amino acid-functionalized diatomite/yttrium alginate hybrid beads as an eco-friendly composite. Chemosphere 336:139233. https://doi.org/10.1016/j.chemosphere.2023.139233

Shim J, Kumar M, Mukherjee S, Goswami R (2019) Sustainable removal of pernicious arsenic and cadmium by a novel composite of MnO2 impregnated alginate beads: a cost-effective approach for wastewater treatment. J Environ Manag 234:8–20. https://doi.org/10.1016/j.jenvman.2018.12.084

Sigdel A et al (2017) Concurrent removal of cadmium and benzene from aqueous solution by powdered activated carbon impregnated alginate beads. Catena 148:101–107. https://doi.org/10.1016/j.catena.2016.06.029

Šillerová H, Komárek M, Liu C, Poch J, Villaescusa I (2015) Biosorbent encapsulation in calcium alginate: effects of process variables on Cr(VI) removal from solutions. Int J Biol Macromol 80:260–270. https://doi.org/10.1016/j.ijbiomac.2015.06.032

Thakur AK, Kumar M (2021) Efficacy of green alginate beads for multi-metal removal from aqueous solution. Case Stud Chem Environ Eng 3:100100. https://doi.org/10.1016/j.cscee.2021.100100

Verma A et al (2020) Graphite modified sodium alginate hydrogel composite for efficient removal of malachite green dye. Int J Biol Macromol 148:1130–1139. https://doi.org/10.1016/j.ijbiomac.2020.01.142

Wang Y, Yao W, Wang Q, Yang Z, Liang L, Chai L (2016) Synthesis of phosphate-embedded calcium alginate beads for Pb(II) and Cd(II) sorption and immobilization in aqueous solutions. Trans Nonferrous Metals Soc China 26(8):2230–2237. https://doi.org/10.1016/S1003-6326(16)64340-6

Wang Z, Wu S, Zhang Y, Miao L, Zhang Y, Wu A (2020) Preparation of modified sodium alginate aerogel and its application in removing lead and cadmium ions in wastewater. Int J Biol Macromol 157:687–694. https://doi.org/10.1016/j.ijbiomac.2019.11.228

Wang D, Zhang J, Li J (2023) Phosphate-functionalized magnetic calcium alginate for the engineering remediation of uranium-contaminated water and soil. Chem Eng J 475:145910. https://doi.org/10.1016/j.cej.2023.145910

Wen S et al (2023) Selective adsorption of uranium (VI) from wastewater using a UiO-66/calcium alginate/hydrothermal carbon composite material. Carbohydr Polym 315:120970. https://doi.org/10.1016/j.carbpol.2023.120970

Wu S, Han C, Xin L, Li M, Long H, Gao X (2023) Synthesis of triethylenetetramine modified sodium alginate/CuS nanocrystal composite for enhanced Cr(VI) removal: performance and mechanism. Int J Biol Macromol 238:124283. https://doi.org/10.1016/j.ijbiomac.2023.124283

Xiao Y et al (2023) Functionalized maghemite nanoparticles for enhanced adsorption of uranium from simulated wastewater and magnetic harvesting. Environ Res 216:114569. https://doi.org/10.1016/j.envres.2022.114569

Yacou C, Altenor S, Carene B, Gaspard S (2018) Chemical structure investigation of tropical Turbinaria turbinata seaweeds and its derived carbon sorbents applied for the removal of hexavalent chromium in water. Algal Res 34:25–36. https://doi.org/10.1016/j.algal.2018.06.014

You Y et al (2019) Sodium alginate templated hydroxyapatite/calcium silicate composite adsorbents for efficient dye removal from polluted water. Int J Biol Macromol 141:1035–1043. https://doi.org/10.1016/j.ijbiomac.2019.09.082

Yu J, Wang J, Jiang Y (2017) Removal of uranium from aqueous solution by alginate beads. Nucl Eng Technol 49(3):534–540. https://doi.org/10.1016/j.net.2016.09.004

Zahakifar F, Keshtkar AR, Talebi M (2021) Synthesis of sodium alginate (SA)/polyvinyl alcohol (PVA)/polyethylene oxide (PEO)/ZSM-5 zeolite hybrid nanostructure adsorbent by casting method for uranium (VI) adsorption from aqueous solutions. Prog Nucl Energy 134. https://doi.org/10.1016/j.pnucene.2021.103642

Zhang Z, Abidi N, Lucia L (2023) Smart superabsorbent alginate/carboxymethyl chitosan composite hydrogel beads as efficient biosorbents for methylene blue dye removal. J Mater Sci Technol 159:81–90. https://doi.org/10.1016/j.jmst.2023.02.045

Zhao D, Shen Z, Shen X (2021) Dual-functional calcium alginate hydrogel beads for disinfection control and removal of dyes in water. Int J Biol Macromol 188:253–262. https://doi.org/10.1016/j.ijbiomac.2021.07.177

Zong Y et al (2023) Preparation of a ternary composite based on water caltrop shell derived biochar and gelatin/alginate for cadmium removal from contaminated water: performances assessment and mechanism insight. Int J Biol Macromol 234:123637. https://doi.org/10.1016/j.ijbiomac.2023.123637

Chapter 7
Use of Pectin for Degradation of Environmental Pollutants

Dipika Panmand, Shankar Khade, Debashis Dutta, and Jyoti Srivastava

Abstract Environmental pollution poses a growing threat to the planet's ecosystems and human health. Various manufacturing industrial chemicals released into the water include dyes, phenolic derivatives, heavy metals, and other industrial waste materials that can act as hazardous pollutants affecting the animals nearby. The pectin-based nanocomposites such as pectin-graphene oxide, pectin-thorium tungsto-molybdate, pectin-cadmium sulfide, pectin-graphene oxide-magnesium ferrite-zinc oxide, etc. have been effective in adsorption or photocatalytic bioremediation pollutants. The dual-responsive pectin-graphene oxide nanocomposites are efficient in the adsorption of heavy metal Cr (III) ions and act as photocatalyst to degrade organic dyes in the water. In this context, using naturally occurring compounds to combat pollutants has gotten a lot of attention. A natural polymer pectin can potentially help to reduce this global issue by assisting the degradation of environmental pollutants. Pectin comes from fruits like citrus and apples. It's been recognized as a good way to fight against pollutants because it's biodegradable and not toxic. The chemical structure of it is both effective as a catalyst and adsorbent for various types of contamination like heavy metals, organic dyes, and industrial waste.

There are three ways that pectin breaks down pollutants: adsorption, coagulation, and precipitation. All these ways eventually separate the pollutant from its water sources. Pectin also has certain functional groups that form complexes with contaminants that assist in removing the pollutants. The nanocomposites made with pectin or modified pectin derivatives help to remove and degrade the pollutants.

Keywords Pollutants · Biodegradation · Pectin · Sustainability · Nanocomposites

D. Panmand · S. Khade (✉) · J. Srivastava
Department of Biotechnology Engineering, School of Engineering, Ajeenkya DY Patil University, Pune, Maharashtra, India

D. Dutta
Department of Food Processing Technology, Mirmadan Mohanlal Government Polytechnic, Plassey, West Bengal, India

A. Guldhe et al. (eds.), *Biomaterials for Environmental Applications: Detection and Remediation*, Environmental Challenges and Solutions,
https://doi.org/10.1007/978-3-032-12701-3_7

7.1 Introduction

Pectin is a soluble fiber found in fruits and vegetables. Galactose is the source of galacturonic acid, the main chemical component of pectin (Gawkowska et al. 2018). Commercially produced citrus fruits are used to make pectin, which is a white to light brown powder (Twinomuhwezi et al. 2020). It is utilized as an edible gelling ingredient, especially in jams and jellies, dessert fillings, medications, and confections, as well as a food stabilizer in fruit juices and milk drinks. It is also a source of nutritional fiber (Bagal-Kestwal et al. 2019).

Pectin Sources:

Pears, apples, guavas, quince, plums, gooseberries, oranges, and other citrus fruits contain high quantities of pectin; soft fruits such as cherries, grapes, and strawberries include trace amounts of the fruit (May 1990; Srivastava and Malviya 2011). Dried citrus peels and apple pomace, both byproducts of juice production, are the primary raw materials used in the manufacturing of pectin (Güzel and Akpınar 2019). A tiny amount of sugar beet pomace is also utilized (Table 7.1).

7.2 Applications of Pectin

7.2.1 Food Industry

In the food business, pectin is typically utilized as an emulsifying, thickening, stabilizing, and gelling ingredient (Brejnholt 2009; Vanitha and Khan 2019). Pectin is frequently utilized in wet and viscous meals because it helps produce hydrogels. Because pectin is widely used in desserts, fruit juices, jams, jellies, and dairy goods, its gelling qualities are well known (May 1990; Christensen 2020).

7.2.2 Biomedical Formulations

Pectin, a polysaccharide with promising health effects, is being investigated as an innovative biomaterials technology for various applications such as drug delivery, gene delivery, wound healing, cholesterol reduction, tissue engineering, membrane

Table 7.1 Typical levels of pectin in various natural sources (Srivastava and Malviya 2011)

Apples	1–1.5%
Apricots	1–1.5%
Cherries	0.4%
Oranges	0.5–3.5%
Carrots	1.4%
Citrus peels	30%

fabrication for contact lens development, artificial corneas, catheters, anticancer activities, and mucosal and gastrointestinal drug delivery (Kedir et al. 2022; Prasathkumar et al. 2023; Abualsoud et al. 2025).

7.2.3 Degradation of Environmental Pollutants

7.2.3.1 Heavy Metal Removal

Pectin's ability to chelate heavy metals like lead, cadmium, and mercury makes it effective in removing these contaminants from wastewater and soil (Zhexenbay et al. 2020). Pectin-based biofilters and adsorbents can immobilize heavy metals, preventing their release into the environment.

7.2.3.2 Organic and Inorganic Pollutant Degradation

Pectin's adsorption capacity allows it to capture organic pollutants, including pesticides, herbicides, and dyes from wastewater. Paulino et al investigated the significant role of pectin hydrogels modified with montmorillonite in the removal of synthetic dye methylene blue and herbicide 2,4-dichlorophenoxyacetic acid (2,4-D) from the wastewater (Fernandes Filho et al. 2023). In addition, the pectin plays pivotal in the removal of heavy metals. Dandan et al formulated a pectin and β-cyclodextrin composite and successfully used in removing the Zn^{2+} and Cu^{2+} by adsorption (Fu et al. 2024). Similarly, Wang et al. developed the hydrogel beads of pectin with carboxy methyl cellulose (CMC) using chitosan as crosslinker and efficiently removed the Cd^{2+}, Pb^{2+}, and Cu^{2+} from the water (Zhang et al. 2025).

7.2.3.3 Microplastic Degradation

Microorganisms produced by pectin can develop enzymes which degrade microplastics in smaller molecules. Han et al investigated the polyethylene and polypropylene plastics degradation using the fungal consortia *Aspergillus niger*, *Cunninghamella elegans* along with pectin-biochar beads which enhanced the degradations (Arshad et al. 2025).

7.3 Structure of Pectin

Pectin has a complicated structure that changes based on where it comes from. All pectins, however, have a-D-galacturonate residues as their common backbone. 1,4-Glycosidic linkages bind these residues together. It's possible that the galacturonate residues are methyl esterified, in which case the carboxyl group (–COOH) is joined to a methyl group (–CH3). Methyl esterification (DM) can range from 0% to 100% in degree (Ropartz and Ralet 2020).

In addition to its galacturonate backbone, pectin can also include side chains of neutral sugars such as arabinose, galactose, and rhamnose. These side chains come in different lengths and compositions, and they can be either branched or unbranched. Pectin's structure is what gives it its special qualities. The gelling ability of pectin is attributed to its galacturonate backbone, with its solubility and viscosity being influenced by its side chains. The characteristics of pectin are likewise impacted by DM. For instance, compared to pectin with a low DM, pectin with a high DM is more soluble and has a poorer gelling ability (Gawkowska et al. 2018).

Pectin is a substance with several uses in the food business due to its versatility. It functions as an emulsifier, stabilizer, thickening, and gelling agent. Certain medicinal and cosmetic items also include pectin.

7.4 Properties of Pectin

Gelling ability When calcium ions are present, pectin can gel. Pectin produces strong, transparent gels that are heat stable (Said et al. 2023).

Thickening ability Pectin can thicken liquids, increasing their viscosity. This property is useful in sauces, jams, and jellies (Ebrahim et al. 2023).

Stabilizing ability Pectin can stabilize emulsions, preventing them from separating. This property is useful in dairy products, such as yogurt and ice cream (Leroux et al. 2003).

Emulsifying ability Pectin can help to emulsify two immiscible liquids, such as oil and water. This property is useful in salad dressings and mayonnaise (Yang et al. 2018). Dapeng et al. investigated that the pectin extracted from the persimmon peel exhibited a significant natural emulsifying property that can be used in the food industries (Jiang et al. 2020).

7.5 Pectin Extraction

Pectin is primarily produced from citrus peels (85%), apple pomace (14%), and beetroot (1%), on an industrial scale (Pereira et al. 2016). However, new studies suggest that pectin can be extracted from a range of food industry by-products, enabling the assessment of agro-industrial waste. Some of the by-products that have recently been studied include the peels of passion fruit, mango, grape marc, jackfruit, kiwi, potato pulp, melon peels, watermelon peels, coffee pulp, cocoa shells, banana peels, pomegranate peels, durian peels, okra pods, pumpkin peels, and papaya peels (Oliveira et al. 2016; Dao et al. 2021). On a large scale, pectin is usually produced by alcoholic precipitation and acid extraction, despite the significant energy and solvent costs. This can be explained by the process's simplicity of operation. The process of acid extraction typically involves the high-temperature hydrolysis of protopectin. Several studies have examined the effects of different acid solvents and extraction parameters (such as pH, temperature, duration, and solid: liquid ratio) on the extraction yield, structure, and physicochemical properties of pectin. Most frequently, high temperatures (60–90 °C) were used for prolonged periods of time (1–6 hours), with the use of sulfuric, hydrochloric, and citric acids, followed by alcoholic precipitation (Abbaszadeh 2008; Shaha et al. 2013).

7.6 Types of Environmental Pollutants

Environmental pollutants are substances or mixtures of substances that are introduced into the environment and cause harm to human health or the natural environment. The chemical characteristics, origins, and effects allow for their classification into multiple categories.

7.6.1 Organic Pollutants

Organic pollutants are substances that contain carbon and are derived from living organisms or their byproducts. They include a wide range of chemicals, such as:

- Pesticides: These are chemicals used to kill pests, such as insects, rodents, and weeds. Some pesticides, such as DDT and PCBs, are persistent and can accumulate in the environment and harm wildlife (Sánchez-Bayo 2011; Boudh and Singh 2018).
- Herbicides: These are chemicals used to kill weeds. Some herbicides, such as glyphosate, are controversial because they may be harmful to human health and the environment (Valavanidis 2018; Caiati et al. 2020).

7.6.2 Heavy Metals

Naturally occurring elements with a high density are called heavy metals. They are present in the air, water, and soil. Lead, mercury, and cadmium are among the heavy metals that are harmful to humans and can result in a number of health issues, such as renal disease, cancer, and damage to the nervous system (Balali-Mood et al. 2021; Jomova et al. 2025).

7.7 Use of Pectin-Based Materials for Pollutant Removal

Heavy metals can now be extracted from wastewater using a variety of techniques, such as membrane filtration, ion exchange, chemical precipitation, electrodeposition, and biosorption. Of them, biosorption appears to be the most adaptable due to its appealing adsorption efficiency, environment friendliness, ease of use, affordability, and accessibility. We explored various agrarian squander such as marine green growth, olive padmese, shelled nut shells, sunfiber stems, citrus shells, and hazelnut shells to extricate overwhelming metals from fluid arrangements. Gri bone squander such as ocean growth, olive swelling, shelled nut trunks, sunfiber stems, citrus shells, and hazelnut shells were inspected to extricate overwhelming metals from the watery arrangement. Adsorption is advanced by biopolymers such as cellulose, chitosan, alginate, and pectin. They may have high adsorption capabilities, biocompatibility, and new adaptability.

Pectin is a complex polymer that is common in plant cell walls and is naturally non-toxic. It can be derived from various natural sources, including apple pomace and citrus peels. Pectin and pectin-rich wastes are promising candidates for biosorbents because of their significant affinity for metal ions, as demonstrated by numerous studies. Nevertheless, there is currently no review article covering the most recent advancements, benefits and drawbacks, guidelines and procedures for using Pectin and its byproducts as absorbents for heavy metals. Because of this, the potential to use pectin and its derivatives as heavy metal adsorbents has not yet been fully realized.

In arrange to help the improvement and application of pectin as an overwhelming metal adsorbent, this considers the current state, impact and impacting components of critical variables, current challenges, and future advancement patterns of pectin-based adsorbents. In expansion, the intelligent between pectin and light metal particles, such as Fe^{2+}, Mg^{2+}, and Ca^{2+}, were compiled for comparison and a more profound comprehension of the overwhelming metal adsorption prepared by pectin and its subsidiaries (Zhang et al. 2020; Ahmad et al. 2023) (Fig. 7.1).

7.7.1 Pectin-Based Composites and Nanoparticles

Pectin-based composites and nanoparticles are a promising class of materials with a wide range of potential applications. Pectin is a natural polysaccharide that is found in fruits and vegetables. It is a non-toxic, biodegradable, and biocompatible material that can be used to form films, hydrogels, and other structures. Nanoparticles are particles that are less than 100 nanometers in size. Their distinct characteristics render them advantageous in numerous applications, including medication administration, food packaging, and sensing.

Pectin-based composites can be formed by incorporating nanoparticles into a pectin matrix. This can be done by physical or chemical methods. Physical

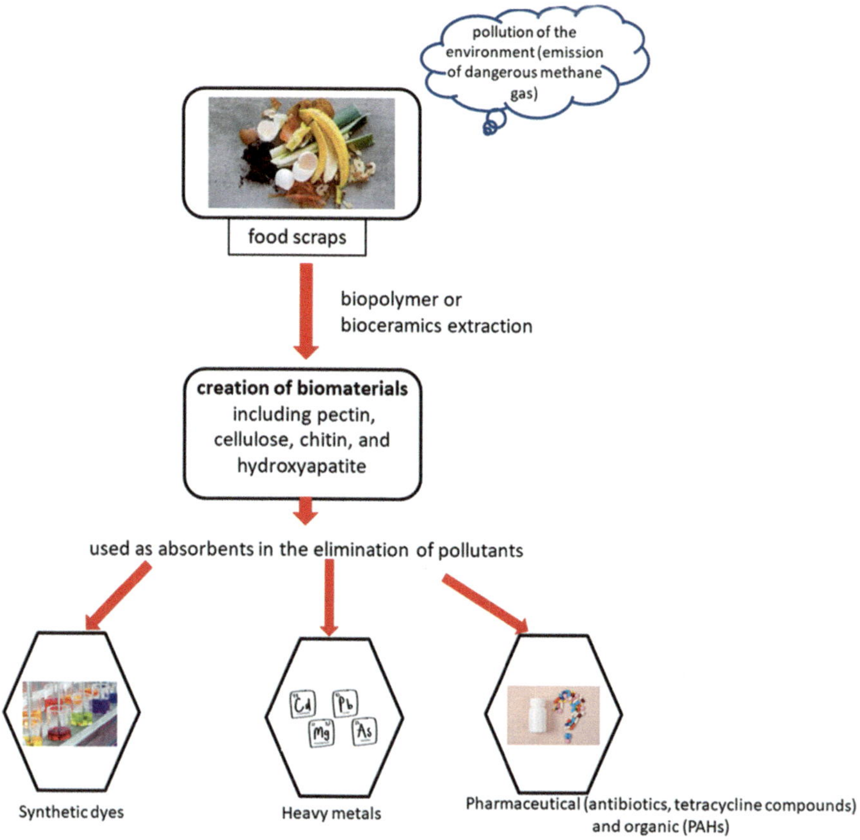

Fig. 7.1 Pectin used for heavy metal degradation

methods, such as mixing or blending, are often used to incorporate nanoparticles into pectin solutions. Chemical methods, such as in situ synthesis, can be used to form nanoparticles directly within the pectin matrix. Pectin-based composites have a number of advantages over traditional materials. They are often stronger, more flexible, and more barrier to gases and liquids. They can also be tailored to have specific properties, such as antibacterial or antioxidant activity. Pectin-based nanoparticles can be used to deliver drugs, target cancer cells, and enhance the properties of foods and packaging materials. They can also be used to develop sensors for detecting environmental contaminants and other substances (Lara-Espinoza et al. 2018).

Here are some specific examples of pectin-based composites and nanoparticles and their applications:

- Pectin-silver nanoparticle composites have been shown to have antibacterial activity. They can be used to develop food packaging materials that prevent the growth of bacteria on food.

- Pectin-zinc oxide nanoparticle composites have been shown to have UV-blocking properties. They can be used to develop sunscreen lotions and protective coatings for materials.
- Pectin-iron nanoparticle composites have been shown to have magnetic properties. They can be used to develop magnetic resonance imaging (MRI) contrast agents.
- Pectin-gold nanoparticle composites have been shown to have catalytic properties. They can be used to develop catalysts for a variety of chemical reactions.

Pectin-based composites and nanoparticles are a promising area of research with a wide range of potential applications.

Removal of Pectin-Bound Heavy Material from Water
There are a number of methods for removing pectin-bound heavy material from water. Some of the most common methods include:

Precipitation This method involves adding a chemical agent to the water to cause the pectin-bound heavy material to precipitate out of solution. Common precipitating agents include aluminum sulfate, iron sulfate, and lime.

Coagulation This method is similar to precipitation, but it involves adding a coagulant aid to help the pectin-bound heavy material form larger particles. Common coagulant aids include alum, polyaluminum chloride, and ferric chloride.

Adsorption This method involves using an adsorbent material to bind to the pectin-bound heavy material and remove it from the water. Common adsorbent materials include activated carbon, zeolites, and biochar.

Membrane filtration This method involves passing the water through a membrane that will filter out the pectin-bound heavy material. Common types of membrane filtration include microfiltration, ultrafiltration, and reverse osmosis.

The choice of method will depend on the specific type of pectin-bound heavy material that is present in the water, as well as the desired level of removal (Table 7.2).

A variety of other cutting-edge technologies are being developed in addition to the foregoing techniques for the extraction of heavy materials bound to pectin from water. These technologies include:

1. *Electrochemical methods*: These methods use electricity to remove heavy metals from water.
2. *Photocatalytic methods*: These methods use light to decompose heavy metals into harmless compounds.
3. *Bioremediation*: This method uses bacteria or other microorganisms to remove heavy metals from water.
4. Although these new technologies are still in their infancy, they could end up being more economical and efficient than conventional approaches.

Table 7.2 Summary of the advantages and disadvantages of pollutant removal methods

Method	Advantages	Disadvantages
Precipitation	Simple, inexpensive	May generate large amounts of sludge
Coagulation	Effective for removing a wide range of heavy metals	Requires the use of coagulant aids
Adsorption	Highly effective for removing specific heavy metals	May require regeneration of the adsorbent
Membrane filtration	Produces high-quality water	Can be expensive

7.7.2 Pectin as a Chelating Agent to Prevent the Toxicity of Heavy Metals

Pectin is a naturally occurring, non-toxic complex polysaccharide found in the cell walls of plants, fruits, and vegetables. Pectin may be found in fruit peels of papaya, bananas, grapefruits, oranges, and pomace from apples, kiwis, and sugar beets. Apple pomace, citrus peels, and sugar beet pulp are the most widely used industrial sources of pectin because they are readily available, inexpensive, and have high pectin contents (15–25% in dried apple pulp, 30–35% in dry citrus peel, and up to 23% in sugar beet). Thus, countries in the European Union produce around half of the world's sugar beet, resulting in an annual waste output of 111.6 million tons. Approximately 15–106 tons of industrial citrus trash are produced per year worldwide. Utilizing fruit and vegetables, we might thereby address the issue of waste accumulation. The functionality of pectin and its potential for use in the food and pharmaceutical sectors are largely determined by its structure. Pectin is mostly made up of D-galacturonic acid units connected by (1 + 4) glycosidic linkages, specifically, homogalacturonan (HG), rhamnogalacturonan-1 (RG-I), and rhamnogalacturonan-II (RG-II). Throughout the whole molecule, there are alternating "smooth" and "hairy" areas. Homogalacturonanas (HGs) are the representative for smooth parts, whereas rhamnogalacturonans are the representative for hairy regions. Pectin can have three different degrees of esterification (DE): high methoxy pectin (DE is greater than 50); low methoxy pectin (DE is below 50); and pectic acid (DE is below 10), depending on the percentage of carboxylic groups esterified by a molecule of methanol (Yu et al. 2022). Pectin can range in molecular weight from 50 to 150 kDa. Pectins have been utilized for years in the food business for gelling, thickening, and stability. These days, their applications include the delivery of drugs to specific colons, the transport of microbes, the delivery of genes, tissue engineering, and the creation of edible films and coatings for food packaging. Pectins have several advantageous qualities that are crucial for preserving human health, including the capacity to detoxify heavy metals and radioactive elements, reduce cholesterol and serum glucose, have an anti-ulcer, prebiotic, and anti-cancer effect, among others. Pectins have been shown in a number of studies to be effective biosorbents for removing metals from wastewater. However, there are certain limitations because of instability and issues with pectin extraction from aqueous solution. From a practical standpoint, using hydrogels based on pectate and chemically modified pectins may be a reasonable option. The oligosaccharides that are produced when pectin is fermented in the gut are essential for cancer treatment, intestinal disease prevention, and immunological defense. Low esterification levels of pectins have been shown to reduce metastasis, induce apoptosis, limit tumor growth, and alter immune responses. Pectin is a prebiotic that is still in its infancy, but it has the potential to improve the gut microbiota by stimulating the growth of probiotics like Lactobacillus acidophilus and Bifidobacterium longum and preventing the growth of Clostridium perfringens, pathogenic E. coli and Helicobacter pylori. Pectins improve the function of the gastrointestinal barrier, protecting the kidneys and liver from the translocation of pro-inflammatory microorganisms and their byproducts. Scientific Zhexenbay & Associates/Procedia Studies on humans and animals have demonstrated pectin's ability to reduce the amount of toxins in the body and pectin's mechanisms for binding heavy metals. One of pectin's important functions is its ability to bind divalent cations, which is particularly relevant given the state of the environment today. The sorption process is dependent on the metal ions attached to non-esterified carboxyl groups or other binding sites. The strength of the sorption processes and the binding capacity are determined by the amount of free carboxyl groups in the pectin structure. It relates to the formation of junction zones between the free carboxyl groups of pectin and heavy metals, as per the "egg-box" paradigm. This indicates that pockets are created when fiber chains are piled together and allowing metals to compound with fibers. The procedure has to do with how esterified something is. The binding mechanism is based on the formation of hydrogen bonds between metal atoms and non-esterified carboxyl groups as well as ionic interactions between metal and oxygen atoms. Pectin travels through the stomach and small intestine without breaking down, and it undergoes significant fermentation in the colon. Low-ester pectins break down more quickly than high-ester pectins. Pectins are employed to carry medications to the colon with low degradation because of this feature. After being absorbed, pectin pieces are transported, changed, surrounded by toxicants, and, by forming insoluble pectinates within the organism, they are expelled through the urine and feces. Maxwell et al. (2012) state that the absorption method of pectin fragments involves both passive absorption and aggressive cell capture activities (Maxwell et al. 2012). Additionally, uptake by lymphoid tissue linked with the gastrointestinal tract may occur. The degree of esterification and the molecular charge of pectin are important factors for passive absorption. It was demonstrated that positively charged, acidic Pectin fragments were unable to penetrate monolayers of Caco- cells. Therefore, only neutral pectin segments are effective. Lower atomic weight pectin (52 kDa) can be effortlessly ingested into the circulatory system amid absorption. Unmodified pectin may be a non-digestible polysaccharide in long polymers of cross-linked chains with a tall molecular weight(Mw). Shorter chain atoms and a moo atomic weight can be produced through adjustment, which incorporates pH and enzymatic adjustment. A number of writers claim that the bloodstream absorbed modified citrus pectin (MCP) with a decreased molecular weight, reduced side-chain structure, and shorter chain molecules very quickly. The pectin source, extraction techniques, cation type, structural variation, esterification degree, pH, and molecular weight all affect how well pectin is absorbed. The ability of pectin to attach to its targets improves as the degree of esterification decreases. Studies have demonstrated that less esterified pectins enter colonic epithelial crypts more deeply than highly esterified ones. The ability to bind heavy metal ions was enhanced by 10% rhamnogalacturonan II and a low degree of esterification, but not by necessary minerals. When the degree of methylation is decreased, Celus et al. observed similar outcomes, which included a larger amount of bound and stronger interactions between the cation and Pectin. According to Liang et al. (2020), ethylenediamine-modified pectins with a high degree of amination have a considerable Pb^{2+} removal efficiency (≥94%) by chelation with amino and acylamino groups and carboxylic group ion exchange (Liang et al. 2020). The majority of the research that is currently available indicates that pectins do not cause the body to release vital minerals. Contrarily, some research findings indicate that pectins may upset the mineral balance and reduce the bioavailability of important elements because of cation exchange, depending on the type of pectin administered and its degree of esterification. Using pectin as a heavy metal detoxifying agent to lower health risks and environmental contamination 557 pectin interactions in the small intestine. Low esterified apple pectin was found in several studies to have a detrimental influence on the calcic minerals' bioavailability. However, sugar beet pectin that is more branched and heavily acetylated can increase the bioaccessibility of minerals. However, during colon microflora's pectin fermentation, bounded minerals may be released, boosting the rate at which minerals are absorbed. However, given that the small intestine is the primary location of nutrient absorption, it is possible that the colon will not absorb free minerals. Furthermore, minerals may be used by the microflora in the colon, making them unavailable to the human body. By using pectins, such as sugar beet pectin, and managing the structure of products containing pectin, this problem could be resolved in a way that benefits the body. Birthing of necessary minerals reduces their bioaccessibility in the small intestine. The pH level is one of the key elements that influences the sorption process. Different pH values are appropriate depending on the kind and structure of pectin. When pH drops below 2, pectins precipitate, which reduces sorption capacity. Pectins become unstable at pH values greater than 8. Accordingly, at pH 2.3 the unmodified pectin removed heavy metals at a rate of 60%, and at pH 4.0 that rate rose to 90%. The optimal pH range for the lead-binding properties of low-esterified pectin was found to be 4–8. A very acidic 15–3.5 in the stomach, 6 in the duodenum and small intestine. 7.4 in the terminal ileum, 5.7 in the caecum, and 6.7 in the rectum is the pH range of the gastrointestinal system, a typical blood pH falls between 7.35 and 7.45. This means that scientific research on humans and animals has confirmed the possibility of heavy metal binding by pectins in the gastrointestinal tract and blood conditions. The cation affects the propensity of pectin to bind heavy metals. The cation's ionic size and degree of electronegativity are significant factors. The most noteworthy liking for divalent cation sorts, especially Cu^{2+} and Pb^{2+}, is shown by the pectins. Based on a bunch try in a watery arrangement for settling overwhelming metals by orange peel adjusted with sodium hydroxide and calcium chloride, the sequence of selectivity was taken after: Cu^{2+} > Zn^{2+} > Pb^{2+}. The comparative selectivity was found by Kartel et al. in a simulant watery arrangement, where $Pb^{2+} > Cu^{2+} > Co^{2+} > Zn^{2+} > Ni^{2+} > Cd^{2+}$ and sugar beet pectin appeared a solid liking for Pb. These findings demonstrate the modified pectin's strong lead-binding affinity. According to the authors, within the first 10 minutes, the sorption process moves quite quickly (Feng and Guo 2012). The metal absorption time was 20 minutes, according to Khotimchenko et al. In an in vivo investigation, children with elevated blood lead levels responded well to PectaSol, a modified citrus pectin containing 3.8% degree of esterification and roughly 10% rhamnogalacturonan II based on the presence of 2-keto-3-deoxy-octatonic acid. The children's blood lead levels dropped (161% average change) and rose (132%) in the urine (132%) after taking 15 g of pectin every day for 24 hours, but no vital minerals were released. The same pectin supplement, PectaSol, was used in a second trial to find out how pectin affected healthy subjects' excretion of toxicants through their urine. Urine arsenic levels dramatically rose (130%, $p < 0.05$) 24 hours after 5 g/daily of pectin was administered. On the sixth day, there was a substantial increase in the excretion of cadmium (150%, $p < 0.05$) and lead (560%, $p < 0.08$) in the urine. These findings might be explained by the essence of rhamnogalacturonan II, which chelates metals. These comes about were plained by the end of the medium-pectin complex by pee and the coordinate chelation of molecular weight pectin containing rhamnogalacturonan II within the blood. A few clinical inquiries about found that 4 g per day was adequate to produce perceptible comes about within the treatment of overwhelming metals in 3–6 months. Since different dairy items are a noteworthy portion of the human count calories on an everyday premise, devouring dairy items containing pectin makes it simple to ingest the previously mentioned amounts of pectin. Consequently, pectin can be taken as a supplement or added to dietary items like dairy products to serve as a safe and effective chelating agent, particularly for young children who are more susceptible. It would be simple to consume such pectin-containing items and be an innovative preventive measure against heavy metals toxicity for ongoing use.

7.8 Conclusion

By controlling gut flora, pectin effectively lowers obesity brought on by environmental pollutants and provides a possible course of treatment for illnesses brought on only by exposure to pollution. Investigating logical writing on pectin authoritative capability shows that it is or of the most secure and successful characteristic chelators. Pectin's frame influences how successful it is as a potential nutraceutical or bioactive dietary polysaccharide. The degree of esterification, homogalacturonan and rhamnogalacturonan II structure, and sum of free carboxyl bunches in pectin are imperative for overwhelming metal chelation. Hence, sugar beet pectin may be utilized in dairy acid products to improve technological product attributes due to its high affinity for binding lead. It can also be used consistently as a pleasant and mild chelating agent. Due to the growing use of pectin worldwide, dairy products were chosen as the pectin matrix. Because of the vast array of advantageous pectin qualities and the mineral- and vitamin-rich composition of milk products, these products may be beneficial to human health and the environment. Using pectin, which has the technological qualities needed for dairy production, is successful at excreting heavy metals from the body without upsetting the mineral balance of the organism and opens up new possibilities for the development of radically new safe food products with distinct functional qualities. These items' numerous health-promoting pectin qualities would also help avoid or even completely eradicate other severe human ailments.

References

Abbaszadeh AH (2008) Pectin and galacturonic acid from citrus wastes. University of Boras, School of Engineering, Boras

Abualsoud BM, Alhomrani M, Alamri AS, Alsanie WF, Baldaniya L, Jyothi SR, Prince PGK, Kashyap A, Nainwal P, Deepak A (2025) Pectin-based composites: a promising approach for tissue engineering and wound healing. Int J Polym Mater Polym Biomater 75:1–28

Ahmad S, Sabir A, Khan SM (2023) Synthesis and characterization of pectin/carboxymethyl cellulose-based hybrid hydrogels for heavy metal ions adsorption. Chem Pap 77(8):4165–4177

Arshad I, Sajid S, Yu Q, Li J, Zveushe OK, Rehman S, Lal MM, Atakpa EO, Chen X, Zhang W (2025) Elucidating polyethylene and polypropylene microplastics degradation mechanisms and metabolic pathways via pectin-biochar mediated fungal consortia. J Hazard Mater 496:139217

Bagal-Kestwal DR, Pan M, Chiang BH (2019) Properties and applications of gelatin, pectin, and carrageenan gels. In: Bio monomers for green polymeric composite materials. Wiley, pp 117–140

Balali-Mood M, Naseri K, Tahergorabi Z, Khazdair MR, Sadeghi M (2021) Toxic mechanisms of five heavy metals: mercury, lead, chromium, cadmium, and arsenic. Front Pharmacol 12:643972

Boudh S, Singh JS (2018) Pesticide contamination: environmental problems and remediation strategies. In: Emerging and eco-friendly approaches for waste management. Springer, pp 245–269

Brejnholt SM (2009) Pectin. In: Food stabilisers, thickeners and gelling agents. Wiley, pp 237–265

Caiati C, Pollice P, Favale S, Lepera ME (2020) The herbicide glyphosate and its apparently controversial effect on human health: an updated clinical perspective. Endocr Metab Immune Disord Drug Targets (Formerly Curr Drug Targets-Immune Endocr Metab Disord) 20(4):489–505

Christensen SH (2020) Pectins. In: Food hydrocolloids. CRC Press, pp 205–230

Dao TAT, Webb HK, Malherbe F (2021) Optimization of pectin extraction from fruit peels by response surface method: conventional versus microwave-assisted heating. Food Hydrocolloids 113:106475

Ebrahim SA, Othman HA, Mosaad MM, Hassabo AG (2023) Eco-friendly natural thickener (pectin) extracted from fruit peels for valuable utilization in textile printing as a thickening agent. Textiles 3(1):26–49

Feng N-C, Guo X-Y (2012) Characterization of adsorptive capacity and mechanisms on adsorption of copper, lead and zinc by modified orange peel. Trans Nonferrous Met Soc China 22(5):1224–1231

Fernandes Filho WB, Agassin STR, Naidek KP, Paulino AT (2023) Pectin hydrogels modified with montmorillonite: case studies for the removal of dyes and herbicides from water. J Environ Chem Eng 11(5):110846

Fu M, Tuo X, Yan X, Li D, Zhu H, Gao S, Han X, Zhou J, Mou D, Xiu J (2024) Adsorption performance and mechanism of pectin modified with β-cyclodextrin for Zn2+ and Cu2+. Int J Biol Macromol 274:133563

Gawkowska D, Cybulska J, Zdunek A (2018) Structure-related gelling of pectins and linking with other natural compounds: a review. Polymers 10(7):762

Güzel M, Akpınar Ö (2019) Valorisation of fruit by-products: production characterization of pectins from fruit peels. Food Bioprod Process 115:126–133

Jiang Y, Xu Y, Li F, Li D, Huang Q (2020) Pectin extracted from persimmon peel: a physicochemical characterization and emulsifying properties evaluation. Food Hydrocoll 101:105561

Jomova K, Alomar SY, Nepovimova E, Kuca K, Valko M (2025) Heavy metals: toxicity and human health effects. Arch Toxicol 99(1):153–209

Kedir WM, Deresa EM, Diriba TF (2022) Pharmaceutical and drug delivery applications of pectin and its modified nanocomposites. Heliyon 8(9):e10654

Lara-Espinoza C, Carvajal-Millán E, Balandrán-Quintana R, López-Franco Y, Rascón-Chu A (2018) Pectin and pectin-based composite materials: beyond food texture. Molecules 23(4):942

Leroux J, Langendorff V, Schick G, Vaishnav V, Mazoyer J (2003) Emulsion stabilizing properties of pectin. Food Hydrocoll 17(4):455–462

Liang R-h, Li Y, Huang L, Wang X-d, Hu X-x, Liu C-m, Chen M-s, Chen J (2020) Pb2+ adsorption by ethylenediamine-modified pectins and their adsorption mechanisms. Carbohydr Polym 234:115911

Maxwell EG, Belshaw NJ, Waldron KW, Morris VJ (2012) Pectin–an emerging new bioactive food polysaccharide. Trends Food Sci Technol 24(2):64–73

May CD (1990) Industrial pectins: sources, production and applications. Carbohydr Polym 12(1):79–99

Oliveira TÍS, Rosa MF, Cavalcante FL, Pereira PHF, Moates GK, Wellner N, Mazzetto SE, Waldron KW, Azeredo HM (2016) Optimization of pectin extraction from banana peels with citric acid by using response surface methodology. Food Chem 198:113–118

Pereira PHF, Oliveira TÍS, Rosa MF, Cavalcante FL, Moates GK, Wellner N, Waldron KW, Azeredo HM (2016) Pectin extraction from pomegranate peels with citric acid. Int J Biol Macromol 88:373–379

Prasathkumar M, Dhrisya C, Lin FH, Sadhasivam S (2023) The design and developments of protein-polysaccharide biomaterials for corneal tissue engineering. Adv Mater Technol 8(15):2300171

Ropartz D, Ralet M-C (2020) Pectin structure. In: Pectin: technological and physiological properties. Springer, pp 17–36

Said NS, Olawuyi IF, Lee WY (2023) Pectin hydrogels: gel-forming behaviors, mechanisms, and food applications. Gels 9(9):732

Sánchez-Bayo F (2011) Impacts of agricultural pesticides on terrestrial ecosystems. In: Ecological impacts of toxic chemicals. Bentham Science Publishers, pp 63–87

Shaha RK, Punichelvana Y, Afandi A (2013) Optimized extraction condition and characterization of pectin from kaffir lime (Citrus hystrix). Res J Agric For Sci ISSN 2320:6063

Srivastava P, Malviya R (2011) Sources of pectin, extraction and its applications in pharmaceutical industry-an overview. Indian J Nat Prod Resour 2(1):10–18

Twinomuhwezi H, Godswill A, Kahunde D (2020) Extraction and characterization of pectin from orange (Citrus sinensis), lemon (Citrus limon) and tangerine (Citrus tangerina). Am J Phys Sci 1(1):17–30

Valavanidis A (2018) Glyphosate, the most widely used herbicide. Sci Rev

Vanitha T, Khan M (2019) Role of pectin in food processing and food packaging. IntechOpen, London

Yang X, Nisar T, Hou Y, Gou X, Sun L, Guo Y (2018) Pomegranate peel pectin can be used as an effective emulsifier. Food Hydrocoll 85:30–38

Yu Y, Cui L, Liu X, Wang Y, Song C, Pak U, Mayo KH, Sun L, Zhou Y (2022) Determining methyl-esterification patterns in plant-derived homogalacturonan pectins. Front Nutr 9:925050

Zhang W, Song J, He Q, Wang H, Lyu W, Feng H, Xiong W, Guo W, Wu J, Chen L (2020) Novel pectin based composite hydrogel derived from grapefruit peel for enhanced Cu (II) removal. J Hazard Mater 384:121445

Zhang L, Qiao M, Zheng H, Vancov T, Antoniadis V, Shaheen SM, Joseph S, Chen C, Shan S, Chen H (2025) Integrating spectroscopic analysis and theoretical calculations to elucidate the adsorption efficiency and mechanisms of Cd, Pb, and Cu using novel carboxymethyl cellulose/pectin-based hydrogel beads. Int J Biol Macromol 305:141028

Zhexenbay N, Akhmetsadykova S, Nabiyeva Z, Kizatova M, Iskakova G (2020) Using pectin as heavy metals detoxification agent to reduce environmental contamination and health risks. Procedia Environ Sci Eng Manag 7(4):551–562

Chapter 8
Biosurfactant as a Potential Strategy for Remediation

W. G. Madusha Lakmali, A. D. Sarangi N. P. Athukorala, E. Sanduni M. Edirisinghe, and M. H. Kasuni T. Rupasinghe

Abstract Several toxic substances have been introduced by humans and are of global concern now. These compounds are hazardous to the environment and human health. With the concern of problems associated with conventional methods to degrade, reduce, or remove these substances, interest in alternative approaches increased in the recent past. Among them, microorganisms and their secreting chemicals have proven a unique role in the degradation and detoxification of pollutants in soil, air, and water. Biosurfactants are surface-active mediators produced by biological systems, mainly by microorganisms. In the sense of bioremediation, they integrate biological processes to degrade, remove, or transform these compounds to less harmful forms of contaminants. Microorganisms contain both hydrophilic and hydrophobic fractions in the composite. According to the type of microorganism, a composite of bioemulsifiers varies. The diversity of biosurfactants makes them an attractive group and has important key roles in biotechnological applications in the way of pharmaceuticals, biodegradation of pollutants, and enhanced oil recovery. Industrial application and eco-friendly use of microbial surfactants have been shown promising mainly due to their biodegradation efficacy, resulting in compounds with low toxicity and high solubility. However, the associated behavior and mode of action of biosurfactants in the bioremediation process have to be further investigated. The current chapter is to entitle the potential applications of biosurfactants as a strategy for remediation.

Keywords Biosurfactant · Contamination · Microorganism · Remediation · Toxic substances

W. G. Madusha Lakmali · A. D. Sarangi N. P. Athukorala (✉)
Postgraduate Institute of Science, University of Peradeniya, Peradeniya, Sri Lanka

Department of Botany, Faculty of Science, University of Peradeniya, Peradeniya, Sri Lanka

E. S. M. Edirisinghe · M. H. Kasuni T. Rupasinghe
Department of Botany, Faculty of Science, University of Peradeniya, Peradeniya, Sri Lanka

A. Guldhe et al. (eds.), *Biomaterials for Environmental Applications: Detection and Remediation*, Environmental Challenges and Solutions,
https://doi.org/10.1007/978-3-032-12701-3_8

8.1 Background

Pollution has significantly altered human perception of life on Earth and their relationship with the environment. Human activities have profoundly impacted the planet, influencing geochemical, biochemical, physiological, population, and community characteristics. Industrial and agricultural sectors extensively utilize numerous harmful chemical substances, which, when released into the environment, become contaminants, leading to pollution (Bhatt et al. 2021). Urbanization and industrialization have exacerbated environmental contamination, introducing pollutants into air, soil, and water, which negatively affect human health, flora, and fauna. Pollution is now a global concern, as ecosystems worldwide, from coastal regions to inland areas, face moderate to severe contamination (Ukaogo et al. 2020). Containing and preventing further pollution remains a major technical challenge.

Pollution originates from various sources, including organic aqueous waste such as pesticides, organic solvents from dry-cleaning, oils from lubricants, automotive fluids, and tars from dye production. Many pollutants result from accidental spills, excessive pesticide use in agriculture, and improper waste disposal, including household and industrial waste. Petroleum hydrocarbons, a major contaminant, arise from crude oil and natural gas leaks, impacting soil properties by altering nitrogen fixation and increasing water retention capacity, leading to soil degradation (Bustamante et al. 2012). Agrochemicals, including pesticides and fertilizers, pose serious environmental risks when improperly stored or overused, causing contamination of soil and groundwater and harming beneficial microorganisms (Qadir et al. 2017). Similarly, polycyclic aromatic hydrocarbons (PAHs), which result from incomplete combustion of organic materials, are persistent carcinogenic pollutants that affect soil structure and microbial diversity (Samanta et al. 2002). Solvents used in various industries contribute to soil and water contamination, leading to reduced soil fertility and groundwater pollution. Heavy metals, such as lead, arsenic, and cadmium, accumulate in the environment through industrial activities, mining, and fertilizers, posing significant health risks and reducing land productivity (Liyanage et al. 2020). Metals do not deteriorate chemically or microbially like organic molecules do. Its accumulation causes health risks to humans and animals through direct contact, ingestion, or its entry into the food chain. They also affect the fertility and productivity of the contaminated land (Lakmali et al. 2022).

The impact of pollution is severe and continues to evolve over time. Toxic chemicals in the environment contribute to health hazards, including cardiovascular diseases, cancer, and respiratory conditions. Pollution-related deaths account for millions globally, with heavy metals and organic compounds contaminating food crops and water sources. Soil degradation caused by pollution reduces agricultural productivity (Münzel et al. 2023), while deforestation and soil erosion release sequestered pollutants, further exacerbating environmental damage (Emenike et al. 2018). Most of the pollutants are persistent and can cause their effect even at very low concentrations (Rastogi and Kumar 2020). Recognizing these threats, global initiatives, such as the European Commission's Action Plan for 2050, aim to reduce pollution to levels that do not harm human or ecosystem health (Ayilara and Babalola 2023).

Remediation of polluted areas is critical for environmental restoration. Advances in detection technologies enable precise identification of contaminants, facilitating effective remediation strategies. Conventional approaches, including excavation, incineration, and chemical treatments, are widely used but often generate secondary pollutants and are costly (Ayilara and Babalola 2023). Alternative methods, such as soil washing, aeration, oxidation, and filtration, are also employed to remove pollutants. However, these techniques may be inefficient or economically unfeasible for large-scale applications. Bioremediation, a sustainable and cost-effective alternative, involves using microorganisms to degrade contaminants in soil and water. This process relies on stimulating microbial activity to break down harmful substances into non-toxic compounds. Bioremediation can be conducted in situ, directly at the contaminated site, or ex situ, where contaminated material is treated elsewhere. Although bioremediation is an effective solution, its success depends on factors such as microbial availability, nutrient levels, temperature, and soil conditions. There is a growing demand for green and non-toxic dispersants for pollutant removal. It is widely recognized as an eco-friendly alternative to conventional remediation methods.

Bioremediation employs the use of living organisms, like microbes or bacteria, in the removal of contaminants, pollutants, and toxins from soil, water, and other environments. Bioremediation techniques include biostimulation, bioaugmentation, and intrinsic bioremediation. In biostimulation, chemicals or nutrients are introduced to activate native microbial populations. Bioaugmentation involves adding specific bacterial strains to accelerate the degradation of contaminants. Intrinsic bioremediation utilizes existing microbial communities to convert toxic substances into harmless compounds. This method has gained global attention due to its efficiency, minimal environmental disruption, and cost-effectiveness. The United States Environmental Protection Agency (EPA) has implemented bioremediation at over 1500 contaminated sites, highlighting its growing importance. As environmental pollution continues to pose significant challenges, adopting sustainable remediation methods like bioremediation will be crucial in ensuring a cleaner and healthier planet for future generations.

8.2 Biological Remediation

Several microbial applications are implemented for removal of contaminants from the environment. These microorganisms and their products are used for removal of contaminants from the environment (Ayilara and Babalola 2023). Table 8.1 depicts the different microorganisms used for remediation of various pollutants.

Table 8.1 Different microorganisms used for removal of pollutants

Species	Pollutant remediated	References
Bacteria		
Aerococcus sp., and *Rhodopseudomonas palustris*	Cadmium, lead, and chromium	Sravya and Sangeetha (2022)
Bacillus licheniformis	Dyes	Mousavi et al. (2021)
Bacillus licheniformis JUG GS2 (MK106145) and *Bacillus sonorensis*	Naphthalene	Rabani et al. (2022)
Bacillus sp. and *Staphylococcus* sp.	Endosulfans	Li et al. (2020)
Bacillus sp., *Rhodopirellula* sp., *Rhodovibrio* sp., *Formosa* sp. *Pseudoalteromonas* sp., and *Agarivorans* sp.	Hydrocarbon	Machado et al. (2019), Dell'Anno et al. (2021)
Bacillus subtilis and *Pseudomonas fluorescence*	Iron and zinc	Siric et al.
Burkholderia sp. and *Myceliophthora thermophila*	N,N-dimethylpphenylenediamine and polycyclic aromatic hydrocarbons	Mohapatra and Phale (2021)
Chlorella sp. and *Spirulina* sp.	Lead, nickel, and dichromate	Geetha et al. (2021)
Cladosporium sp., *Paecilomyces* sp., *A. fumigatus*, *Paecilomyces* sp., and *Terichoderma* sp	Cadmium	Unuofin et al. (2021)
Dehalococcoides sp.	Chloroethenes	Dutta et al. (2022)
E. coli	Hexavalent chromium	Mohamed et al. (2020)
Fusarium sp., *Corynebacterium propinquum*, *P. aeruginosa*, and *Alcaligenes odorans*	Oils	Pande et al. (2020)
K. oxytoca, *B. firmus*, *B. macerans*, and *Staphylococcus aureus*	Vat dyes	Sangkharak et al. (2020)
Lysinibacillus sphaericus CBAM5	Lead, cobalt, copper crude oil, and chromium	Kharangate-Lad and D'Souza (2021)
Microbacterium sp., *Micrococcus* sp., *Bacillus* sp., and *Shigella* sp.	Uranium and arsenic	Bhakat et al. (2019)
Oscillatoria laete-virens, *Arthrospira platensis*, *Pseudochlorococcum typicum*, and *Spirogyra insignis*	Lead	Song et al. (2019)
P. cepacia, *B. coagulans*, *B. cereus*, and *Serratia ficaria*	Diesel oil	Miri et al. (2022)
Pseudomonas aeruginosa	Crude oil	Mukjang et al. (2022)
Ralstonia sp., *Microbacterium* sp., *Pseudomonas* sp., and *Acinetobacter* sp.	Aromatic hydrocarbons	Basu et al. (2018)

(continued)

Table 8.1 (continued)

Species	Pollutant remediated	References
Saccharomyces cerevisiae and *Cunninghamella elegans*	Heavy metals and mercury	Duc et al. (2021)
Fungi		
Aspergillus sp. and *Penicillium* sp.	Aliphatic hydrocarbons, polycyclic aromatic hydrocarbons, and chlorophenols	Liu et al. (2018), Li et al. (2020)
Phanerochaete chrysosporium	N-heterocyclic explosives, benzene, xylene, ethylbenzene, toluene, and organochlorines	Singh et al. (2020)
Phanerochaete chrysosporium	4,4-Dibromodiphenyl ether	Sen et al. (2019)
Saccharomyces cerevisiae, *Aspergillus* sp.	Arsenic	Verma and Kuila (2019), Mohd et al. (2019)
Trichoderma sp., *Penicillium* sp., and *Aspergillus* sp.	Cobalt and copper	Dusengemungu et al. (2020)
Algae		
Microcystis aeruginosa, *Chlamydomonas reinhardtii*, and *S. almeriensis*	Arsenic	Saavedra et al. (2018), Wang et al. (2018)
Fucus vesiculosus	Zinc	Brinza et al. (2020)
Chlorococcum humicola	Iron	Chugh et al. (2022)
Chlorella sp., *Isochrysis galbana*, and *Phaeodactylum tricornutum*	Phenol	Wu et al. (2022)
F. vesiculosus	Chromium, nickel, cadmium, and lead	Moreira et al. (2019)
Cystoseria indicant	Nickel and cadmium	Moreira et al. (2019)
Chlamydomonas reinhardtii	Chromium and cadmium	Nowicka et al. (2020)
Chlorococcum humicola	Cobalt	Chugh et al. (2022)

8.3 What Is a Surfactant?

Surfactants are mixtures of water and oil that form foam at the water-oil interface, providing foaming, emulsifying, detergency, and dispersing properties (Saenz-Marta et al. 2015). While surfactants can exist as simple monomers, they can also form more complex polymers. These compounds are capable of reducing the surface tension of liquids or the interfaces between different liquids (Rebello et al. 2014). Typically, surfactant molecules have two distinct regions: a hydrophilic head and a hydrophobic tail, as shown in Fig. 8.1. Based on the nature of the head group, surfactants are classified as anionic, cationic, nonionic, or zwitterionic (containing both cationic and anionic groups). The hydrophilic head is water-soluble and can include carbohydrate, amino acid, cyclic protein peptide, phosphate, or alcohol groups. The hydrophobic tail, on the other hand, consists of hydrocarbons, which

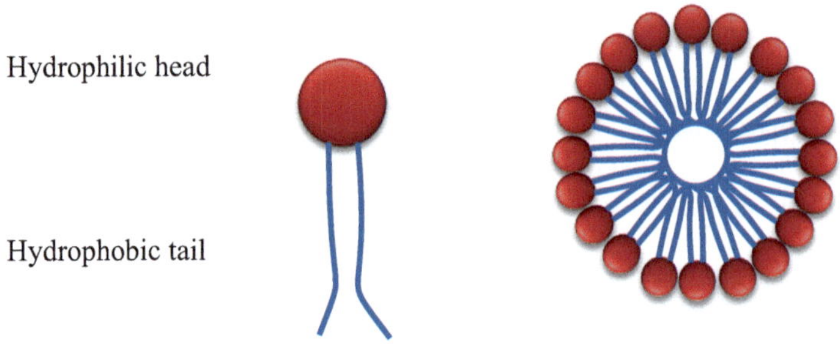

Fig. 8.1 Micelle formation of biosurfactant

may be long-chain fatty acids, α-alkyl-β-hydroxy fatty acids, or hydroxyl fatty acids. These tails typically range from C8 to C22 alkyl chains or alkyl aryls (Rahman and Gakpe 2008).

8.4 Structures and Role of Biosurfactant

Biosurfactants or natural surfactants, a term derived from the biological surface-active agents, can be produced extracellularly by a wide range of microorganisms including bacteria, fungi, and yeast. Microbial compounds, which exhibit particularly high surface activity and emulsifying activity, are classified as biosurfactants (Cameotra and Makkar 2010). Basically, the microbial surfactants are considered as complex molecules with a diverse range of chemical structures (Georgiou et al. 1992). Unlike chemically synthesized surfactants, which are categorized based on their polar group, biosurfactants are primarily classified by their chemical contents/composition and microbial origin such as fatty acids, antibiotics, peptides, glycolipids, phospholipids, and lipopeptides. Biosurfactants are versatile amphiphilic compounds containing both lipophilic and hydrophilic groups, enabling them to reduce surface and interfacial tension in water-hydrocarbon mixtures (Cameotra and Makkar 2010). It is also made up of a hydrophobic and hydrophilic component. The polar hydrophilic part of biosurfactant can be an amino acid, peptides, anions, cations, mono-, di-, or polysaccharides, and/or a phosphate group. Nonpolar hydrophobic portion may comprise a long chain unsaturated and saturated fatty acid constituents (Jahan et al. 2020). It preferentially partitions at the interface between fluid phases having different degrees of polarity and hydrogen bonding as oil and water or air and water interface. This property enables them to locate at the cell envelop and make non-soluble substrates bioavailable to the cells. Thus, they have a strong influence over interfacial rheological behavior and mass transfer (Zahed et al. 2022).

Microbial biosurfactants fall into two primary categories as high molecular weight (HMW) polymeric compounds, including polysaccharides, proteins, lipoproteins, and lipopolysaccharides, and low molecular weight (LMW) compounds like lipopeptides and glycolipids (Patel et al. 2021). HMW biosurfactants excel at adhering firmly to surfaces and acting as bioemulsifiers. LMW biosurfactants, such as rhamnolipids and sophorolipids, are disaccharides combined with long-chain acetylated fatty acids or hydroxyl fatty acids, effectively reducing surface and interfacial tension (McClements and Gumus 2016). However, according to a classification proposed by Neu (1996), the term biosurfactant denotes low molecular weight microbial surfactant while high molecular weight polymers can be defined as bioemulsifiers. These bioemulsifiers do not lower the surface tension to a great extent as biosurfactant due to the presence of amphiphilic and polyphilic polymers are more effective in stabilizing oil in water emulsion (Neu 1996).

8.5 Different Types of Biosurfactants

Numerous types of biosurfactants have been studied for their potential in remediation, particularly in enhancing the degradation and removal of pollutants. Unlike synthetic surfactants, microbial biosurfactants are classified based on their chemical composition and the microorganisms that produce them (Sachdev and Cameotra 2013). The major groups of microbial biosurfactants include glycolipids, lipopeptides, neutral lipids, phospholipids, fatty acids, polymeric biosurfactants, and particulate surfactants. These biosurfactants, which are generally low in molecular weight, increase the surface area of hydrophobic, water-insoluble substrates. Additionally, they reduce pollutant toxicity by forming complexes with pollutants, facilitating their interaction with the cell surface and altering pollutant uptake (Cameotra and Makkar 2010). In most cases, microbial biosurfactants are derived from diverse sources and exist as either anionic or neutral compounds (Fig. 8.2).

8.5.1 Glycolipid Biosurfactants

The primary category of biosurfactants includes glycolipids which are composed of sugars linked to long-chain aliphatic or hydroxyaliphatic acids, connected through ether or ester bonds. Glycolipids, which are conjugates of carbohydrates and fatty acids linked by either ether or ester groups, are among the most well-known biosurfactants (Mnif et al. 2018). Glycolipids are further divided into three categories based on the attachment of carbohydrate group as rhamnolipids, sophorolipids, and trehalolipids. These glycolipids are produced by various microorganisms and exhibit distinct properties and applications in different industries (Ayilara and Babalola 2023). These compounds help in mobilization of microorganism by reducing the interfacial tension of environment while providing better location for the growth, reproduction, and colonization of the microorganisms.

One of the most researched biosurfactants is rhamnolipid which is characterized by one or two molecules of rhamnose linked to one or two molecules of

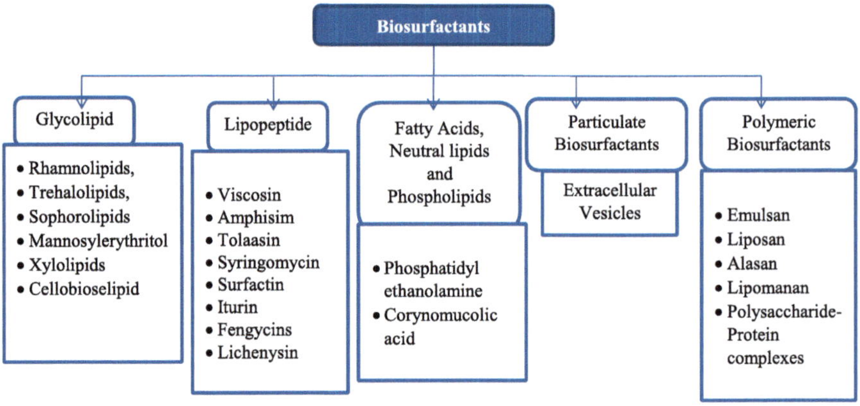

Fig. 8.2 Classification chart of microbial biosurfactants based on their chemical composition and the producing microorganisms

hydroxydecanoic acid. These compounds consist of sugar groups, typically one or two rhamnose molecules, linked to one or two fatty acid chains composed mainly of 10–14 carbon atoms. This molecular arrangement results in the formation of mono or di-rhamnolipid molecules (Gautam and Tyagi 2006). Numerous bacteria, including *Pseudomonas aeruginosa*, generate them. They are extensively studied and are the principal glycolipids primarily produced by the bacterium *Pseudomonas aeruginosa* and they hold significant promise for a range of environmental, industrial, and bioremediation applications (Costa and Ferrão 2010). Rhamnolipids have been utilized to improve the bioremediation of hydrocarbon-contaminated soils because they are efficient in solubilizing hydrophobic molecules (Nguyen et al. 2008). Rhamnolipids are particularly noteworthy among biosurfactants due to their extensive research and application (Müller et al. 2012).

Sophorolipids consist of a dimeric carbohydrate called sophorose linked to a long-chain hydroxyl fatty acid through a glycosidic bond. These glycolipids produced by yeasts such as *Candida bombicola*, *Torulopsis bombicola*, *T. petrophilum*, and *T. apicola* are the primary producers of sophorolipids (Shekhar et al. 2015). These biosurfactants have applications in oil bioremediation (Elshafie et al. 2015). While purified sophorolipids are highly surface-active and less water-soluble, they exhibit strong cytotoxic effects. However, they are not efficient emulsifying agents. They have shown promise in enhancing the bioremediation of petroleum hydrocarbons and heavy metals in contaminated soils (Van Bogaert et al. 2007).

Trehalolipids are a class of glycolipids characterized by the presence of disaccharide trehalose linked to long-chain aliphatic or hydroxyaliphatic acids through an ester group. These compounds are produced by various species of fungi, yeasts, and bacteria, including *Mycobacterium*, *Nocardia*, and *Corynebacterium*. The specific properties of trehalolipids can vary among different organisms, primarily in terms of the size and structure of the mycolic acid component, the number of carbon atoms present, and the degree of unsaturation (Banat et al. 2010). For instance,

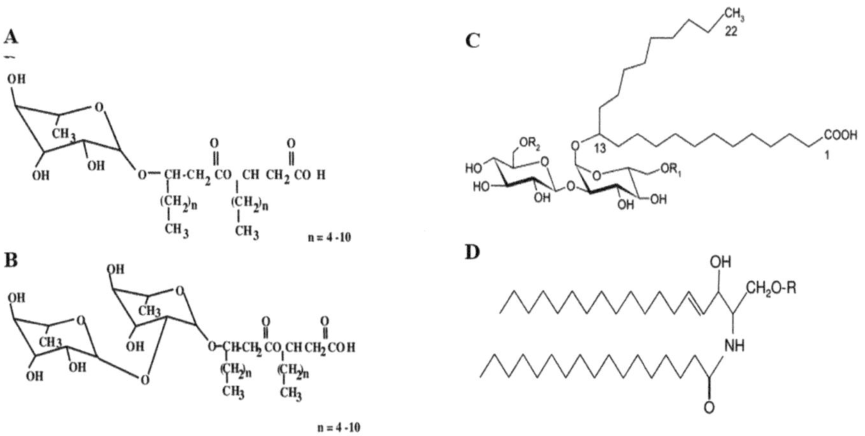

Fig. 8.3 Chemical structure of (**a**) monorhamnolipids, (**b**) dirhamnolipd, (**c**) sophorolipids, (**d**) trehalolipids

Rhodococcus erythropolis is known to produce trehalose dimycolate, and this particular trehalolipid has been studied for its ability to reduce surface and interfacial tension in culture broths (Kuyukina and Ivshina 2019). They have been studied for their potential to enhance the biodegradation of hydrocarbons in contaminated soils (Kuyukina et al. 2015).

Mannosylerythritols are glycolipids containing mannose, erythritol, and two fatty acid chains. They center around the disaccharide mannosylerythritol, with fatty acids linked to mannose through ester bonds (Sajadi Bami et al. 2022).

Xylolipids are glycolipid biosurfactants with a xylose head and fatty acid tails. They are primarily produced by lactic acid bacteria, although some yeast species, *Pichia caribbica* can synthesize them (Jameel and Haider 2021). Xylolipids can reduce surface tension effectively (Joshi-Navare et al. 2014).

Cellobioselipids are glycolipid biosurfactants produced by microbes, *Cryptococcus humicola* JCM 1461. They are usually a mixture of different acylated low molecular weight D-glucolipids, linked to hydroxyl palmitic acid via ω-hydroxyl groups (Sajadi Bami et al. 2022) (Fig. 8.3).

8.5.2 *Lipopeptide Biosurfactants*

Lipopeptides and lipoproteins are compounds consisting of lipids attached to polypeptide chains and have demonstrated antimicrobial properties against various microorganisms, including bacteria, algae, fungi, and viruses (Vijayakumar and Saravanan 2015). These diverse biosurfactants play critical roles in various applications, including bioremediation and antimicrobial activities. These molecules often have lower critical micelle concentration (CMC) values compared to synthetic surfactants, enhancing their effectiveness in various applications (Johnson et al. 2021).

Viscosin is a lipopeptide which has a peptide moiety consisting of nine amino acids. Amphisin is a group that harbors 11 amino acids in the peptide moiety. Tolassin is a group of lipopeptides that has a length of 15–25 amino acids containing unusual 2,3-dehydro-2-aminobutyric acid and homoserine consisting of the lipid tail of 3-hydroxydecanoic acid or 3-hydroxyctanoic acid (Raaijmakers et al. 2006).

In the category of lipopeptides, Arthrofactin is produced by *Arthrobacter*, *Actinomyces*, and *Streptomyces* (Mnif and Ghribi 2015). It is notable for its ability to reduce the surface tension of water significantly and has applications in biofilm control and various industries (Banat et al. 2000).

Syringomycin is a group of lipopetide that has nine amino acids containing 2,4-diaminobutyric acid and a C-terminal chlorinated thereonine residue with the 3-hydroxydecanoic acid tail. *Pseudomonas aeruginosa* has been studied for syringomycin production (Chauhan et al. 2023).

Iturin is a lipopetide family that consists of six variants as iturin A and C, bacillomycin D, F, L, and mycosubtilin. The structural linkage consists of a heptapeptide attached to an amino fatty acid chain. These iturin, produced by *Bacillus subtilis*, exhibits both antifungal and antibacterial properties, remaining active under various conditions (Wan et al. 2022).

Surfactin is an effective biosurfactant with low toxicity, which is produced by *Bacillus subtilis* (De França et al. 2015). Its amphiphilic nature allows it to function in both hydrophobic and hydrophilic environments. They found that the surfactins are able to reduce the surface tension and interfacial tension of water. The inactivation of herpes and retrovirus was also observed with surfactin, which has been investigated for its ability to enhance the bioremediation of hydrocarbon-contaminated soils. It has strong emulsification properties and can enhance microbial degradation of pollutants (Seydlová and Svobodová 2008). A remarkable number of cyclic lipopetides, including decapeptide anti-toxins (gramicidins) and lipopeptide anti-toxins (polymyxins), are delivered. These molecules consist of a lipid linked to a polypeptide chain.

Fengycin, another cyclic lipopeptide, consists of decapeptides attached to a linear chain of β-fatty acid (Mantil et al. 2019). It interacts readily with lipid bilayers and has antimicrobial effects (González-Jaramillo et al. 2017). It is primarily produced by *Bacillus* species and closely related to it is plipastatin, which differs in the Tyr position (Cortés-Camargo et al. 2021).

Lichenysin is also a biosurfactant produced by *Bacillus licheniformis*, is known for its synergistic, and exhibits excellent stability under extreme conditions, including temperature, pH, and salt levels. It reduces both surface and interfacial tension of water and holds promise for various applications (McInerney et al. 1990). They have high structural and physicochemical similarities with surfactin. The difference is the presence of a glutaminyl residue in position 1 of the peptide sequence in place of glutamic acid in surfactin. Lichenysin is a better chelating agent having high association constants with Ca^{2+} and Mg^{2+} increased by a factor of 4 and 16 respectively. Lichenysin has a role in medical applications that *Acinetobacter* spp. 1-N, phosphatidyl ethanolamine-rich vesicles are produced, forming optically clear microemulsions of alkanes in water. These vesicles are capable of forming optically clear micro-emulsions of alkanes in water (Mclnerney et al. 1990).

8.5.3 Fatty Acids, Phospholipid, and Neutral Lipids Biosurfactants

Microorganisms produce large quantities of fatty acid and phospholipid biosurfactants when grown in media rich in n-alkanes. The length of the hydrocarbon chain in their structures correlates with proportion of hydrophilic and lipophilic moieties of the surfactants (Satpute et al. 2010). Phosphatidyl ethanolamine-containing vesicles are secreted by *Acinetobacter* sp. which forms optically clear micro-emulsions of alkane-water. In addition, phosphatidyl ethanolamine produced by *Rhodococcus erythropolis* grown on alkane lowers the interfacial tension of mixture of hexadecane and water. Some of the most popular biosurfactants fall into phospholipid category. *Corynebacterium lepus* produces phospholipid biosurfactants (Rajkumari et al. 2017). Gautam and Tyagi reported that the deficiency of the phospholipid protein complex is a major cause of respiratory failure in prematurely born children. They also suggested that isolating and cloning the responsible genes could be used for their fermentative production (Gautam and Tyagi 2006).

8.5.4 Polymeric Biosurfactants

Within the polymeric biosurfactants subcategory, compounds such as emulsan, liposan, alasan, lipomanan, and other polysaccharide protein complexes have been extensively studied.

Emulsan is recognized as a highly effective bioemulsifier for stabilizing hydrocarbon-water mixtures, demonstrating efficiency even at extremely low concentrations ranging from 0.001% to 0.01% (Hatha et al. 2007). It plays multifaceted roles in various microbial environments, including biofilms, flocs, and biological sludge, resulting in both adverse and advantageous outcomes. In the accumulation of microorganisms, plants, algae, or small animals where it is not wanted on surfaces, these polymeric substances including emulsan contribute to increased friction resistance and alterations in surface properties such as hydrophobicity, roughness, and color (Qi and Christopher 2019). Emulsan and biodispersan produced by *Acinetobacter calcoaceticus* consist of a heteropolysaccharide moiety covalently linked to fatty acids (Sekhon et al. 2012).

Liposan is another polymeric biosurfactant, an extracellular water-soluble emulsifier produced by *Candida lipolytica*. It comprises 83% carbohydrates and 17% protein (Rahman and Gakpe 2008). Liposan finds applications in various industries.

Emulsan and liposan exhibit diverse and influential roles in microbial environments, influencing factors such as biofouling, metal biocorrosion, bioweathering, and pollutant sorption.

A similar glycoprotein complex is also synthesized by *Yarrowia lipolytica*. Other examples include mannan-lipid protein from *Candida tropicalis* and protein PA from *Pseudomonas aeruginosa*. *Saccharomyces cerevisiae* produces mannoproteins, which consist of 44% mannose and 17% protein (Shekhar et al. 2015).

8.5.5 Particulate Biosurfactants

The particulate biosurfactants such as extracellular membrane vesicles partition oil–water mixture and form micro-emulsion (at the interface), which facilitates uptake of alkane by microorganisms (Rocha e Silva et al. 2019). These substances are involved in microbial processes where hydrocarbons are taken up by microbial cells. They create extracellular membrane vesicles that partition to form microemulsions. These microemulsions play a crucial role in the uptake of alkane compounds by microbial cells. A classic example includes vesicles produced by *Acinetobacter* sp. strain HO1-N. These vesicles typically have a diameter ranging from 20 to 50 nanometers and a buoyant density of approximately 1.158 cubic grams per cubic centimeter. They consist of lipopolysaccharides, protein, and phospholipids (Vijayakumar and Saravanan 2015).

Some researchers have explored using mixtures of different biosurfactants to improve their overall effectiveness. Combining different types of biosurfactants can provide a broader range of pollutant solubilization capabilities (Cameotra and Makkar 2010).

8.6 Microbial Surfactant Production

Microbial surfactants are produced intracellularly or extracellularly by microorganisms in the presence of water-soluble and/or water-insoluble substrates via de novo pathway and/or assembly from other substrates with consequent variations in structure (Satpute et al. 2010). The production of these biomolecules is dependent on proper selection of microbial strains, substrate type, and fermentation technology (Yaraguppi et al. 2020).

The biosurfactant production affected by several bioprocess parameters including carbon and nitrogen sources, temperature, pH, metal ions, agitation speed, etc. influenced the growth of microorganisms and surfactants production. Each of those parameters interacts with each other and affects the kinetics of microbial surfactant production (Yaraguppi et al. 2020). The various nutritional and physicochemical parameters influence the yield of microbial surfactants.

The carbon sources are incorporated in fermentation media for microbial surfactant production. These carbon sources can be carbohydrates, fats & oils and hydrocarbon groups which are classified as water-soluble (e.g., glucose, sucrose, glycerol) and water insoluble (such as crude oils, vegetable oils) (Prabhu and Phale 2003). Microorganisms primarily utilize glucose due to its efficient metabolism through the glycolytic pathway, which facilitates the production of desired metabolites (Nurfarahin et al. 2018). These carbon sources serve as essential building blocks for synthesizing the hydrophilic and hydrophobic components of biosurfactants. Vegetable oil and coconut oil were reported as promising substrates for biosurfactant production (Adetunji and Olaniran 2019).

Nitrogen sources form another most essential nutrient for microbial growth and production of biosurfactants (De Almeida et al. 2016). Various nitrogen sources such as $NaNO_3$, $(NH4)_2SO_4$, and CH_4, N_2O are employed for biosurfactant production (Abdel-Mawgoud et al. 2010). *Pseudomonas aeruginosa* has been investigated and resulted $NaNO_3$ was more effective when compared to other nitrogen sources, resulting in higher rhamnolipid production (Santa Anna et al. 2002). Enhanced secretion of sophorose lipids by *Torulopsis bombicola* and *Candida bombicola* has been observed when yeast extract and urea are used as nitrogen sources. Furthermore, addition of amino acids (such as aspartic acid, asparagine, glycine, and glutamic acid) in growth media enhances microbial biosurfactant production (Rodríguez-López et al. 2020).

Physicochemical parameters such as temperature, pH, metal ions, oxygen requirements, and agitation speed play a crucial role in influencing the growth and/or metabolic activity of microorganisms for biosurfactant production. Notably, maximum rhamnolipid production by *Pseudomonas* sp. has been recorded at an optimal pH range of 6.0–6.5. Additionally, the ideal incubation temperature for biosurfactant secretion varies depending on the physiological characteristics of the microorganism. As a result, microorganisms are cultivated at different temperatures to achieve high biosurfactant yields. Furthermore, oxygen supply and agitation speed play a crucial role in influencing the kinetics of biosurfactant production by *Acinetobacter calcoaceticus* RAG-1. Moreover, metal ions act as a key cofactor for the enzyme-catalyzed biosynthetic pathways of biosurfactant production (Abdel-Aziz et al. 2017). However, an excessive concentration of these metal ions can be harmful to microorganisms. Furthermore, incubation period plays a crucial role in microbial surfactant production since the biomolecules are produced at different time intervals (Rodríguez-López et al. 2020).

8.7 Mechanisms Involving the Remediation Using Biosurfactants

Biosurfactants play a crucial role in enhancing the microbial degradation process, particularly when it comes to pollutants that are not readily available to degrading bacteria due to their low solubility. Particulate biosurfactants play a significant role in microbial processes, particularly in the uptake of hydrocarbons by microbial cells. They form extracellular membrane vesicles that partition to create microemulsions (Vijayakumar and Saravanan 2015). These molecules improve the bioavailability of hydrophobic contaminants to microbes by increasing their solubility. This enhancement of the biodegradation process is primarily achieved through two mechanisms: including the micelle formation and increasing the availability of pollutant by emulsification and dispersion.

Biosurfactants interact with the phase boundary between two phases in a heterogeneous system due to their amphipathic nature, which allows molecules with both hydrophilic and hydrophobic components to preferentially accumulate at interfaces between fluid phases with varying polarity and hydrogen bonding, such as oil-water or air-water interfaces. The most effective biosurfactants can reduce both surface tension and interfacial tension. Through micelle formation, biosurfactants lower surface and interfacial tension, enhancing the solubility and bioavailability of hydrophobic organic compounds. Micelles are stable structures where the hydrophobic portions of surfactants orient themselves toward the hydrophobic environment, while the hydrophilic portions interact with the aqueous phase. This structure disperses and solubilizes hydrophobic contaminants in the aqueous solvent, enhancing their availability for microbial consumption. Micelles can also aid in the absorption of substances by microbial cells. Biosurfactant activity is influenced by the concentration of surface-active compounds until the critical micelle concentration (CMC) is reached. The CMC is commonly used as an indicator of surfactant efficiency. At concentrations above the CMC, biosurfactant molecules associate to form micelles biosurfactants form micelles, bilayers, and vesicles in the hydrophilic environment (Fig. 8.4). Efficient biosurfactants have a low CMC, which means that less biosurfactant is required to decrease the surface tension (Jahan et al. 2020).

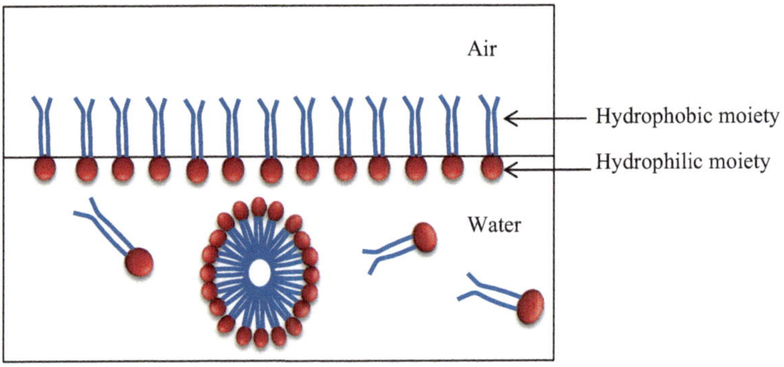

Fig. 8.4 Biosurfactant in water and air interface

Biosurfactants act as emulsifiers for hydrophobic compounds, making them more soluble and accessible to microbial attack. Pollutant remediation using biosurfactants can occur below and above the critical micelle concentration (CMC). Below CMC, biosurfactants alter the system's affinity for water, promoting pollutant release. Above CMC, contaminants are incorporated into micelles, facilitating their partition into the aqueous phase (Ali et al. 2016).

Micelle formation plays a crucial role in the formation of microemulsions which are clear and stable liquid mixtures of water and oil domains separated by monolayer or aggregates of biosurfactants (Jahan et al. 2020). The critical micelle concentration (CMC) of biosurfactants, which indicates their efficiency, ranges from 1 to 2000 mg/L. Their interfacial tension (oil/water) and surface tension are typically around 1 mN/m and 30 mN/m, respectively. In principle, a lower CMC denotes greater efficacy of the surfactant and, from the economic standpoint, favors its use in industrial processes (Abdel-Aziz et al. 2017).

Regarding heavy metal remediation, anionic biosurfactants exhibit a strong affinity for cationic heavy metals like Zinc, Lead, Chromium, and Copper. They form complexes with these metals, and these biosurfactant-metal complexes can then detach from soil surfaces and form micelles. This property makes biosurfactants useful for the remediation of heavy metal pollution. Smaller micelles are especially advantageous as they diffuse more effectively in soil, increasing the contact area with heavy metals and thus improving the efficiency of bioremediation. Biosurfactants enhance biodegradation by improving solubility and facilitating the transfer of hydrophobic contaminants through micelle formation. Biosurfactants form complexes with these metals and increasing their mobility in the soil, ultimately aiding in their removal (Fig. 8.5).

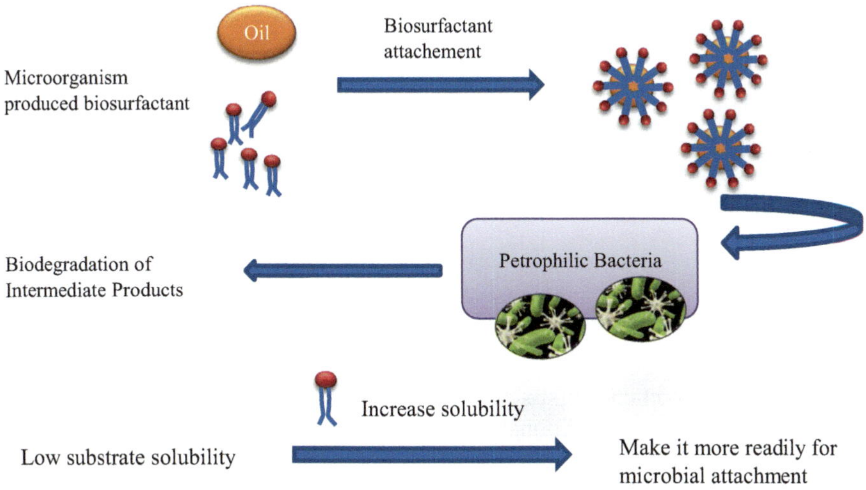

Fig. 8.5 Mechanisms of microbial degradation of the hydrophobic compound with the aid of biosurfactants

The extent of microbial adhesion to hydrocarbons is directly related to the ability of these microorganisms to thrive in environments where hydrocarbons or other hydrophobic substrates are present. Biosurfactants, when bound to microbial cell walls, render the cell surface more hydrophobic. Microorganisms can use biosurfactants to regulate their cell-surface properties, allowing them to attach to or detach from surfaces as needed.

The permeability of microbial cell membranes can be changed by biosurfactants. This can improve nutrient intake and make it easier for metabolic waste products to be excreted, both of which are critical for the overall microbial activity and breakdown of pollutants. Large droplets of hydrophobic contaminants can be broken down into smaller, easier-to-manage droplets by biosurfactants. This accelerates the deterioration process and expands the surface area open to microbial attack. In some cases, biosurfactants can reduce the toxicity of pollutants by promoting their transformation into less harmful compounds through microbial metabolism. This is particularly important for pollutants that are initially toxic to microorganisms (Pitocchi et al. 2020). There are two common mechanisms that reduce the harmful effects by dry or fresh biomass adsorption or absorption.

Synergistic effects may arise from mixing microorganisms that produce biosurfactants with those that break down pollutants. By enhancing pollutant accessibility, the biosurfactants created by one particular group of microbes might be advantageous to the whole microbial population. Nowadays, some studies have focused on using microbial consortia that naturally produce biosurfactants, instead of using a specific biosurfactant where these consortia can have synergistic effects on pollutant degradation (Pitocchi et al. 2020).

8.8 Application of Biosurfactant as Bioremediation Process in Industries

Biosurfactants offer a greener alternative, being biodegradable, non-toxic, and produced through eco-friendly processes. As industries seek sustainable solutions, biosurfactants could find applications in sectors such as agriculture, textiles, pharmaceuticals, and personal care.

8.8.1 Detergent Industry

The detergent market encompasses personal care, home cleaning, and heavy industrial cleaning products. Traditionally, surfactant compounds used in these products are derived from the petrochemical industry. For example, the disinfectants widely used during the COVID-19 pandemic in personal care and home care contain surfactants derived from crude oil, posing challenges for biodegradation and potential harm to aquatic environments (Ambaye et al. 2021). As a result, there's a growing interest in replacing these synthetic surfactants with bio-based alternatives. One approach is to develop biodegradable detergents based on straight-chain organic compounds, which facilitate efficient microbial degradation (Ambaye et al. 2021). To create more sustainable detergents, synthetic surfactants are being substituted with "green" surfactants, including biosurfactants, preferably those effective at low temperatures and in hard water (Md 2012).

Biosurfactants offer several valuable properties in this context, including their emulsifying capacity, essential for detergent effectiveness. Additionally, biosurfactants exhibit properties similar to commercial detergents and find applications in the detergent and laundry industries. For instance, Sarubbo et al. reported that a lipopeptide from B. subtilis SPB1 outperformed commercial detergents in removing vegetable oil and coffee stains (Sarubbo et al. 2022). Studies found that surfactin from B. subtilis HSO121 could replace chemical surfactants while offering benefits like low toxicity, no irritation, excellent emulsifying activity, wetting capacity, high compatibility, stability, biodegradability, and substantial foaming capacity. Moreover, a study involving the biosurfactant from *Ochrobactrum intermedium* MZV101 demonstrated its remarkable ability to remove oil from fabric (Zarinviarsagh et al. 2017). Another study by Liley et al. explored a mixture of ternary surfactants and biosurfactants, highlighting that the addition of rhamnolipids increased tolerance to temperature reduction, making them suitable for detergent formulations at lower temperatures (Liley et al. 2017).

In summary, the shift toward sustainable detergents involves replacing synthetic surfactants with eco-friendly alternatives like biosurfactants. These biosurfactants not only exhibit excellent cleaning properties but also offer advantages such as biodegradability and low toxicity, making them attractive for various applications in the detergent industry.

8.8.2 Nanotechnology

Biosurfactants have significant potential in nanotechnology, particularly in the synthesis of nanoparticles. They can serve as both reducing agents and stabilizers, with a notable application being the synthesis of silver nanoparticles. This approach offers a "green" alternative to traditional chemical methods, which often involve high pressure, temperature, energy consumption, and the generation of toxic byproducts. Biosurfactants enable an eco-friendly, energy-efficient process without the production of harmful compounds (Sarubbo et al. 2022).

Certain microorganisms, like *Bacillus subtilis*, possess the capability to produce gold and silver nanoparticles, both intracellularly and extracellularly, with reported biological activities. Biosurfactants derived from bacteria such as *Pseudomonas aeruginosa* and *Brevibacterium casei* have shown promise in stabilizing nanoparticles in microemulsions and preventing the formation of nanoparticle aggregates, maintaining uniform morphology for extended periods (Curstedt et al. 2013). Silver nanoparticles synthesized by microorganisms have potential applications in oil recovery, where they reduce surface tension and viscosity, and exhibit antimicrobial properties. However, further exploration is needed to expand the use of these nano-biotechnology applications in bioremediation (Sarubbo et al. 2022).

8.8.3 Agriculture

Biosurfactants find diverse applications in agriculture, particularly in the substitution of synthetic surfactants within pesticide and agrochemical formulations. This shift supports the expansion of "green chemistry" in agriculture, addressing concerns about the environmental and human health impacts associated with excessive chemical compound usage (Fenibo et al. 2019). Notably, rhamnolipid and lipopeptide biosurfactants contribute to enhanced soil quality, crucial for crop production, by acting as bioremediation agents for soils, surface water, groundwater, and waste streams contaminated with hydrophobic organic compounds like polycyclic aromatic hydrocarbons and metals (Sarubbo et al. 2022).

Due to their antimicrobial properties, biosurfactants from lipid classes such as mannosylerythritol, rhamnolipids, and lipopeptides serve as biopesticides to control various pests, pathogens, phytopathogenic fungi, and weeds. For example, lipopeptides produced by *Pseudomonas putida* and *Pseudomonas fluorescens* can lyse *Phytophthora capsici* zoospores and inhibit the growth of phytopathogens like *Pythium ultimum*, *Fusarium oxysporum*, and *Phytophthora cryptogea* (Thavasi et al. 2014). Other lipopeptides and glycolipids have shown promise in inhibiting aphids, mosquitoes, and toxins produced by fungi like Aspergillus parasiticum in peanut, cotton, and corn crops, safeguarding plants from infections and pest infestations (Wołejko et al. 2020).

Additionally, biosurfactants and biosurfactant-producing microorganisms can serve as plant nutrients (carbon sources) and facilitate the absorption of essential substances for plant growth. They efficiently distribute metals and micronutrients in the soil, form biofilms on roots, and protect plants from harmful substances. These applications demonstrate the versatility of biosurfactants in promoting sustainable and environmentally friendly agricultural practices (Manga et al. 2021).

8.8.4 Enhanced Oil Recovery and Energy Industry

Biosurfactants are eco-friendly alternatives for bioremediation and oil cleanup. In the energy sector, biosurfactants hold promise for enhanced oil recovery (EOR). By lowering the interfacial tension between oil and water, biosurfactants increase the mobility of trapped oil in reservoirs, making extraction more efficient. As conventional oil reserves become scarcer, the application of biosurfactants in EOR could become increasingly vital, ensuring a more sustainable utilization of existing resources (Mnif and Ghribi 2015). They enhance biodegradation in contaminated soils, reduce handling efforts, and offer low toxicity.

8.8.5 Bioremediation and Environmental Cleanup

Biosurfactants are useful tools for bioremediation because they have special qualities. They have the capacity to make hydrophobic contaminants more soluble, allowing microbial oxidation and breakdown. This site might be used to clean up oil spills and polluted water and soil. Biosurfactants might make a substantial contribution to environmental restoration efforts by increasing the effectiveness of microbial decomposition (Makkar and Rockne 2003).

The soil is a biologically active, porous medium which functions as a source of nutrients and water as well as a filtering and decomposition medium for harmful pollutants. Additionally, it aids in the cycle of carbon and other components throughout the entire ecosystem. Soil contamination is a type of land degradation brought by the spillage, migration, or burying of hazardous contaminants to the soil ecosystem from untreated industrial effluents, local waste, manufacturing wastes, agrochemicals, and other man-made substances or activities. In addition to harming the soil, toxins in it also have a negative impact on the health of humans, plants, and animals. It can be exposed to polluted soil through inhaling, swallowing, touching, or eating food that has been cultivated there. Most metropolitan areas, as well as old industrial and mining sites, are affected by soil pollution (Ambaye et al. 2021).

When a pollutant enters the soil, it can be adsorbed or carried away by the wind and runoff, or leached by infiltration water and enter to groundwater. Agriculture residues, byproducts, air pollutants, irrigation, floods, unintentional oil spills, improper management of municipal waste and sewage, heavy metals, and hydrocarbon deposition are some of the main contributors of soil pollution (Sales da Silva et al. 2020).

Biosurfactants contribute to the effectiveness of bioremediation in soil pollution scenarios (Filonov et al. 2012). Many soil pollutants, such as pesticides, hydrocarbons, and heavy hydrophobic organic compounds, have a strong affinity to soil particles hence they are insoluble in water. This reduces their availability for microbial degradation. Amphiphilic qualities refer to the presence of both hydrophilic (water-attracting) and hydrophobic (water-repellent) components in biosurfactants. These hydrophobic pollutants can be successfully solubilized and emulsified by them, resulting in micelles or emulsions that increase the contaminants' accessibility to microorganisms for destruction.

In addition, surfactant promotes microbial growth in the soil. Many microbes may use biosurfactants as a source of carbon and energy. Certain bacteria can promote both their own development as well as the growth of other microbes that degrade pollutants by creating biosurfactants. In polluted soils, this cooperative action may hasten the biodegradation process. Surfactin produced by species like *B. licheniformis* and *B. subtilis* achieved removal rates exceeding 85% in soil

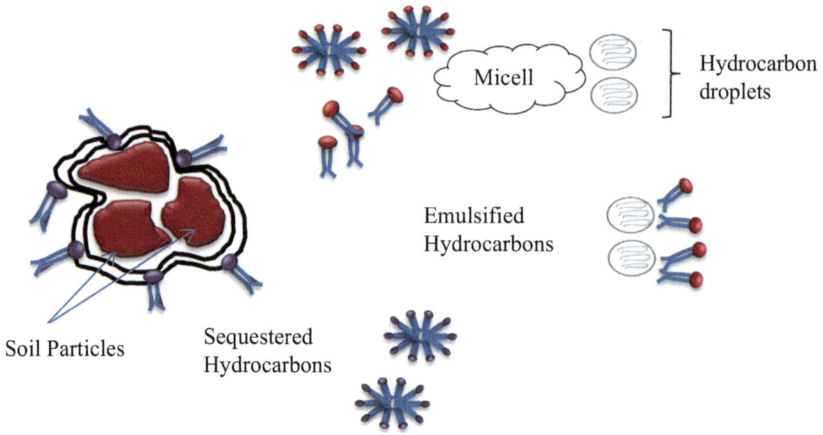

Fig. 8.6 Hydrocarbon and biosurfactants interaction with soil during the bioremediation process

remediation (Mulligan 2005). Other biosurfactants from *Bacillus*, *Pseudomonas*, and *Candida* species have shown success in soil cleanup (Fig. 8.6).

8.8.6 Heavy Metal Removal

In the context of bio-corrosion of metals, these polymeric biosurfactants demonstrate their capability to bind metal ions, which can exacerbate the corrosion process. In addition, bio-weathering processes, these substances play a role through their complexion properties, aiding in the dissolution of minerals (Bhakat et al. 2019).

Extracellular polymeric substances (EPSs), including polymeric biosurfactants which are produced by several microorganisms, serve as sorption sites for pollutants such as heavy metal ions and organic molecules, potentially mitigating environmental contamination. In terms of classification, chemically synthesized surfactants are typically categorized based on their polarity. In contrast, biosurfactants are generally classified according to their microbial origin and chemical composition (Fig. 8.7).

8.9 Role of Biotechnology in Biosurfactants-Based Bioremediation Process

The environment is now facing a highly critical situation due to contamination trapping on the earth as a result of rapid urbanization and industrialization. Thus, there is a requirement for developed biotechnological technology to remove the contaminated materials. Now with recent advancements in genetic technologies, bioremediation processes have been developed to protect the natural environment through

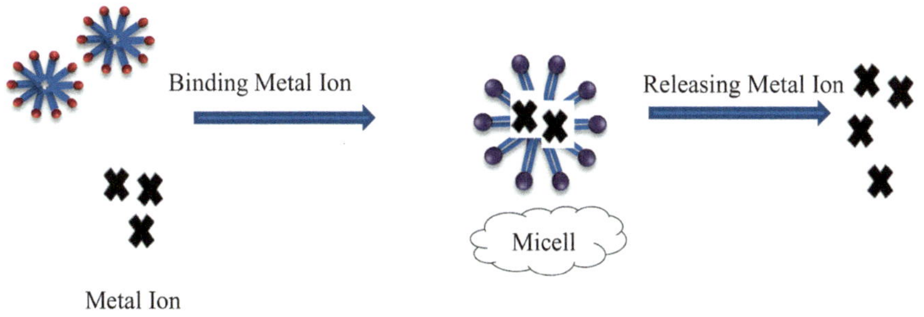

Fig. 8.7 The removal of potentially toxic elements through the mediation of biosurfactants

the recycling of waste materials. The biotechnological approaches play a prominent role in the production of biosurfactants primarily through genetic modification techniques, focusing on the optimization of bacteria and microbial populations to increase surfactant production yield of surfactants and their surface-active properties. Genetic modification involves altering the genetic material of an organism to achieve specific characteristics or traits, such as increased biosurfactant production. The primary goal of genetic modification in this context is to enhance the production and surface activity of biosurfactants. Bacteria and other microorganisms are targeted for optimization through genetic modification methods. There are only a few studies that have reported genetic modification of wild *Bacillus* sp. strains to increase biosurfactant production. Scientific studies have reported that such a biosurfactant obtained through genetic modification is surfactin, which is produced by engineered *Bacillus* species (Adetunji et al. 2021).

Some biotechnological approaches including metagenomics, single-cell sequencing, next-generation sequencing (NGS), microarray technology, metatranscriptomics, metaproteomics, and bioinformatics are novel biotechnological approaches that are applied in the bioremediation process (Mohanta et al. 2015).

Metagenomics approaches offer great possibility to find unknown sequences or functions from the environment. Microbes that are living in already polluted site are used for the isolation of genomic DNA from polluted environments (sludge, waste water, sediment, or soil) due to the presence of pollutant degradation genomes on their genome. The extracted DNA is sequenced using NGS. Metagenomic allows for the identification of functional genes related to biosurfactant production within the microbial community. These genes are associated with surfactant synthesis pathways relevant to bioremediation. Metagenomics data can guide the optimization and engineering of microbial communities to enhance biosurfactant production (Schultz and Rosado 2020).

Single-cell sequencing is a technology that allows the genetic information of individual microbial cells to be analyzed separately. This approach provides collective data for entire microbial communities, and single-cell sequencing dissects the genetic content of single microorganisms. Environmental samples are collected from contaminated sites. Single cells are isolated from these samples using specialized techniques, such as fluorescence-activated cell sorting (FACS) or microfluidic devices. The isolated single cells are subjected to whole-genome amplification. The amplified DNA from individual cells is subjected to high-throughput sequencing techniques, such as Next-Generation Sequencing (NGS). Researchers can identify genes and pathways associated with biosurfactant production, surfactant properties, and other relevant functions (Slatko et al. 2018).

Next-Generation Sequencing (NGS) is a revolutionary technology that has transformed the field of biosurfactants-based bioremediation by providing comprehensive insights into the genetic composition, diversity, and functional potential of microbial communities. NGS allows for the identification and profiling of microorganisms within the community that have the genetic potential to produce biosurfactants. Specific genes and pathways associated with biosurfactant production can be detected. NGS allows for the profiling of functional genes related to biosurfactant production and degradation of contaminants. This information helps in understanding the metabolic potential of the microbial community (Schultz and Rosado 2020).

It can be used to analyze how the genes related to biosurfactant production, degradation of contaminants, and other relevant metabolic pathways are being expressed. This information can help assess the efficiency and effectiveness of the bioremediation process. By comparing gene expression profiles between biosurfactant-producing and non-producing strains, we can pinpoint the specific genes and pathways involved in biosurfactant synthesis (Ndlovu 2017). This can help researchers understand how biosurfactants facilitate the breakdown of pollutants by enhancing microbial activity and solubilizing hydrophobic compounds. Microarrays can aid in the design of genetically engineered microorganisms optimized for biosurfactant production and bioremediation. By manipulating the expression of specific genes involved in biosurfactant production or pollutant degradation, researchers can create strains with enhanced capabilities.

Metatranscriptomics allows researchers to identify which microorganisms within the environmental sample are actively producing biosurfactants. By analyzing the RNA transcripts associated with biosurfactant genes, it's possible to pinpoint the specific organisms responsible for biosurfactant production. It provides real-time information about the expression levels of genes involved in biosurfactant production. This data can help researchers assess when and to what extent biosurfactants are being produced during the bioremediation process. Metatranscriptomics can help assess the environmental impact of biosurfactant-based bioremediation. Metaproteomics can help identify the composition of the microbial community responsible for biosurfactant production and contaminant degradation (Manga et al. 2021). By analyzing the proteins expressed by different microorganisms, researchers can gain insights into which species are actively contributing to the bioremediation process. Researchers can detect and quantify proteins involved in biosurfactant synthesis, confirming whether biosurfactants are being actively produced during bioremediation. This information can be used to determine which proteins are highly expressed during bioremediation, indicating their importance in the process. Metaproteomic data from different environmental samples can be compared to assess changes in protein expression patterns. This helps researchers understand how microbial communities adapt and evolve over the course of the bioremediation process. By identifying key proteins and pathways, researchers can design interventions to enhance biosurfactant production or pollutant degradation (Slatko et al. 2018).

Bioinformatics can be applied to biosurfactant-based bioremediation to enhance the understanding, selection, and optimization of biosurfactant-producing microorganisms and their associated biodegradation processes. Bioinformatics tools can analyze the genomes of microorganisms to identify potential biosurfactant-producing strains (Slatko et al. 2018). Genomic data can reveal genes and pathways involved in biosurfactant production, allowing researchers to screen and select strains with high biosurfactant-producing potential. By annotating the functional genes responsible for biosurfactant production, bioinformatics can help identify the specific enzymes and metabolic pathways involved. This information aids in the targeted manipulation and optimization of biosurfactant production in selected strains. Bioinformatics allows researchers to construct metabolic pathways and networks associated with biosurfactant production. This knowledge can be used to enhance biosurfactant production through genetic engineering or metabolic engineering approaches (Reis et al. 2013).

Bioinformatics tools can predict the structural properties of biosurfactants, including their surface-active properties and interaction with pollutants. This information aids in understanding the way biosurfactants enhance the solubility and biodegradation of contaminants. Bioinformatics can assist in identifying and characterizing the metabolic pathways involved in the biodegradation of specific pollutants facilitated by biosurfactants. This knowledge helps in the design of tailored bioremediation strategies. Bioinformatics can help design monitoring systems to track the performance of biosurfactant-based bioremediation processes. Bioinformatics databases can provide valuable information on biosurfactants, their production pathways, and their role in bioremediation (Reis et al. 2013). These databases can aid in the selection of biosurfactant-producing strains and relevant biodegradation pathways.

8.10 Advantages and Disadvantages of Biosurfactants Use as a Potential Strategy for Remediation

Biosurfactants are bio-based extensive applications in various industries. These eco-friendly alternatives offer several benefits compared to chemical surfactants. Biosurfactants offer several advantages and disadvantages when used for bioremediation. Understanding these pros and cons is essential for making informed decisions about the application of biosurfactants in environmental cleanup efforts. Biosurfactants can increase the solubility of hydrophobic pollutants, making them more accessible to microorganisms for degradation.

Emulsification and Dispersion: Biosurfactants can break down larger pollutant aggregates into smaller droplets, increasing the surface area available for microbial activity. Biosurfactants improve the availability of pollutants to microbial communities, facilitating faster and more efficient biodegradation. Biosurfactants can be synthesized from abundant raw materials and even from industrial waste. Some biosurfactants can serve as a nutrient source for pollutant-degrading microorganisms, enhancing their growth and activity (Akbari et al. 2018). Biosurfactants can work in synergy with microbial consortia, enhancing the overall efficiency of bioremediation processes. Biosurfactants are often biodegradable and have lower toxicity compared to synthetic surfactants, reducing the risk of introducing additional pollutants. Biosurfactants have shown effectiveness in remediating various types of contaminants, including hydrocarbons, heavy metals, and organic compounds (Usman et al. 2016). Biosurfactants are derived from natural sources and can be produced using renewable resources, aligning with environmentally friendly practices. Therefore, according to the purpose different biosurfactants can be selected and tailored to specific contaminants and environmental conditions, allowing for customized bioremediation strategies.

8.11 Benefits of Biosurfactants Compared to Synthetic Surfactants in Remediation

In comparison to synthetic surfactants, biosurfactants are often biodegradable and have lower environmental impacts. This makes them a more sustainable choice for enhancing bioremediation processes. Biosurfactants exhibit anionic and cationic properties that reduce the toxicity of harmful chemicals in both aquatic and terrestrial environments when compared to synthetic surfactants (Vijayakumar and Saravanan 2015).

It's important to note that while biosurfactants offer several advantages in soil bioremediation, their effectiveness can vary based on factors such as the type of contaminant, the microbial community present, environmental conditions, and the specific biosurfactant-producing organisms used. Research is ongoing to optimize the use of biosurfactants in various bioremediation strategies and to understand their interactions with different soil matrices and pollutants (Curstedt et al. 2013).

Biosurfactants can act as anti-biofilm agents and prevent the adhesion of contaminants exhibiting anti-adhesive properties. This helps in the formation of microbial biofilms, which enhance the interaction of active biosurfactants with hydrophobic surfaces, facilitating pollutant removal where these properties could not be observed in physicochemical conventional remediation strategies (Banat et al. 2010). Biosurfactants improve the solubility and bioavailability of toxic substances, aiding in their removal where these properties could not be observed in physicochemical conventional remediation strategies.

8.12 Disadvantages of Biosurfactants for Bioremediation

Though there are several advantages that are promising with biosurfactants, there are such limitations to them and their usage in the remediation process. Biosurfactants have specificity of their action where different biosurfactants have varying degrees of effectiveness against different types of pollutants. Not all biosurfactants are equally efficient in solubilizing or dispersing all types of contaminants (Kashif et al. 2022). The selection of substrate plays a vital role in influencing the yield and efficiency of the final product. Consequently, selecting the proper biosurfactant for a particular contaminant might be difficult. Basic considerations, especially regarding agro-industrial substrates, should take into account factors such as the availability of raw materials, transportation costs, and the necessary pre-treatment before the production process (Mohanty et al. 2021). Introducing biosurfactant-producing microorganisms may compete with indigenous microbial populations, affecting the microbial balance in the environment.

The production of biosurfactants can sometimes be costly, particularly when considering large-scale applications. Downstream processing plays a major role in the overall expenses of biotechnology products. A diverse range of analytical techniques is employed to isolate, purify, and characterize various biosurfactants. Further, high production costs hinder market competitiveness of biosurfactants. Thus, the cost-effectiveness of using biosurfactants compared to other remediation methods needs to be carefully evaluated.

Biosurfactant production and activity can be influenced by environmental factors such as temperature, pH, and salinity. The effectiveness of biosurfactants can vary based on factors such as the specific biosurfactant used, pollutant type, microbial community, and site conditions. Introduction of biosurfactants could have unintended ecological consequences, affecting non-target organisms or ecosystem dynamics. Therefore, condition optimization and controlling biological parameters are bit of difficulties. Moreover, bioremediation processes, even with the assistance of biosurfactants, may require a significant amount of time to achieve desired pollutant reductions (Kashif et al. 2022). The use of biosurfactants in bioremediation may require regulatory approvals, especially if the site is sensitive or poses risks to human health or the environment. While biosurfactant production and application can work well in laboratory settings, scaling up the process to industrial or field conditions can present challenges (Reis et al. 2013).

8.13 Future Prospects Related to Biosurfactants

The long-term effects of biosurfactant application on ecosystem health and stability are not always well understood, and more research is needed in this area. In addition, a substantial amount of research has been conducted to increase the productivity and yield of biosurfactants in the upstream production process. These challenges are opportunities for research groups that should be thoroughly examined and addressed (Fracchia et al. 2012).

The development of biotechnology has the potential to influence how biosurfactants are used in the future. Scientists may create custom biosurfactants with particular qualities suited to a variety of applications by genetically altering microbial strains and optimizing them. The incorporation of biosurfactants into established businesses may be sped up by this modification, which could result in more effective and economical manufacturing techniques (De Almeida et al. 2016).

The increasing demand for eco-friendly and biodegradable products creates a fertile ground for the growth of the biosurfactant market. As consumer preferences shift toward sustainable options, industries are likely to invest in research and development, as well as production technologies, driving innovation and driving down production costs. This could lead to a virtuous cycle of increasing adoption and technological refinement.

Biosurfactants have great potential for influencing a more sustainable and effective future across numerous sectors because of their environmentally friendly nature and adaptable qualities (Fracchia et al. 2012). Biosurfactants are set to play a crucial role as the globe continues to struggle with environmental issues and the demand for greener alternatives. Biosurfactants provide a glimpse of a more harmonious interaction between business, technology, and the environment by encouraging innovation, promoting environmentally friendly practices, and promoting sustainable growth. Unlocking their full potential and taking advantage of what they have to offer will depend on continued research, investment, and cooperation.

8.14 Conclusion

In summary, microorganisms and plants are remarkable for their ability to produce various types of surfactants, particularly fatty acids, phospholipids, and neutral lipids, during growth on n-alkanes, with potential applications in medicine, industry, and beyond. Despite these limitations, biosurfactants remain promising tools for enhancing bioremediation efforts. To overcome these limitations, a thorough understanding of the specific site conditions, pollutant types, and the potential effects of biosurfactants is essential when designing and implementing bioremediation strategies. In summary, biosurfactants offer a promising avenue for enhancing bioremediation efforts, but their use should be carefully considered within the context of site-specific conditions, contaminant types, and regulatory requirements. It's important to conduct thorough testing, monitoring, and assessment to ensure the success and safety of biosurfactant-based bioremediation strategies.

References

Abdel-Aziz SM, Elsoud MMA, Anise AAH (2017) Microbial biosynthesis: a repertory of vital natural products. In: Food biosynthesis. Elsevier, pp 25–54

Abdel-Mawgoud AM, Lépine F, Déziel E (2010) Rhamnolipids: diversity of structures, microbial origins and roles. Appl Microbiol Biotechnol. Springer 86:1323–1336

Adetunji AI, Olaniran AO (2021) Production and potential biotechnological applications of microbial surfactants: an overview. Saudi J Biol Sci 28(1):669–679

Adetunji CO et al (2021) Strain improvement methodology and genetic engineering that could lead to an increase in the production of biosurfactants. In: Green sustainable process for chemical and environmental engineering and science. Elsevier, pp 299–315

Akbari S et al (2018) Biosurfactants—a new frontier for social and environmental safety: a mini review. Biotechnol Res Innov. Elsevier 2(1):81–90

Ali I et al (2016) Hydroxychavicol: a phytochemical targeting cutaneous fungal infections. Sci Rep. Nature Publishing Group 6:1–20. https://doi.org/10.1038/srep37867

Ambaye TG et al (2021) Preparation, characterization and application of biosurfactant in various industries: a critical review on progress, challenges and perspectives. Environ Technol Innov. Elsevier 24:102090

Ayilara MS, Babalola OO (2023) Bioremediation of environmental wastes: the role of microorganisms. Front Agron. Frontiers Media S.A 5:1183691. https://doi.org/10.3389/fagro.2023.1183691

Banat IM, Makkar RS, Cameotra SS (2000) Potential commercial applications of microbial surfactants. Appl Microbiol Biotechnol. Springer 53:495–508

Banat IM et al (2010) Microbial biosurfactants production, applications and future potential. Appl Microbiol Biotechnol. Springer 87:427–444

Basu S et al (2018) Engineering PGPMOs through gene editing and systems biology: a solution for phytoremediation? Trends Biotechnol. Elsevier 36(5):499–510

Bhakat K, Chakraborty A, Islam E (2019) Characterization of arsenic oxidation and uranium bioremediation potential of arsenic resistant bacteria isolated from uranium ore. Environ Sci Pollut Res. Springer 26:12907–12919

Bhatt P et al (2021) New insights into the degradation of synthetic pollutants in contaminated environments. Chemosphere. Elsevier 268:128827

Brinza L et al (2020) Baltic Fucus vesiculosus as potential bio-sorbent for Zn removal: mechanism insight. Chemosphere. Elsevier 238:124652

Bustamante M, Duran N, Diez MC (2012) Biosurfactants are useful tools for the bioremediation of contaminated soil: a review. J Soil Sci Plant Nutr. SciELO Chile 12(4):667–687

Cameotra SS, Makkar RS (2010) Biosurfactant-enhanced bioremediation of hydrophobic pollutants. Pure Appl Chem 82:97–116. https://doi.org/10.1351/PAC-CON-09-02-10

Chauhan V et al (2023) Pseudomonas lipopeptide: an excellent biomedical agent. MedComm Biomater Appl. Wiley Online Library 2(1):e27

Chugh M et al (2022) Algal bioremediation of heavy metals: an insight into removal mechanisms, recovery of by-products, challenges, and future opportunities. Energy Nexus. Elsevier 7:100129

Cortés-Camargo S et al (2021) Biosurfactant production by Bacillus tequilensis ZSB10: structural characterization, physicochemical, and antifungal properties. J Surfactant Deterg. Wiley Online Library 24(5):773–782

Costa I, Ferrão P (2010) A case study of industrial symbiosis development using a middle-out approach. J Clean Prod. Elsevier 18(10–11):984–992

Curstedt T, Calkovska A, Johansson J (2013) New generation synthetic surfactants. Neonatology. S. Karger AG 103(4):327–330

De Almeida DG et al (2016) Biosurfactants: promising molecules for petroleum biotechnology advances. Front Microbiol. Frontiers Media SA 7:1718

De França ÍWL et al (2015) Production of a biosurfactant by Bacillus subtilis ICA56 aiming bioremediation of impacted soils. Catal Today. Elsevier 255:10–15
Dell'Anno F et al (2021) Highly contaminated marine sediments can host rare bacterial taxa potentially useful for bioremediation. Front Microbiol. Frontiers Media SA 12:584850
Duc HD, Hung NV, Oanh NT (2021) Anaerobic degradation of endosulfans by a mixed culture of Pseudomonas sp. and Staphylococcus sp. Appl Biochem Microbiol. Springer 57(3):327–334
Dusengemungu L et al (2020) Recent advances in biosorption of copper and cobalt by filamentous fungi. Front Microbiol. Frontiers Media SA 11:582016
Dutta N et al (2022) Efficacy of emerging technologies in addressing reductive dechlorination for environmental bioremediation: a review. J Hazard Mater Lett. Elsevier 3:100065
Elshafie AE et al (2015) Sophorolipids production by Candida bombicola ATCC 22214 and its potential application in microbial enhanced oil recovery. Front Microbiol. Frontiers Media SA 6:1324
Emenike CU et al (2018) Biotransformation and removal of heavy metals: a review of phytoremediation and microbial remediation assessment on contaminated soil. Environ Rev. NRC Research Press 26(2):156–168
Fenibo EO et al (2019) Microbial surfactants: the next generation multifunctional biomolecules for applications in the petroleum industry and its associated environmental remediation. Microorganisms. MDPI 7(11):581
Filonov A et al (2012) Oil-spill bioremediation, using a commercial biopreparation "MicroBak" and a consortium of plasmid-bearing strains "V&O" with associated plants. In: Introduction to enhanced oil recovery (EOR) processes and bioremediation of oil-contaminated sites. IntechOpen, pp 291–318
Fracchia L et al (2012) Biosurfactants and bioemulsifiers biomedical and related applications–present status and future potentials. In: Biomedical science, engineering and technology, vol 14. InTech, Rijeka, pp 1–49
Gautam KK, Tyagi VK (2006) Microbial surfactants: a review. J Oleo Sci. Japan Oil Chemists' Society 55(4):155–166
Geetha N et al (2021) Insights into nanomycoremediation: secretomics and mycogenic biopolymer nanocomposites for heavy metal detoxification. J Hazard Mater. Elsevier 409:124541
Georgiou G, Lin S-C, Sharma MM (1992) Surface–active compounds from microorganisms. Bio/Technology. Nature Publishing Group US New York 10(1):60–65
González-Jaramillo LM et al (2017) Antimycotic activity of fengycin C biosurfactant and its interaction with phosphatidylcholine model membranes. Colloids Surf B Biointerfaces. Elsevier 156:114–122
Hatha AAM, Edward G, Pattanathu Rahman KSM (2007) Microbial biosurfactants–review. J Mar Atmos Res 3:1–17
Jahan R et al (2020) Biosurfactants, natural alternatives to synthetic surfactants: physicochemical properties and applications. Adv Colloid Interf Sci. Elsevier 275:102061
Jameel AA, Haider NH (2021) Study the antimicrobial and antiadhesive activity of purified biosurfactant produced from Lactobacillus plantarum against pathogenic bacteria. Iraqi J Agric Sci 52(5):1194–1206
Johnson P et al (2021) Effect of synthetic surfactants on the environment and the potential for substitution by biosurfactants. Adv Colloid Interf Sci. Elsevier 288:102340
Joshi-Navare K, Singh PK, Prabhune AA (2014) New yeast isolate Pichia caribbica synthesizes xylolipid biosurfactant with enhanced functionality. Eur J Lipid Sci Technol. Wiley Online Library 116(8):1070–1079
Kashif A et al (2022) Current advances in the classification, production, properties and applications of microbial biosurfactants – a critical review. Adv Colloid Interf Sci. Elsevier 306:102718
Kharangate-Lad A, D'Souza NC (2021) Current approaches in bioremediation of toxic contaminants by application of microbial cells; biosurfactants and bioemulsifiers of microbial origin. In: Rhizobiont in bioremediation of hazardous waste. Springer, pp 217–263

Kuyukina MS, Ivshina IB (2019) Bioremediation of contaminated environments using Rhodococcus. In: Biology of Rhodococcus. Springer, pp 231–270
Kuyukina MS et al (2015) Trehalolipid biosurfactants from nonpathogenic Rhodococcus actinobacteria with diverse immunomodulatory activities. New Biotechnol. Elsevier 32(6):559–568
Lakmali WGM, Athukorala ADSNP, Jayasundera KB (2022) Investigation of Pb (II) bioremediation potential of algae and cyanobacteria strains isolated from polluted water. Water Sci Eng. Elsevier 15(3):237–246
Li Q, Liu J, Gadd GM (2020) Fungal bioremediation of soil co-contaminated with petroleum hydrocarbons and toxic metals. Appl Microbiol Biotechnol. Springer 104:8999–9008
Liley JR et al (2017) Self-assembly in dilute mixtures of non-ionic and anionic surfactants and rhamnolipd biosurfactants. J Colloid Interface Sci. Elsevier 487:493–503
Liu Z et al (2018) Effects of rhamnolipids on the removal of 2,4,2,4-tetrabrominated biphenyl ether (BDE-47) by Phanerochaete chrysosporium analyzed with a combined approach of experiments and molecular docking. Chemosphere. Elsevier 210:922–930
Liyanage LMM et al (2020) Application of live Chlorococcum aquaticum biomass for the removal of Pb (II) from aqueous solutions. J Appl Phycol. Springer 32:4069–4080
Machado LF et al (2019) Tracking mangrove oil bioremediation approaches and bacterial diversity at different depths in an in situ mesocosms system. Front Microbiol. Frontiers Media SA 10:2107
Makkar RS, Rockne KJ (2003) Comparison of synthetic surfactants and biosurfactants in enhancing biodegradation of polycyclic aromatic hydrocarbons. Environ Toxicol Chem. Wiley Online Library 22(10):2280–2292
Manga EB et al (2021) Biosurfactants: opportunities for the development of a sustainable future. Curr Opin Colloid Interface Sci. Elsevier 56:101514
Mantil E, Crippin T, Avis TJ (2019) Domain redistribution within ergosterol-containing model membranes in the presence of the antimicrobial compound fengycin. Biochim Biophys Acta. Elsevier 1861(4):738–747
McClements DJ, Gumus CE (2016) Natural emulsifiers—biosurfactants, phospholipids, biopolymers, and colloidal particles: molecular and physicochemical basis of functional performance. Adv Colloid Interf Sci. Elsevier 234:3–26
McInerney MJ, Javaheri M, Nagle DP Jr (1990) Properties of the biosurfactant produced by Bacillus licheniformis strain JF-2. J Ind Microbiol Biotechnol. Oxford University Press 5(2–3):95–101
Md F (2012) Biosurfactant: production and application. J Pet Environ Biotechnol 3(4):124
Miri S et al (2022) Biodegradation of p-xylene—a comparison of three psychrophilic Pseudomonas strains through the lens of gene expression. Environ Sci Pollut Res.. Springer 29(15):21465–21479
Mnif I, Ghribi D (2015) Review lipopeptides biosurfactants: mean classes and new insights for industrial, biomedical, and environmental applications. Pept Sci. Wiley Online Library 104(3):129–147
Mnif I, Ellouz-Chaabouni S, Ghribi D (2018) Glycolipid biosurfactants, main classes, functional properties and related potential applications in environmental biotechnology. J Polym Environ. Springer 26:2192–2206
Mohamed MSM et al (2020) Reduction of chromium-VI by chromium-resistant Escherichia coli FACU: a prospective bacterium for bioremediation. Folia Microbiol. Springer 65:687–696
Mohanta TK, Mohanta YK, Mohanta N (2015) Role of biotechnology in bioremediation. In: Handbook of research on uncovering new methods for ecosystem management through bioremediation. IGI Global, pp 399–432
Mohanty SS et al (2021) A critical review on various feedstocks as sustainable substrates for biosurfactants production: a way towards cleaner production. Microb Cell Fact. BioMed Central 20(1):1–13
Mohapatra B, Phale PS (2021) Microbial degradation of naphthalene and substituted naphthalenes: metabolic diversity and genomic insight for bioremediation. Front Bioeng Biotechnol. Frontiers Media SA 9:602445

Mohd S et al (2019) Fungal mediated biotransformation reduces toxicity of arsenic to soil dwelling microorganism and plant. Ecotoxicol Environ Saf. Elsevier 176:108–118
Moreira VR et al (2019) Simultaneous biosorption of Cd (II), Ni (II) and Pb (II) onto a brown macroalgae Fucus vesiculosus: mono-and multi-component isotherms, kinetics and thermodynamics. J Environ Manage. Elsevier 251:109587
Mousavi SM, Hashemi SA, Iman Moezzi SM, Ravan N, Gholami A, Lai CW, Chiang WH, Omidifar N, Yousefi K and Behbudi G (2021) Recent advances in enzymes for the bioremediation of pollutants. Biochemistry research international, 2021(1), p.5599204
Mukjang N et al (2022) Bacterial communities associated with crude oil bioremediation through composting approaches with indigenous bacterial isolate. Life. MDPI 12(11):1712
Müller MM et al (2012) Rhamnolipids—next generation surfactants? J Biotechnol. Elsevier 162(4):366–380
Mulligan CN (2005) Environmental applications for biosurfactants. Environ Pollut. Elsevier 133(2):183–198
Münzel T et al (2023) Soil and water pollution and human health: what should cardiologists worry about? Cardiovasc Res. Oxford University Press 119:440–449. https://doi.org/10.1093/cvr/cvac082
Ndlovu T (2017) Bioprospecting for novel biosurfactants and biosurfactant producing bacteria in wastewater. Stellenbosch University, Stellenbosch
Neu TR (1996) Significance of bacterial surface-active compounds in interaction of bacteria with interfaces. Microbiol Rev, American Society for Microbiology 60(1):151–166
Nguyen TT et al (2008) Rhamnolipid biosurfactant mixtures for environmental remediation. Water Res. Elsevier 42(6–7):1735–1743
Nowicka B et al (2020) The inhibitor-evoked shortage of tocopherol and plastoquinol is compensated by other antioxidant mechanisms in Chlamydomonas reinhardtii exposed to toxic concentrations of cadmium and chromium ions. Ecotoxicol Environ Saf. Elsevier 191:110241
Nurfarahin AH, Mohamed MS, Phang LY (2018) Culture medium development for microbial-derived surfactants production—an overview. Molecules. MDPI 23(5):1049
Pande V et al (2020) Bioremediation: an emerging effective approach towards environment restoration. Environ Sustain. Springer 3:91–103
Patel P et al (2021) Microbial biosurfactants in management of organic waste. In: Sustainable environmental clean-up. Elsevier, pp 211–230
Pitocchi R et al (2020) Cerato-platanins from marine fungi as effective protein biosurfactants and bioemulsifiers. Int J Mol Sci.. MDPI 21(8):2913
Prabhu Y, Phale PS (2003) Biodegradation of phenanthrene by Pseudomonas sp. strain PP2: novel metabolic pathway, role of biosurfactant and cell surface hydrophobicity in hydrocarbon assimilation. Appl Microbiol Biotechnol. Springer 61:342–351
Qadir A, Hashmi MZ, Mahmood A (2017) Xenobiotics, types, and mode of action. In: Xenobiotics in the soil environment: monitoring, toxicity and management. Springer, pp 1–7
Qi L, Christopher GF (2019) Role of flagella, type IV pili, biosurfactants, and extracellular polymeric substance polysaccharides on the formation of pellicles by Pseudomonas aeruginosa. Langmuir. ACS Publications 35(15):5294–5304
Raaijmakers JM, De Bruijn I, De Kock MJD (2006) Cyclic lipopeptide production by plant-associated Pseudomonas spp.: diversity, activity, biosynthesis, and regulation. Mol Plant-Microbe Interact. American Phytopathological Society 19(7):699–710
Rabani MS et al (2022) Characterization and identification of naphthalene degrading bacteria isolated from petroleum contaminated sites and their possible use in bioremediation. Polycycl Aromat Compd. Taylor & Francis 42(3):978–989
Rahman PKSM, Gakpe E (2008) Production, characterisation and applications of biosurfactants-review. Biotechnology. Asian Network for Scientific Information 7:360–370
Rajkumari J et al (2017) Facile green synthesis of baicalein fabricated gold nanoparticles and their antibiofilm activity against Pseudomonas aeruginosa PAO1. Microb Pathog. Elsevier 107:261–269

Rastogi S, Kumar R (2020) Remediation of heavy metals using non-conventional adsorbents and biosurfactant-producing bacteria. In: Environmental degradation: causes and remediation strategies. Agro Environ Media – Agriculture and Ennvironmental Science Academy, Haridwar, pp 133–153. https://doi.org/10.26832/aesa-2020-edcrs-010
Rebello S et al (2014) Surfactants: toxicity, remediation and green surfactants. Environ Chem Lett. Springer Verlag 12:275–287. https://doi.org/10.1007/s10311-014-0466-2
Reis RS et al (2013) Biosurfactants: production and applications. In: Biodegradation-life of science, vol 2. InTech, Rijeka, pp 31–61
Rocha e Silva NMP et al (2019) Natural surfactants and their applications for heavy oil removal in industry. Sep Purif Rev. Taylor & Francis 48(4):267–281
Rodríguez-López L et al (2020) Extraction, separation and characterization of lipopeptides and phospholipids from corn steep water. Sep Purif Technol. Elsevier 248:117076
Saavedra R et al (2018) Comparative uptake study of arsenic, boron, copper, manganese and zinc from water by different green microalgae. Bioresour Technol. Elsevier 263:49–57
Sachdev DP, Cameotra SS (2013) Biosurfactants in agriculture. Appl Microbiol Biotechnol. Springer 97:1005–1016
Sáenz-Marta CI, de Lourdes Ballinas-Casarrubias M, Rivera-Chavira BE and Nevárez-Moorillón GV (2015) Biosurfactants as useful tools in bioremediation. IntechOpen
Sajadi Bami M et al (2022) Biosurfactants aided bioremediation mechanisms: a mini-review. Soil Sediment Contam Int J. Taylor & Francis 31(7):801–817
Sales da Silva IG et al (2020) Soil bioremediation: overview of technologies and trends. Energies. MDPI 13(18):4664
Samanta SK, Singh OV, Jain RK (2002) Polycyclic aromatic hydrocarbons: environmental pollution and bioremediation. Trends Biotechnol. Elsevier 20(6):243–248
Sangkharak K et al (2020) The degradation of phenanthrene, pyrene, and fluoranthene and its conversion into medium-chain-length polyhydroxyalkanoate by novel polycyclic aromatic hydrocarbon-degrading bacteria. Curr Microbiol. Springer 77:897–909
Santa Anna LM et al (2002) Production of biosurfactants from Pseudomonas aeruginosa PA 1 isolated in oil environments. Braz J Chem Eng. SciELO Brasil 19:159–166
Sarubbo LA et al (2022) Biosurfactants: production, properties, applications, trends, and general perspectives. Biochem Eng J. Elsevier 181:108377
Satpute SK et al (2010) Biosurfactants, bioemulsifiers and exopolysaccharides from marine microorganisms. Biotechnol Adv. Elsevier 28(4):436–450
Schultz J, Rosado AS (2020) Extreme environments: a source of biosurfactants for biotechnological applications. Extremophiles. Springer 24:189–206
Sekhon KK, Khanna S, Cameotra SS (2012) Biosurfactant production and potential correlation with esterase activity. J Pet Environ Biotechnol. OMICS Publishing Group 3(7):133
Sen SK et al (2019) Pilot-scale evaluation of bio-decolorization and biodegradation of reactive textile wastewater: an impact on its use in irrigation of wheat crop. Water Resour Ind. Elsevier 21:100106
Seydlová G, Svobodová J (2008) Review of surfactin chemical properties and the potential biomedical applications. Cent Eur J Med. Springer 3:123–133
Shekhar S, Sundaramanickam A, Balasubramanian T (2015) Biosurfactant producing microbes and their potential applications: a review. Crit Rev Environ Sci Technol. Taylor & Francis 45(14):1522–1554
Singh RK et al (2020) Fungi as potential candidates for bioremediation. In: Abatement of environmental pollutants. Elsevier, pp 177–191
Slatko BE, Gardner AF, Ausubel FM (2018) Overview of next-generation sequencing technologies. Curr Protoc Mol Biol.. Wiley Online Library 122(1):e59
Song J et al (2019) Purification and characteristics of an aflatoxin B1 degradation enzyme isolated from Pseudomonas aeruginosa. FEMS Microbiol Lett.. Oxford University Press 366(5):fnz034
Sravya K, Sangeetha S (2022) Feasibility study on bioremediation techniques to contaminated soils. Mater Today Proc. Elsevier 51:2556–2560

Thavasi R, Marchant R, Banat IM (2014) 15 Biosurfactant applications in agriculture. In: Biosurfactants: production and utilization—processes, technologies, and economics, vol 159. CRC Press, p 313
Ukaogo PO, Ewuzie U, Onwuka CV (2020) Environmental pollution: causes, effects, and the remedies. In: Microorganisms for sustainable environment and health. Elsevier, pp 419–429
Unuofin JO, Falade AO, Aladekoyi OJ (2021) Applications of microbial laccases in bioremediation of environmental pollutants: potential issues, challenges, and prospects. In: Bioremediation for environmental sustainability. Elsevier, pp 519–540
Usman MM et al (2016) Application of biosurfactants in environmental biotechnology; remediation of oil and heavy metal. AIMS Bioeng. Aims Press 3(3):289–304
Van Bogaert INA et al (2007) Microbial production and application of sophorolipids. Appl Microbiol Biotechnol. Springer 76:23–34
Verma S, Kuila A (2019) Bioremediation of heavy metals by microbial process. Environ Technol Innov. Elsevier 14:100369
Vijayakumar S, Saravanan V (2015) Biosurfactants-types, sources and applications. Res J Microbiol 10(5):181–192
Wan C et al (2022) Iturin: cyclic lipopeptide with multifunction biological potential. Crit Rev Food Sci Nutr. Taylor & Francis 62(29):7976–7988
Wang Z et al (2018) Biokinetics of arsenate accumulation and release in Microcystis aeruginosa regulated by common environmental factors: practical implications for enhanced bioremediation. J Clean Prod. Elsevier 199:112–120
Wołejko E et al (2020) Soil biological activity as an indicator of soil pollution with pesticides – a review. Appl Soil Ecol. Elsevier 147:103356
Wu P et al (2022) Bioremediation of phenolic pollutants by algae-current status and challenges. Bioresour Technol.. Elsevier 350:126930
Yaraguppi DA et al (2020) Response surface methodology-based optimization of biosurfactant production from isolated Bacillus aryabhattai strain ZDY2. J Pet Explor Prod Technol. Springer 10:2483–2498
Zahed MA et al (2022) Biosurfactant, a green and effective solution for bioremediation of petroleum hydrocarbons in the aquatic environment. Discov Water. Springer Science and Business Media LLC 2(1):5. https://doi.org/10.1007/s43832-022-00013-x
Zarinviarsagh M, Ebrahimipour G, Sadeghi H (2017) Lipase and biosurfactant from Ochrobactrum intermedium strain MZV101 isolated by washing powder for detergent application. Lipids Health Dis. Springer 16:1–13

Chapter 9
Utilization of Biochar for Remediation

Geetanjali Mallick, Hemant Mittal, and Jaya Lakkakula

Abstract Various human activities, such as indiscriminate littering and the discharge of industrial and domestic waste into water bodies, have significantly worsened environmental contamination. The presence of various pollutants has culminated in significant adverse effects, including soil degradation and water pollution. With environmental problems on the rise, biochar has emerged as an innovative and sustainable solution. This stable, carbon-rich material is produced by heating various types of organic matter—such as agricultural leftovers, animal manure, or microbial sources—without oxygen. Because biochar can be made from such diverse feedstocks, thoroughly analyzing its characteristics is vital for determining how well it can work in environmental clean-up efforts. Advanced analytical methodologies are routinely utilized to evaluate its physicochemical properties, which directly influence its adsorption capacity, nutrient retention, and overall performance in soil amendment and carbon sequestration strategies. Evaluations have shown that biochar has a highly porous structure, which enhances its ability to adsorb and remove pollutants from contaminated environments. By optimizing these conditions, biochar can be used as an effective and sustainable material for mitigating environmental contamination. Its application in pollution control demonstrates its potential as an eco-friendly approach to addressing environmental challenges.

G. Mallick
Amity Institute of Biotechnology, Amity University Maharashtra,
Mumbai, Maharashtra, India

H. Mittal
Mechanical Engineering Department, Khalifa University of Science and Technology,
Abu Dhabi, United Arab Emirates

J. Lakkakula (✉)
Amity Institute of Biotechnology, Amity University Maharashtra, Mumbai-Pune Expresway,
Panvel, Maharashtra, India

Centre for Computational Biology and Translational Research,
Amity University Maharashtra, Mumbai-Pune Expressway, Panvel, India
e-mail: jrlakkakula@mum.amity.edu

A. Guldhe et al. (eds.), *Biomaterials for Environmental Applications: Detection and Remediation*, Environmental Challenges and Solutions,
https://doi.org/10.1007/978-3-032-12701-3_9

Keywords Biochar · Environmental pollution · Soil remediation · Wastewater treatment · Pharmaceutical wastes · Adsorption

9.1 Introduction

Environmental pollution is an urgent global challenge, significantly impacting ecosystems and human health. Human activities such as industrialization, urbanization, and improper waste disposal have intensified environmental contamination, leading to adverse effects such as biodiversity loss, soil degradation, water pollution, and air contamination (Keerthana et al. 2023). Addressing such environmental hazards requires strategic interventions, one of which is "environmental remediation," a process aimed at removing and neutralizing pollutants. These contaminants include greenhouse gases, toxic chemicals, heavy metals, oil spills, radioactive substances, plastic waste, and pharmaceutical residues, all of which stem from activities such as mining, drilling, deforestation, industrial processing, agriculture, and manufacturing. One of the most promising and sustainable solutions for environmental remediation is biochar (Alkharabsheh et al. 2021).

Biochar is a remarkably stable, carbon-rich material created by heating organic matter—like plant waste or manure—in a low-oxygen setting, usually at temperatures between 300 °C and 1000 °C. This process, known as pyrolysis, helps lock carbon into a solid form, making biochar useful for a variety of environmental applications. The feedstock for biochar production encompasses a diverse range of organic materials, which include agricultural residues, forest byproducts, sewage sludge, animal waste, and municipal solid waste. Key characteristics such as how porous biochar is, how much surface area it offers, the types of chemical groups it contains, and its pH level all play important roles in how well it can absorb substances and function in different uses. These properties are especially important for improving soil health and storing carbon, making biochar valuable for environmental solutions. These properties are further influenced by various factors and any modifications made before or after the pyrolysis process (Tisserant and Cherubini 2019).

One of the major usages of biochar is soil remediation. Biochar enhances soil structure, increases microbial activity, and improves fertility, ultimately benefiting plant growth and agricultural productivity. Biochar is also widely employed in water treatment due to its superior adsorption properties because it effectively removes contaminants, which include heavy metals (lead, cadmium, and arsenic), organic pollutants, pesticides, and pharmaceutical residues. Biochar has been successfully applied in treating industrial effluents from dye manufacturing, battery production, dairy processing, municipal wastewater, agricultural runoff, and stormwater. The efficiency of biochar in water treatment is highly determined by its surface charge, pore structure, and functional group interactions with target pollutants.

A particularly concerning issue in environmental pollution is the increasing prevalence of pharmaceutical contaminants. The widespread production and use of pharmaceuticals in hospitals, healthcare centers, and veterinary facilities have led to the continuous discharge of pharmaceutical residues into water bodies. These contaminants primarily enter the environment through human excretion, improper disposal, and wastewater discharges from pharmaceutical industries. Common pharmaceutical pollutants include antibiotics, analgesics, hormones, and personal care products, all of which pose risks to aquatic ecosystems and contribute to antimicrobial resistance (Kang et al. 2022).

The efficacy of biochar in adsorbing the contaminants from pharmaceutical is contingent upon several factors, including its physicochemical characteristics, the specific pharmaceutical agents involved, and the prevailing environmental parameters such as temperature and pH levels. Empirical studies indicated that biochar produced from biomass sources like wood and manure exhibits superior adsorption capabilities for pharmaceutical pollutants, attributed to its enhanced surface area. Nonetheless, certain biochar variants have shown variable performance in adsorption experiments, highlighting a critical need for further investigation to refine and optimize their applications in wastewater treatment.

Beyond its established functions in soil and water remediation, biochar presents promising applications in air quality management, carbon sequestration, and waste mitigation. As a stable carbon sink, biochar effectively reduces greenhouse gas emissions by sequestering carbon in a persistent format over prolonged timescales. Nonetheless, the widespread adoption of biochar is hindered by a range of challenges that need to be addressed for large-scale implementation.

In conclusion, biochar offers a sustainable and effective approach to environmental remediation, providing solutions for soil and water contamination, the removal of pharmaceutical pollutants, air pollution control, and carbon sequestration. Its versatility, cost-effectiveness, and stability make it a valuable tool for tackling various environmental challenges. This chapter further explores the diverse applications of biochar, detailing its interactions with pollutants and its potential contribution to a cleaner environment (Table 9.1).

Table 9.1 Applications of biochar in the environment

Characteristics	Soil remediation	Water remediation	Pharmaceutical waste
Source of pollutants	Overuse of pesticides in agriculture Intensive industrial activities Improper waste management and disposal	Sewage Agricultural and oil pollution Industrial waste The burning of fossil fuels River and marine dumping	Factories and production facilities Hospitals, clinics, and long-term care centers Companies producing personal care products Veterinary clinics and animal care centers
Types of pollutants	Heavy metals, polycyclic aromatic hydrocarbons, industrial wastes, and pesticides.	Insecticides, herbicides, food waste, livestock pollutants, VOCs, heavy metals, and chemical waste	Solid pharmaceutical wastes (ex: medicine), liquid pharmaceutical waste (ex: sludge)
Methods used to treat the contamination	Excavation and extraction of pollution, bioremediation or phytoremediation, and mycoremediation.	Biological, chemical, and physical treatment technologies.	Sorption, abiotic, and biotic degradation.

9.2 Biochar for Environmental Applications

9.2.1 *Biochar Application for Soil Remediation*

Contaminated soil refers to the presence of hazardous substances that pose significant risks to living organisms and disrupt ecological balance. This contamination often results from pollutants affecting the soil directly or infiltrating nearby water sources. Natural ecosystems are degraded by environmental contaminants, and severe health risks, including potential long-term illnesses, are posed to human populations.

Soil remediation encompasses a range of techniques designed to remove or neutralize pollutants, making the soil safer for the environment and human use. The primary motive of these remediation strategies is to eliminate or transform harmful substances into less toxic forms. Common soil contaminants include heavy metals, pesticides, cyanide, and other hazardous chemicals. Eliminating these pollutants is crucial in preventing their migration into groundwater, surface water, agricultural produce, and natural vegetation (Fig. 9.1).

Biochar has gained attention as an effective remediation strategy for heavy metal contamination in soils, particularly in its ability to immobilize hazardous elements. Bousdra et al. conducted an in-depth investigation into how well biochar made from different natural sources—like sugarcane bagasse, bamboo, and rice straw—can work. Their study specifically looked at how effective these types of biochar are at trapping heavy metals such as lead, cadmium, and copper in polluted soils. The

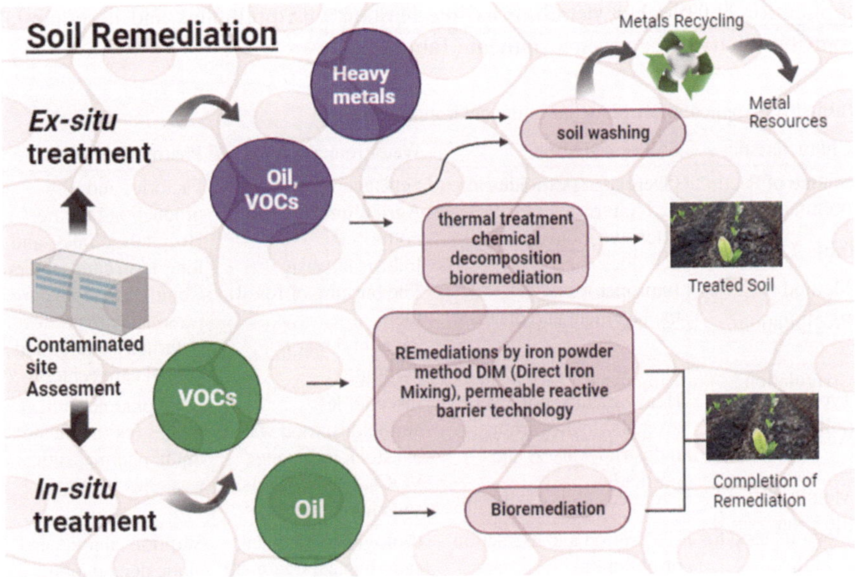

Fig. 9.1 Soil remediation approaches for contaminated sites

findings offer valuable perspectives on how biochar could be put to use in cleaning up contaminated environments. In terms of lead immobilization efficiency, sugarcane bagasse biochar outperformed all other studied materials. Notably, the assessment of bamboo biochar included two distinct particle sizes, applied at rates of 0%, 1%, and 5%. Rice straw biochar underwent analysis under comparable conditions to evaluate its effectiveness. Additionally, garden waste biochar was pyrolyzed at 400 °C (GB1) and 600 °C (GB2) and tested; the BCR method was further utilized to evaluate the immobilization of Cd and Cu, confirming biochar's broad applicability in heavy metal remediation. The study concluded that sugarcane-derived biochar exhibited the highest remediation potential, followed by paulownia, garden waste, bamboo, and rice straw biochar. The critical role of biochar as an effective material for mitigating heavy metal contamination present in soils is highlighted by these findings, paving the way for further research into optimizing its applications for environmental remediation (Bousdra et al. 2023).

Tackling pollution from organic contaminants is an increasingly serious concern, and biochar has become well-regarded as an effective, eco-friendly option for restoring soil health in a sustainable way. A study by Xiang et al. aimed to evaluate the efficacy of biochar produced from both bacterial and plant feedstocks in the degradation and immobilization of organic contaminants. The research explored various mechanisms through which biochar can interact with these pollutants, contributing to enhanced remediation outcomes. The study showed that the free radicals produced by biochar react with molecular oxygen, triggering the formation of hydroxyl radicals (·OH) or activating compounds like peroxydisulfate ($S_2O_8^{2-}$) and hydrogen peroxide (H_2O_2). These interactions result in the generation of various reactive oxygen species, including hydroxyl, sulfate, and superoxide radicals. Additionally, biochar facilitates the immobilization of pollutants through various mechanisms, and its highly porous surface structure enhances biochar's ability to support microbial communities. It also facilitates electron transfer between microbes and pollutants, thereby influencing microbial metabolism and the breakdown of organic contaminants in the soil. In conclusion, the findings underscore the multifaceted role of biochar in sustainable soil remediation by leveraging its adsorption capacity, free radical generation, and microbial interactions. Biochar presented a promising approach for mitigating organic pollution, restoring soil quality, and supporting long-term environmental sustainability (Xiang et al. 2022).

A study by Burachevskaya et al. investigated biochar derived from different agricultural feedstocks, including rice husk, wood, and sunflower husk, and evaluated its elemental composition, structural characteristics, and adsorption potential. The elemental analysis of the biochars revealed carbon (C) content ranging from 70.4% to 77.3%, nitrogen (N) between 2.1% and 2.4%, hydrogen (H) from 3.0% to 4.8%, and oxygen (O) within 7.3% to 11.8%. XRF analysis demonstrated that rice husk biochar had the highest silica (SiO_2) content at 47.52%, while sunflower husk biochar exhibited high levels of phosphorus (P_2O_5) at 20.96% and potassium (K_2O) at 24.33%. Additionally, wood biochar was characterized by a significantly high calcium oxide (CaO) concentration of 26.30%. Structural characterization using X-ray diffraction (XRD) identified prominent peaks in sunflower husk biochar at

$2\theta = 24.40°$, $30.57°$, and $42.67°$, indicating its crystalline nature. FTIR spectroscopy of wood biochar revealed a strong absorption peak in the range of 3100–3600 cm^{-1}, corresponding to hydroxyl (–OH) functional groups. SEM analysis determined the particle size distribution of biochar to be between 1 and 10 μm, highlighting its porous structure. Thermogravimetric analysis (TGA) of sunflower husk biochar indicated significant mass loss between 105–110 °C, initially because of the desorption of moisture from its surface. With its diverse physicochemical properties and strong heavy metal adsorption capabilities, biochar offers a sustainable solution for mitigating soil contamination while promoting resource-efficient waste management (Burachevskaya et al. 2023).

Mazarji et al. assessed the efficiency of biochar and its nanocomposites for the remediation of heavy metal-laden soils. The biochar was synthesized via the pyrolysis of biomass, and this research concentrated on the adsorption and extraction capabilities for multiple hazardous heavy metals, which are known for their persistence and resistance to biodegradation in the environment. The study found that the composite material developed had a surface area of 38.08 m^2/g, an average particle size of about 0.12 μm, and a saturation magnetization of 0.24 emu/mg. It also highlighted that the ability of biochar and its nanocomposites to remove heavy metals depends greatly on factors like soil pH and the overall makeup of the soil. The importance of further research to refine the application of biochar and optimize its effectiveness in sustainable soil decontamination strategies is highlighted by the findings. Overall, the study suggests that biochar could be a valuable tool in addressing heavy metal contamination in soils, but additional investigations are needed to maximize its potential benefits (Mazarji et al. 2021).

Biochar derived from wood waste has been recognized as a viable and sustainable approach for enhancing soil quality and remediation. In their research, the potential was investigated by Papageorgiou et al. employing four robust analytical methodologies: scenario analysis, (Material and Energy Flow Analysis) MEFA, (Life Cycle Assessment) LCA, and (Substance Flow Analysis) SFA. These methods provided a comprehensive evaluation of biochar's efficiency in remediating contaminated soil and its role in energy recovery. The integration of LCA with MEFA and SFA highlighted the environmental benefits of biochar-based remediation over conventional landfilling. The MEFA results showed that fuel consumption was significantly reduced through on-site remediation. It was determined that on-site remediation was the most effective approach. Additionally, SFA demonstrated lower polycyclic aromatic hydrocarbon (PAH) leaching from remediated soil compared to landfilled soil. A comparative analysis of scenarios showed reduced leaching of Cu, Hg, Ni, and Zn from remediated soil, whereas As, Ba, Cd, Cr, Co, Mo, Pb, and V exhibited higher leaching levels. These findings reinforce the potential of biochar as an environmentally friendly alternative for soil remediation, emphasizing the need for further optimization to maximize its efficacy and minimize heavy metal leaching (Papageorgiou et al. 2021) (Fig. 9.2).

Niedzinski et al. explored the pyrolysis process for biochar production, where organic waste was thermally decomposed at various temperatures (200–900 °C).

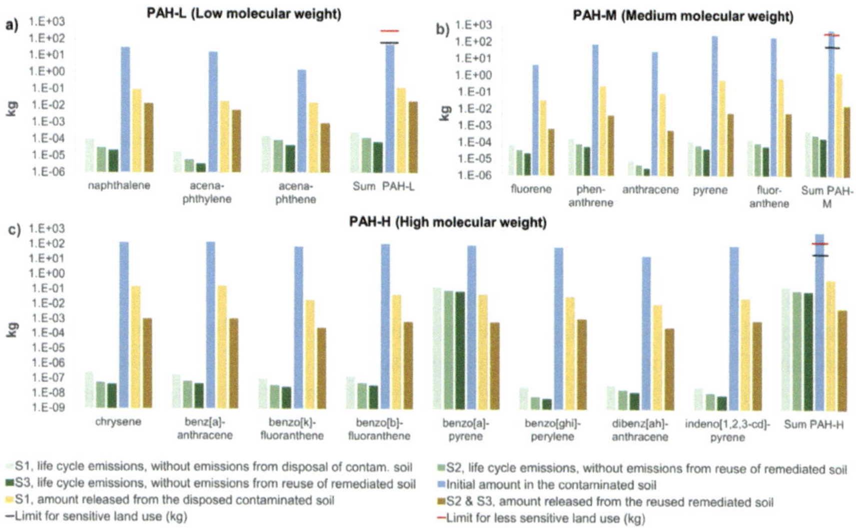

Fig. 9.2 Results of the SFA for PAH (in logarithmic scale). The figure also displays the Swedish guideline values for contaminated soil for sensitive or less sensitive land uses, scaled to the functional unit of the study (i.e., 19,583 kg of contaminated soil with a density of 1600 kg/m^3) (Papageorgiou et al. 2021)

The study analyzed the resulting biochar based on its pH and elemental composition, classifying its effectiveness in soil remediation. The findings revealed that biochar substrates varied in their efficiency, following the order: manure > wood > lignin-cellulosic matter > green plant parts. Manure-derived biochar demonstrated the highest effectiveness due to its rich nutrient composition. The study highlighted biochar's significant sorption capacity, demonstrating its efficacy in immobilizing a range of organic and inorganic contaminants that are present in soil, thus minimizing their bioavailability and potential toxicity. An analysis of its porous structure, especially the activated carbon component, revealed a clearly organized network of pores. These were categorized into three types based on size: micropores smaller than 2 nm, mesopores ranging from 2 to 50 nm, and macropores larger than 50 nm. These findings underscore biochar's potential as a cost-effective agent for soil remediation. Its capability to sequester harmful compounds while enhancing soil quality positions biochar as a promising strategy for promoting sustainable agriculture and environmental stewardship (Niedziński et al. 2023).

Blenis et al. conducted a study on biochar created through the combustion of biomass in an oxygen-restricted environment, focusing on its role in addressing soil pollution. The findings emphasized biochar's capacity to stabilize heavy metals, effectively reducing their solubility and bioavailability, which in turn lessens their negative impact on the environment. Moreover, biochar is adept at eliminating various organic contaminants, such as antibiotics, fuels, oils, pesticides, and persistent organic substances, from soil. The study also highlighted the importance of both pre- and post-treatments in enhancing the selectivity and efficiency of biochar, making it better suited for specific remediation tasks. These treatments enable the modification of feedstocks that might otherwise be deemed unsuitable, thus expanding the variety of waste materials that can be transformed into biochar for environmental use. In conclusion, the research emphasizes the promise of engineered biochar as a flexible and effective approach to combating soil contamination. It presents sustainable advantages not only for environmental remediation but also for agricultural practices, showcasing its potential to contribute significantly to environmental restoration efforts (Blenis et al. 2023).

Biochar, synthesized from a range of biomass feedstocks via pyrolysis under oxygen-limited conditions at temperatures not exceeding 700 °C, has demonstrated considerable efficacy in mitigating environmental pollution linked to organic contaminants. The adsorption capabilities against a range of pollutants, including pesticides, fertilizers, antibiotics, polycyclic aromatic hydrocarbons, polychlorinated biphenyls, and heavy metals, were evaluated by the research conducted by Wang et al. The investigation also examined the influence of biochar on soil health, utilizing biological indicators such as microbial functional activities and community structure to assess its impact. These factors help determine biochar's influence on soil restoration and ecosystem balance. One crucial aspect of biochar's effectiveness in environmental remediation is its stability in the soil. Biochar consists of two forms of carbon: stable carbon (highly resistant to mineralization) and unstable carbon (low resistance to mineralization). The proportion of these components influences the long-term sustainability of biochar's remediation effects. In conclusion, biochar derived from different feedstocks presents a promising solution for soil decontamination by enhancing microbial activity and stabilizing pollutants (Wang et al. 2022).

The efficacy of biochar in removing arsenic, copper, lead, nickel, and cadmium from polluted soils was examined in a study by Golding et al. The research emphasized the distinctive physicochemical properties of biochar, including its high degradation resistance, negatively charged surfaces, and extensive internal surface area. These attributes allow biochar to effectively adsorb contaminants, thereby limiting their uptake into the soil and decreasing pollution levels. Beyond heavy metal removal, biochar also enhances soil microbial communities. The proliferation of beneficial microorganisms, such as mycorrhizae, is promoted, thereby improving nitrogen and phosphorus cycling and increasing nutrient availability. Additionally, soil pH is influenced by biochar, creating a more favorable environment for growth of the plant and soil health restoration. In conclusion, the implementation of biochar is recognized as a sustainable and efficient approach to soil remediation and by immobilizing toxic metals and fostering microbial activity, it mitigates environmental pollution and enhances soil fertility, making it a valuable asset in ecological restoration and sustainable agriculture (Golding 2021).

Luque et al. examined the potential of biochar-based slow-release phosphorus (P) fertilizers for remediating lead (Pb)-contaminated soil. The biochar used was derived from rice husks and analyzed through X-ray fluorescence (XRF), revealing a significant concentration of silicon oxides. This particular composition enhances the biochar's capacity for adsorption, positioning it as a promising option for soil remediation. The porosity of the biochar was further evaluated using Brunauer–Emmett–Teller analysis, which identified two types of pores: micropores (approximately 2 nm) and larger pores. Among the four biochar samples produced, the highest pore volume of 0.354 cm^3/g at the 2 nm size was demonstrated by the one activated at 500 °C, facilitating efficient phosphorus and lead exchange within the soil environment. Additionally, FT-IR was conducted to identify the superficial functional groups responsible for interactions with contaminants and nutrients. Overall, the study demonstrated that soil nutrient availability is improved by biochar-based slow-release P fertilizer, while heavy metals like Pb are effectively immobilized. This dual-function approach establishes biochar as an innovative and sustainable solution for enhancing soil quality and promoting environmental remediation (Paula et al. 2022).

Zahed et al. explored the impact of biochar production conditions and its effectiveness in treating soil pollutants derived from petroleum hydrocarbons. Their findings indicated that lower temperatures and selective, homogeneous heating resulted in high-quality biochar, enhancing its remediation potential. Petroleum hydrocarbons in soil are classified into saturated hydrocarbons (SHC), aromatic hydrocarbons (AHC), and non-hydrocarbon compounds (NHC). The research indicated that SHCs possess comparatively low boiling points, enabling them to slowly volatilize or undergo degradation through photodegradation processes. Conversely, NHCs were identified as highly persistent, with high fusion and boiling points, making them difficult to remove and increasing their environmental toxicity and mutagenicity. The study emphasized the effectiveness of a biochar-microbe composite in breaking down persistent polycyclic aromatic hydrocarbons (PAHs) like phenanthrene, fluoranthene, and pyrene. Molecular analysis revealed that the inclusion of biochar significantly enriched the microbial communities capable of degrading PAHs, thereby further improving soil remediation efforts. Overall, the findings underscored the effectiveness of biochar as an efficient material for addressing the petroleum-related soil contamination. By enhancing microbial activity and selectively adsorbing pollutants, biochar offers a viable strategy for long-term environmental restoration (Zahed et al. 2021).

Tang et al. explored the effectiveness of low-temperature pyrolytic biochar for remediating soil contaminated with sulfonamide-Cu(II). Their research underscored the benefits of biochar aging, especially at elevated temperatures, in improving the adsorption of sulfapyridine (SPY) in amended soils. A detailed analysis of the SPY sorption mechanism indicated that hydrogen bonding primarily drove the adsorption process. The enhancement of SPY adsorption was also attributed to π-π electron donor-acceptor (EDA) interactions and the occupation of micropores, both of which contributed to improving the overall effectiveness of the remediation process. When introduced into the soil, a natural aging process is experienced by biochar that alters its physicochemical characteristics. This transformation enhances its ability to adsorb and immobilize pollutants, effectively separating contaminants from water and soil. It was concluded in the study that sulfonamide-Cu(II) contaminated soils in tropical regions can be effectively and sustainably remediated using low-temperature pyrolytic biochar. Its ability to undergo natural aging while maintaining strong sorption properties makes it a preferred choice for long-term soil restoration and pollution mitigation efforts (Tang et al. 2023).

Palansooriya et al. employed machine learning techniques to forecast the effect of biochar on the immobilization of heavy metals (HMs) in soil, emphasizing the significant influence of pH in this process. The study showed that higher soil pH levels help deprotonate acidic functional groups on the biochar surface, which boosts the adsorption of positively charged heavy metals and improves their immobilization. Interestingly, unlike some previous research, this study found that factors like the surface area of biochar and the pyrolysis temperature didn't significantly influence heavy metal immobilization. However, it was observed that biochar made at higher pyrolysis temperatures typically had a larger surface area, which could enhance its adsorption ability. The research also noted that when pyrolysis temperatures exceeded 450 °C, the number of key functional groups—such as carboxyl, carbonyl, hydroxyl, and methoxyl—on the biochar surface decreased, which might lower its capacity to adsorb contaminants. By integrating machine learning into biochar research, this study provides valuable insights into optimizing biochar properties for more effective soil remediation. These findings pave the way for data-driven strategies to enhance biochar applications in environmental management and pollution control (Palansooriya et al. 2022).

9.2.2 Water Remediation

Water remediation involves the process of cleaning, disinfecting, and restoring areas affected by water damage. This procedure is essential for mitigating the impact of water-related issues. Prompt action is crucial, as untreated water damage can lead to severe consequences, including the proliferation of mold, bacteria, and viruses in damp environments.

The primary step in water remediation is to identify and stop the source of contamination while removing accumulated water. Water extraction is typically carried out using pumps and vacuums to eliminate standing water effectively. Additionally, smaller items can be compressed to remove excess moisture. The extent of water damage determines the appropriate extraction method to be used. By addressing water damage swiftly and employing suitable remediation techniques, the risk of long-term structural damage and health hazards can be significantly minimized, ensuring a safer and healthier environment (Fig. 9.3).

Qui et al. explored the production of biochar as an effective method for eliminating contaminants from both soil and water. This biochar, derived from a range of carbon-rich organic waste feedstocks, plays a key role in promoting sustainable waste management. The pristine biochar exhibited considerable effectiveness in purifying water by adsorbing pollutants efficiently. In a controlled experiment, 3 g/L of biochar was applied at a solution pH of 6.5 and a temperature of 45 °C over 8 h, resulting in removal capacities of 67.41 mg/g for tannic acid and 62.89 mg/g for phenol. Modified biochar produced from organic waste demonstrated effective

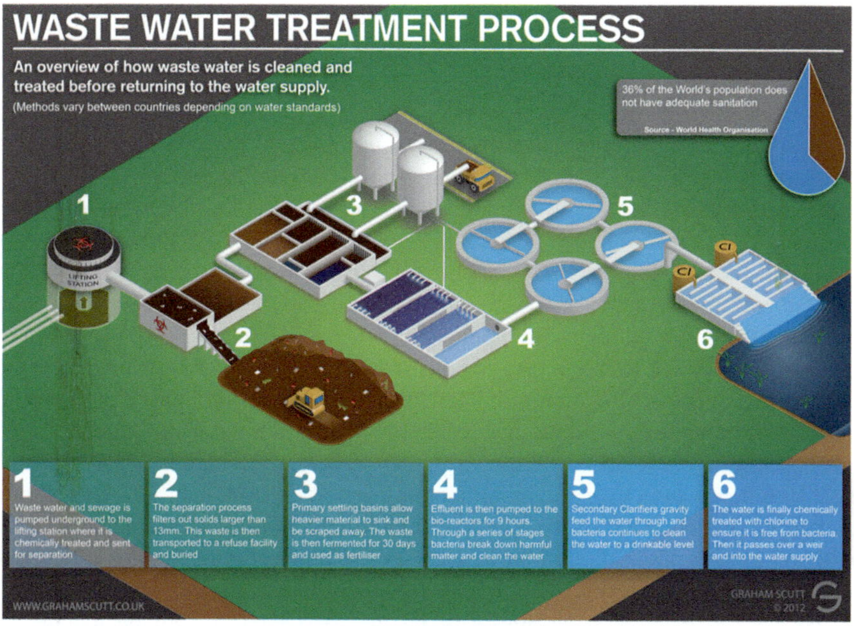

Fig. 9.3 Wastewater treatment process

adsorption of pharmaceutical contaminants such as naproxen, diclofenac, and triclosan from wastewater. The adsorption capacities varied, with the highest removal efficiency recorded for naproxen at 127 mg/g, followed by triclosan at 113 mg/g and diclofenac at 92.7 mg/g, largely due to differences in hydrophobicity. Interestingly, biochar synthesized at lower temperatures exhibited the greatest efficiency in removing contaminants, achieving rates of up to 70%. These results highlight the adaptability of biochar in environmental remediation, providing a sustainable and efficient means of addressing water pollution (Qiu et al. 2022).

Chen et al. investigated the effectiveness of magnetic biochar, sourced from waste wood, in removing arsenic (As), a recognized carcinogen, from wastewater. The study also assessed the improvement of arsenic (As(III)) removal efficiency through the use of $FeCl_3$ and $KMnO_4$, which enhanced the adsorption properties of the modified biochar. The BET analyzer was used to analyze the pore size distribution of average pore sizes 21.7 Å for PB and 46.3 Å for 1FeC and 1Fe2MnC were recorded, confirming their classification as mesoporous materials, defined by pore sizes ranging from 2 to 50 nm. Characteristic peaks near 3400 cm^{-1} were identified through Fourier transform infrared analysis, which correspond to the stretching vibrations of –OH and N–H, indicating the presence of amine functional groups. Additionally, the study assessed the impact of pH variations (ranging from 3.0 to 11.0) on removal efficiency, with experiments conducted at different concentrations. Overall, the effectiveness of modified magnetic biochar as a sustainable and efficient method for arsenic removal was emphasized, presenting a promising solution for wastewater treatment and environmental remediation (Chen et al. 2021).

Stylianou et al. explored the efficacy of biochar, generated from organic waste feedstocks, in removing antibiotic contaminants from water. The research involved naturally drying the biomass feedstocks at approximately 35 °C before converting them into biochar aimed at decontaminating seven different antibiotics. Sorption is a key factor that affects how antibiotics behave and interact in the environment, particularly in terms of their availability and activity. In this study, researchers examined three distinct types of biochar: one made from biosolids, another from cattle manure (referred to as MDB), and the third from used coffee grounds. The evaluation was carried out at two application rates: a low dose (1 g L^{-1}) and a high dose (10 g L^{-1}). The sorption experiments indicated that all types of biochar at the lower dose successfully removed over 70% of specific antibiotics, with some achieving complete removal efficiency for tetracycline (TET), erythromycin (ERY), and clarithromycin (CLA). Additionally, the manure-derived biochar (MDB) exhibited significant effectiveness in eliminating ampicillin (AMP). These findings underscore the potential of biochar as an eco-friendly and sustainable solution for mitigating antibiotic pollution in water bodies, highlighting its role in enhancing water quality and reducing environmental risks (Stylianou et al. 2021).

Zhao et al. examined the ability of biochar derived from bovine manure to remove tetracycline (TC) from wastewater. The study focused on analyzing the physicochemical properties of biochar derived from bovine manure, produced at two pyrolysis temperatures: 500 °C (BC-500) and 700 °C (BC-700). Key aspects of the evaluation included determining the point of zero charge (pHpzc) and identifying the surface functional groups present in each biochar sample. The results demonstrated that BC-500 achieved a TC removal efficiency of 95.31% after 12 h, whereas BC-700 exhibited an impressive removal rate of 99.70% within just 4 h. SEM analysis revealed that BC-500 had a complex interconnected pore structure with a few small pores, likely formed due to the escape of small molecules during pyrolysis. In contrast, BC-700 displayed a more intricate mesh-like structure with a rougher surface, contributing to its superior adsorption capacity. This study underscores the effectiveness of bovine manure-derived biochar as a sustainable and efficient adsorbent for antibiotic removal from wastewater, highlighting its potential for environmental remediation and water purification applications (Zhao et al. 2021).

The use of biochar as a functional material was studied by Gasim et al., further the biochar applications were further evaluated in various advanced treatment processes, these processes were assessed based on their effectiveness in removing organic pollutants, mechanisms of action, tolerance to different water matrices, reusability, and economic feasibility. However, direct comparisons between these methods were challenging due to their distinct mechanisms. The findings revealed that in processes such as photocatalysis, PS activation, hydrogen peroxide (H_2O_2) activation, and O_3 activation, organic pollutants must first undergo adsorption onto biochar before degradation reactions can proceed. Despite its potential, the study noted that biochar exhibited limited durability of active sites during advanced oxidation processes (AOPs), leading to slow pollutant removal and low efficiency.

Additionally, interactions between biochar, pollutants, and oxidants were significantly influenced by variations in water pH, posing a challenge for real wastewater treatment applications. This study highlights biochar's potential and limitations in environmental remediation, emphasizing the need for further modifications to enhance its stability, efficiency, and applicability in practical wastewater treatment scenarios (Gasim et al. 2022).

Praveen et al. explored the synthesis of biochar derived from agricultural biomass for the removal of dyes from contaminated water and for these various raw materials, including rice husks, coconut shells, groundnut shells, wood waste, sunflower seed husks, cauliflower leaves, and orange peels, were utilized in the biochar production process. The biomass was subjected to pyrolysis at different temperatures in an oxygen-free environment, which promoted its transformation into biochar. The study demonstrated that the adsorption efficiency of the biochar was greatly affected by its physicochemical properties and important experimental variables. It was noted that the dye removal rate was initially high, successfully decreasing contamination levels within the first few hours. However, as the biochar dosage increased, the removal rate gradually slowed, indicating a saturation effect. The findings suggest that optimizing biochar properties and operational parameters could enhance its efficiency, making it a promising solution for mitigating water pollution caused by industrial dye contaminants (Praveen et al. 2022).

Osman et al. investigated the generation of biochar for wastewater treatment, specifically focusing on the removal of pollutants, and their findings emphasized the role of biochar as a suitable biofilm habitat for nitrifying bacteria, which significantly enhanced contaminant removal. A 67% reduction in ammonia and a 68% decrease in total suspended particles were demonstrated in the results, highlighting its efficiency. Additionally, biochar enhanced plant growth by optimizing pH to 6.8, leading to a daily increase of 0.0562% in lettuce growth. A related study confirmed the effectiveness of bamboo-derived biochar in improving water quality in tilapia aquaponics by reducing turbidity and suspended particles. Chromium adsorption can be attributed to the electrostatic properties, high porosity, and functional groups of biochar. As a result, removal rates of 98%, 99.2%, and 99.8% were achieved in wastewater, saltwater, and tap water, respectively and further the Biochar's potential as a sustainable and efficient material for wastewater treatment and environmental remediation was highlighted by these findings (Osman et al. 2022).

The potential of biochar-based composites for remediating polluted wastewater was investigated by Issaka et al., with a focus on removing agrochemicals, antibiotics, and heavy metals. The study demonstrated that both pyrolyzed and raw biomass-derived biochar effectively removed these contaminants from water, highlighting its role as a sustainable and efficient adsorbent. Among the various remediation techniques, adsorption using biochar was found to be highly effective due to its numerous advantages, including a simple design, cost efficiency, low energy requirements, etc. However, the study also revealed that the adsorption efficiency of biochar decreased with increasing temperature, potentially limiting its performance under certain environmental conditions. Despite its effectiveness in contaminant removal, the researchers emphasized a critical concern—the disposal of contaminant-loaded biochar. If not managed properly, the used biochar could become an environmental hazard, leading to secondary pollution. Therefore, appropriate disposal or regeneration strategies must be developed to ensure sustainable application, and these findings underscore the potential of biochar-based composites as a viable solution for wastewater treatment while also highlighting the need for responsible management of spent biochar to prevent environmental risks (Issaka et al. 2022).

Ahmed et al. investigated the potential of *Citrullus lanatus* L. seeds in synthesizing a novel $MnFe_2O_4$-biochar for use as an adsorbent in efficiently removing uranium (VI) from wastewater. The study aimed to create an environmentally friendly and cost-effective method for wastewater treatment by utilizing agricultural waste. The adsorption studies showed that the results fit well with the pseudo-second-order kinetic model, with an equilibrium adsorption capacity (qe) of 15.12 mg/g and a strong correlation coefficient (Ahmed et al. 2021) of 0.96. The process also followed the Langmuir isotherm model, revealing a maximum adsorption capacity (qmax) of 27.61 mg/g. The highest adsorption efficiency was observed at 318 K, suggesting that the adsorption performance improves with temperature. Overall, the findings highlight the potential of biochar as an eco-friendly and highly efficient material for removing uranium from the environment. These findings represent a significant advancement toward green and low-cost strategies for wastewater remediation, highlighting the critical role of employing natural resources in pollution control. The study provides valuable insights into employing agricultural waste for environmental remediation, encouraging further exploration of sustainable practices in wastewater treatment (Ahmed et al. 2021) (Fig. 9.4).

A study on porous-structured biochar materials for sustainable water treatment was conducted by Li et al., with an emphasis placed on their potential in removing pollutants from industrial wastewater. However, biochar derived from direct biomass pyrolysis often exhibits low surface functionality, limited active groups, and minimal porosity, ranging between 0.016 and 0.083 cm^3/g, with a surface area generally below 150 m^2/g. The study focused on the treatment of three major industrial wastewater sources: dyeing, pharmaceutical, and mining wastewater. One of the most hazardous xanthene dye pollutants, Rhodamine B, was effectively removed using different biochar sources. Cement waste-derived biochar, through adsorption at a pH of 4.3, achieved a remarkable adsorption capacity of 531.836 mg/g. Additionally, biochar synthesized from *Musa paradisiaca* demonstrated adsorption capacities of 7.003 mg/g and 6.878 mg/g at a pH of 4. This research underscores the significance of biochar in wastewater treatment, particularly in adsorbing toxic industrial pollutants. By optimizing biochar properties, its efficiency in pollutant removal can be significantly enhanced, paving the way for sustainable and eco-friendly water purification technologies (Li et al. 2023).

The research by Al-Nemr et al. explored the potential of biochar-NH_2 as an environmentally friendly adsorbent for the removal of Cu(II) ions from water which was derived from watermelon peels. Differential thermal analysis (DTA) was conducted on both the watermelon peel waste (WWP) and the biochar-NH_2, revealing six distinct thermal peaks at temperatures of 769, 178, 251, 292, 426, and 872 °C. Notably, two significant degradation peaks were observed at 767 and 477 °C. The

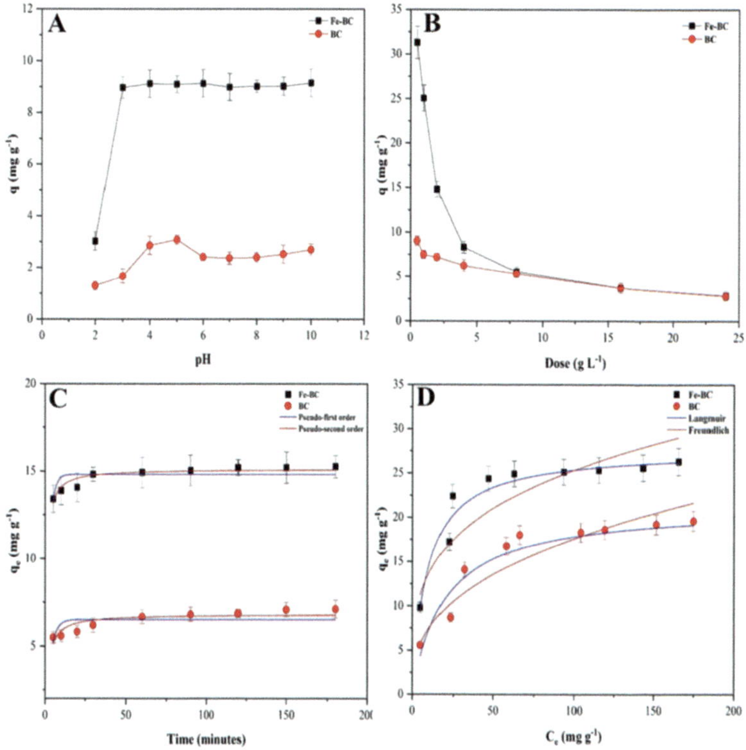

Fig. 9.4 U(VI) adsorption isotherms showing the influence of experimental parameters for BC and MF-BC: (**a**) impact of solution pH, where C0 = 30 mg L^{-1}, m/v = 1 g L^{-1}, t = 120 min and T = 25 °C; (**b**) effect of different sorbent doses, where C0 = 30 mg L^{-1}, pH = 4.0, t = 120 min and T = 25 °C; (**c**) effect of different contact times, where C0 = 30 mg L^{-1}, pH = 4.0, m/v = 1 g L^{-1}, t = 120 min and T = 25 °C; (**d**) effect of uranium concentration at equilibrium, where pH = 4.0, m/v = 1 g L^{-1}, t = 120 min, and T = 25 °C (Ahmed et al. 2021)

BET-specific surface area of the biochar measured 12.26 m^2/g, and its total pore volume was 0.012 cm^3/g, indicating its porous structure. The study performed determined the sorption capacity of the biochar using the Langmuir isotherm model, estimating it to be 140.85 mg/g, which highlights a strong affinity for Cu(II) ions. This investigation illustrates the effectiveness of biochar derived from watermelon peels as a cost-effective and sustainable option for the removal of heavy metals. The notable adsorption capacity underlines its potential as a viable method for addressing water pollution and enhancing green remediation practices. Through this research, watermelon-peel-derived biochar is positioned as a promising solution for effective heavy metal removal, contributing to the advancement of environmentally friendly water treatment strategies (El-Nemr et al. 2022).

The transformation of sewage sludge into biochar for effective wastewater remediation was explored by Gopinath et al. The study emphasized thermochemical techniques, particularly hydrothermal carbonization and pyrolysis, as the primary methods for biochar synthesis. The study highlighted the various applications in wastewater treatment, emphasizing the role of biochar as an effective adsorbent for removing a range of pollutants, including emerging contaminants, dyes, heavy metals, etc., from both synthetic and real wastewater. In addition to its adsorption capabilities, researchers also identified biochar as a catalyst and a support medium in advanced oxidation processes, which further improves pollutant degradation. Furthermore, they investigated the potential use of biochar as an electrode material, paving the way for its application in electrochemical treatment methods. This research emphasizes the importance of transforming sewage sludge into biochar. The multifunctionality of biochar not only enhances pollutant removal but also contributes to environmental sustainability by repurposing waste into valuable remediation materials (Gopinath et al. 2021).

Visiy et al. examined how applying biochar in domestic wastewater treatment influenced the growth of lettuce (*Lactuca sativa*). The biochar, derived from corn cobs, underwent significant changes in its physicochemical properties after wastewater filtration. A notable increase in bulk density was observed, rising from 0.243 g/cm^3 to 0.381 g/cm^3, indicating a reduction in total porosity. This decrease in porosity contributed to a lower bulking density of less than 0.38 g/cm^3. The study further revealed that wastewater filtration resulted in a decline in the ash content of the biochar, leading to a reduction in pH from 7.8 to 5. Additionally, there was a marked decrease in exchangeable cations, with calcium, potassium, magnesium, and sodium concentrations dropping by 84.3%, 74.8%, 43.3%, and 23.4%, respectively. The total phosphorus concentration also declined by 51.4%. This study underscores the transformative effects of biochar in wastewater treatment and highlights the importance of evaluating its impact on soil nutrient dynamics for sustainable agricultural applications (Visiy et al. 2022).

9.2.3 Pharmaceutical

The environment poses significant risks to human and ecological health by the discharge of pharmaceutical waste, necessitating the development of advanced remediation strategies. Traditional removal techniques are often non-selective and may lead to secondary contamination, highlighting the need for more efficient alternatives. Among the emerging solutions, promising adsorbents have gained attention by molecularly imprinted polymers (MIPs).

Various multi-step polymerization techniques are used to synthesize MIPs, which have demonstrated strong adsorption capabilities at the laboratory scale. The adsorption efficiency of multi-templated MIPs is influenced by factors such as the design and operational capacity of wastewater treatment plants, pharmaceutical consumption trends, and others (Fig. 9.5).

The increasing presence of antibiotics in the environment necessitates efficient remediation strategies, and biochar-assisted graphene-based materials have emerged as promising solutions. Ashraf et al. explored the synthesis and application of these

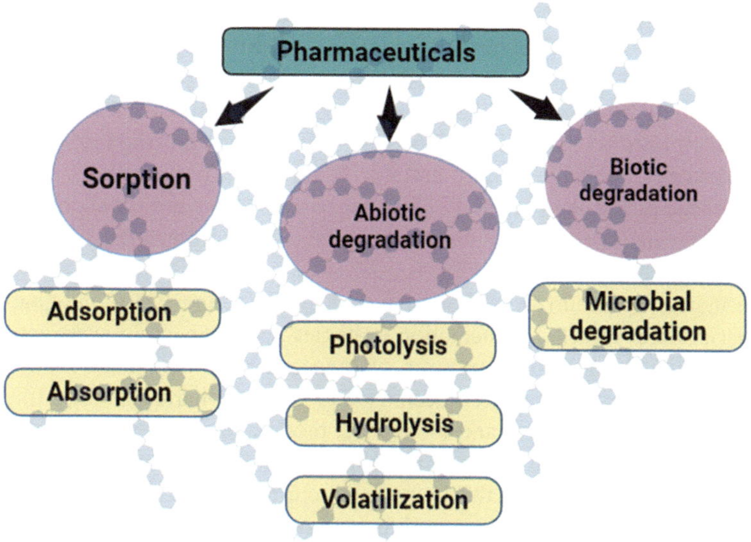

Fig. 9.5 Environmental fate of pharmaceuticals

advanced materials for antibiotic removal, leveraging the unique properties of graphene and biochar composites. Natural graphite serves as a precursor for large-scale production of transparent graphene thin films with high conductivity. The incorporation of graphite carbon enhances photocatalytic activity by improving pollutant adsorption on photocatalyst surfaces. Multiple mechanisms, such as π–π electron donor-acceptor interactions, hydrophobic forces, hydrogen bonding, and covalent-π interactions, facilitate antibiotic adsorption, all of which enhance the overall effectiveness of the process. The combination of biochar and graphene leads to structural and morphological changes that promote non-covalent linkages via various interaction mechanisms. These synergistic effects enhance the overall efficiency of antibiotic removal. The potential of biochar-assisted graphene materials in wastewater treatment is highlighted in this study, presenting an eco-friendly and efficient solution for mitigating antibiotic contamination and minimizing environmental hazards (Ashraf et al. 2022).

The widespread contamination of water bodies with antibiotic pollutants necessitates efficient remediation strategies. Katiyar et al. investigated innovative biochar for the effective removal of antibiotics from aqueous environments. Since antibiotics possess dissociation constants (pKa) within the pH of 1.5–9.5, their charge and solubility vary according to environmental conditions. The study emphasized the crucial role of pH in influencing the adsorption efficiency of biochar, which researchers produced at temperatures ranging from 300 to 900 °C in an oxygen-limited environment. Various biochar dosages (1–10 g/L) were tested for removing antibiotics such as tetracycline, trimethoprim, erythromycin, clarithromycin, ampicillin, ofloxacin, and sulfamethoxazole. The efficiency of removal was significantly impacted by pH, with optimal performance observed between pH 5 and 11. Furthermore, an innovative approach combining waste-fungus-chaff-biochar with *Herbaspirillum huttiense* was explored to purify antibiotic pollution while simultaneously binding heavy metals like copper and zinc. The combined material successfully immobilized 85.5% of Cu and 64.4% of Zn while achieving removal rates of 41.9% for oxytetracycline (OTC) and 40.7% for enrofloxacin (ENR). This study highlights the promising potential of biochar-based approaches for antibiotic remediation, offering a sustainable solution to mitigate environmental contamination (Katiyar et al. 2022).

The increasing occurrence of pharmaceutical compounds (PCs) in aquatic systems necessitates effective remediation strategies. Chauhan et al. investigated how biochar-mediated adsorption removes pharmaceutical contaminants from water. While activated carbon has often been used to address pharmaceuticals like ketoprofen, ibuprofen, naproxen, and diclofenac, biochar has emerged as a promising alternative. The adsorption mechanism primarily involves π–π interactions between the π-electrons of biochar and the aromatic structures of these pharmaceuticals. The resulting pristine biochar, underscoring its potential for wastewater treatment, exhibited adsorption capacities of 7.72 mg/g and 4.99 mg/g, respectively. However, the study raised concerns regarding the stability of biochar during prolonged use in wastewater systems. The instability in its aromatic condensation degree may lead to carbon leaching, thereby increasing the carbon content in treated effluent. Despite these challenges, biochar remains a promising sustainable solution for pharmaceutical wastewater treatment. Further optimization in biochar modification could enhance its stability and long-term efficacy in environmental remediation (Chauhan et al. 2022).

The increasing concern over pharmaceutical contaminants in water systems has driven the development of innovative adsorption techniques using biochar. A multi-component mixture of active pharmaceutical compounds was examined by Puga et al. using biochars derived from grapevine cane, holm tree, and eucalyptus pruning residues. It was demonstrated by the study that venlafaxine (VLX) and trazodone (TRZ) were effectively removed from aqueous solutions by these biochars. Kinetic analysis indicated that the adsorption of FLX and TRZ followed a pseudo-first-order kinetic model, with a rate constant (k1) of 4.23 min^{-1} for both. The experimentally determined uptake capacities aligned with equilibrium values predicted by the kinetic models, confirming the reliability of the adsorption process. The occurrence of organic and inorganic impurities in the system may gradually lead to a reduction in column efficiency. The potential of agroforestry-derived biochar as a cost-effective for the removal of pharmaceuticals from wastewater is underscored by this study, with further optimization of biochar characteristics and adsorption conditions enhancing its applications in environmental remediation (Puga et al. 2022) (Fig. 9.6).

Badiger et al. explored how biochar is applied in sulfate radical-based advanced oxidation processes to remove pharmaceutical contaminants. These pollutants, which primarily enter water systems through hospital waste and human excretion, pose significant risks to both aquatic ecosystems and human health. The study

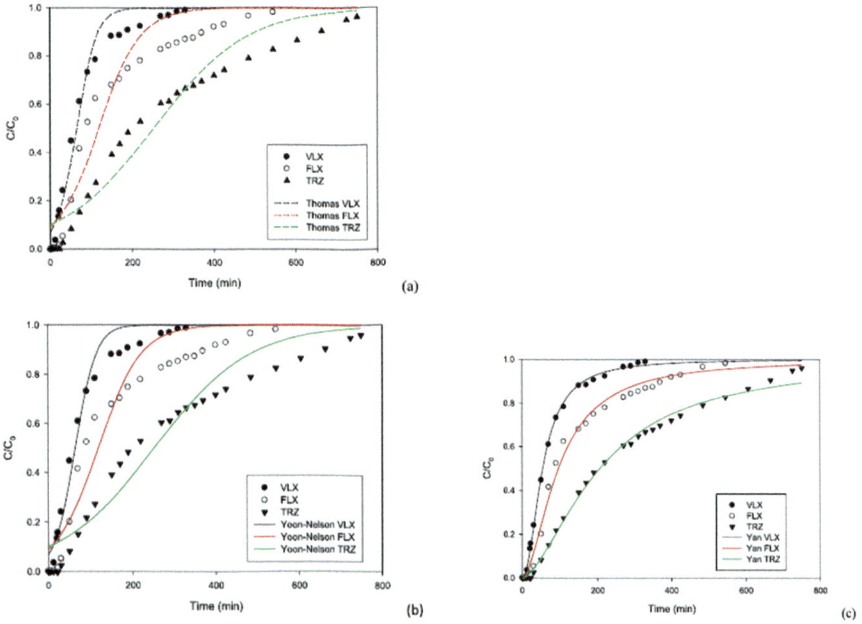

Fig. 9.6 Breakthrough curves for the adsorption of the pharmaceutical mixture in spiked synthetic WWTP. The lines represent the fitting to the models: (**a**) Thomas, (**b**) Yoon-Nelson, and (**c**) Yan (Puga et al. 2022)

underscored the growing importance of functional biochars as sustainable materials for pollutant removal, noting that various modification techniques, such as acid-base treatment and metal loading, significantly enhance their surface properties. A notable case involved treating bone-derived biochar with H_2SO_4, which increased its surface area by 83% due to the removal of inorganic components like $CaCO_3$ and hydroxyapatite, creating additional active adsorption sites. Engineered biochars have shown high efficiency in eliminating pharmaceuticals from aqueous solutions, establishing them as promising candidates for wastewater treatment. This study underscored the potential of biochar-based SR-AOPs as a sustainable and effective approach to mitigating pharmaceutical pollution, contributing to the advancement of environmentally friendly water purification technologies (Badiger and Nidheesh 2023).

The growing concern over pharmaceutical contaminants in water has led to the exploration of advanced adsorbents for effective remediation by researchers. The potential of N-doped and activated porous biochar, derived from cocoa shell, was investigated by Zanli et al. for the removal of norfloxacin (NOR) from aqueous solutions. Various bio-based adsorbents, including wood, sawdust, bamboo, coconut, and reed stalks, have been utilized for eliminating pharmaceutical and personal care products such as norfloxacin, acetaminophen, and tetracycline. However, the study found that conventional biochar exhibited limited efficiency in NOR removal.

To address this limitation, modifications were made to enhance the surface functionality and porosity of the pristine biochar. A novel one-step synthesis approach was introduced, incorporating nitrogen doping with urea and activation using $NaHCO_3$. Furthermore, the potential of engineered biochar as an efficient and excellent adsorbent for pharmaceutical contaminants was noted in the study, contributing to the development of environmentally friendly wastewater treatment strategies (Guy Laurent Zanli et al. 2022).

As antibiotic and antibiotic resistance gene levels rise in agricultural waste, developing effective mitigation strategies becomes critical. Tong et al. investigated how biochar influences antibiotic degradation and variations in resistance genes during the co-composting of the pig manure and corn straw. The composting experiments were carried out outdoors with two treatment groups: a control group (CK) and a biochar-amended group (CK + Biochar). During the initial 14 days, the removal efficiency of tylosin was significantly enhanced in the biochar-amended compost, with a reduction rate of 94.44% compared to 77.08% in the control. However, concentrations of danofloxacin and enrofloxacin were increased in both treatments, with values in the biochar treatment being elevated from 23.49 to 32.32 μg/kg and from 68.06 to 146.32 μg/kg, respectively. It was found that microbial activity was a crucial factor influencing changes in antibiotic concentrations. By the end of the composting period, the removal rates of sulfadiazine, sulfamethazine, ofloxacin, ciprofloxacin, difloxacin, and lincomycin were higher in the biochar-treated group (68.32%–87.24%) compared to the control (38.21%–77.10%). This study highlighted the efficacy of biochar in enhancing antibiotic degradation during composting, offering a sustainable approach to mitigating antibiotic contamination in agricultural waste management (Tong et al. 2022).

In another study, Curiel et al. investigated thermochemical waste conversion, biochar adsorption, and electro-assisted regeneration as sustainable strategies for removing pharmaceuticals. The study included the characterization of biochar produced from rice bran under various conditions to evaluate its adsorption potential. BET surface area analysis revealed that the surface area increased when the pyrolysis temperature was raised from 300 °C to 500 °C accordingly, with notable differences being observed between BC500, BC750, and their autoclave-treated counterparts (BC500A, BC750A). To understand the adsorption mechanism, isotherm and kinetic studies were conducted on BC500A, which was found to exhibit a slightly acidic nature (pHPZC 5.2). The study also examined the behavior of fluoxetine (FLX), which has a pKa of 9.8, meaning its amine functional group is ionized below this value, resulting in a positively charged species in solution. Given that the experimental pH was maintained between 5.2 and 9.8, electrostatic interactions played a key role in adsorption. The potential of biochar as an efficient and sustainable adsorbent is highlighted by these findings, paving the way for environmentally responsible pharmaceutical waste remediation (Escudero-Curiel et al. 2021).

The increasing presence of antibiotic pollutants in wastewater necessitates effective and sustainable removal strategies. Ajala et al. investigated the potential of biomass-derived adsorbents for eliminating antibiotics, focusing on amoxicillin removal using palm bark-based biochar. Biomass-based adsorbents, sourced from agricultural residues, contain functional groups, which enhance their ability to capture emerging contaminants, including pharmaceuticals. In their study, adsorption experiments were conducted with varying conditions: adsorbent dosages ranging from 0.5 to 5 g/L, initial amoxicillin concentrations between 10 and 100 mg/L, and contact times spanning 10 to 180 min. The experiment achieved a 98.1% removal efficiency for amoxicillin within 90 min using a 3 g/L adsorbent dose at an initial antibiotic concentration of 10 mg/L at 25 °C. Additionally, researchers observed that increasing the adsorbent dose from 0.5 to 3.0 g/L significantly improved the removal efficiency, and these findings emphasize the potential of biomass-derived adsorbents as a sustainable and cost-effective solution for wastewater treatment. By harnessing agricultural residues, this approach offers a sustainable method for mitigating pharmaceutical contamination in aquatic environments (Aderemi Ajala et al. 2023).

As pharmaceutical pollutants in water bodies are becoming a growing concern, the effectiveness of biochar in their removal through adsorption and persulfate-based advanced oxidation processes (AOPs) was investigated by Kang et al. The critical role of free radical pathways in degrading pharmaceutical contaminants was emphasized by the study, particularly regarding the removal of tetracycline, which was facilitated by persulfate-based AOPs through external energy excitation and heterogeneous catalytic processes. It was demonstrated that sulfamethoxazole was effectively removed by biochar, primarily through adsorption, with mechanisms, hydrogen bonding, and hydrophobic effects being leveraged. Notably, SMX exhibited poor solubility at higher pH levels, enhancing its adsorption on the hydrophobic biochar surface. Additionally, other antibiotics, including acetaminophen, cephalexin, and levofloxacin, were evaluated through a comparative analysis of adsorption versus persulfate-based AOPs. The findings indicated that increasing the dosage of tetracycline significantly improved pollutant removal, with the rate rising from 21% to 49%. Moreover, the catalytic degradation efficiency was enhanced from 75% to 88%. This study underscores biochar's potential as a versatile and sustainable material for pharmaceutical wastewater treatment. Its dual functionality—adsorption and catalytic degradation—offers a promising approach to mitigating pharmaceutical contamination and improving water quality (Kang et al. 2022).

9.3 Conclusion

Biochar is increasingly being recognized as a sustainable and effective tool for environmental remediation, with significant potential demonstrated in soil restoration, water purification, and pharmaceutical waste management. Its distinct physicochemical properties have contributed to its strong adsorption capabilities. Various pollutants are effectively captured by biochar, while improvements in soil health and microbial activity have made it suitable for large-scale environmental applications.

In the field of water treatment, biochar has been utilized for the removal of pollutants from industrial wastewater, agricultural runoff, and municipal effluents. Its adsorption efficiency is influenced by several factors and continuous research is being conducted to optimize its performance in different environmental settings. Additionally, the increasing prevalence of pharmaceutical contaminants has necessitated the exploration of biochar-based approaches in wastewater treatment, with studies demonstrating its effectiveness in removing antibiotics, analgesics, and hormones from water sources.

Beyond soil and water remediation, biochar has been employed in air pollution control and carbon sequestration; however, large-scale implementation has been hindered by challenges such as production costs, variability in adsorption performance, and scalability. Efforts are being made to enhance biochar's efficiency through modifications and innovative applications, further strengthening its role as an environmentally friendly solution for pollution control.

Overall, biochar has been established as a promising strategy for environmental management, aligning with global sustainability efforts and providing an economical approach to pollution mitigation across various sectors.

References

Aderemi Ajala O, Oluwaseun Akinnawo S, Bamisaye A, Tunrayo Adedipe D, Oluwabunmi Adesina M, Abiodun Okon-Akan O, Adewumi Adebusuyi T, Titi Ojedokun A, Adesina Adegoke K, Solomon Bello O, Olugbenga SB, Solomon Bello is Professor of Physical O (2023) Adsorptive removal of antibiotic pollutants from wastewater using biomass/biochar-based adsorbents. RSC Adv 13(7):4678–4712

Ahmed W, Mehmood S, Núñez-Delgado A, Ali S, Qaswar M, Khan ZH, Ying H, Chen DY (2021) Utilization of Citrullus lanatus L. seeds to synthesize a novel MnFe2O4-biochar adsorbent for the removal of U(VI) from wastewater: insights and comparison between modified and raw biochar. Sci Total Environ 771. https://doi.org/10.1016/j.scitotenv.2021.144955

Alkharabsheh, HM, Seleiman, MF, Battaglia, ML, Shami, A, Jalal, RS, Alhammad, BA, Almutairi, KF, & Al-Saif, AM (2021) Biochar and its broad impacts in soil quality and fertility, nutrient leaching and crop productivity: A review. In Agronomy 11(5). https://doi.org/10.3390/agronomy11050993

Ashraf A, Liu G, Arif M, Mian MM, Rashid A, Yousaf B, Khawar MI, Riaz L, Safeer R (2022) Insights into the synthesis and application of biochar-assisted graphene-based materials in antibiotic remediation. J Clean Prod 361:132211. Elsevier Ltd. https://doi.org/10.1016/j.jclepro.2022.132211

Badiger SM, Nidheesh PV (2023) Applications of biochar in sulfate radical-based advanced oxidation processes for the removal of pharmaceuticals and personal care products. Water Sci Technol. https://doi.org/10.2166/wst.2023.069

Blenis N, Hue N, Maaz TMC, Kantar M (2023) Biochar production, modification, and its uses in soil remediation: a review. Sustainability 15(4). MDPI. https://doi.org/10.3390/su15043442

Bousdra T, Papadimou SG, Golia EE (2023) The use of biochar in the remediation of Pb, Cd, and Cu-contaminated soils. The impact of biochar feedstock and preparation conditions on its remediation capacity. Land 12(2). https://doi.org/10.3390/land12020383

Burachevskaya M, Minkina T, Bauer T, Lobzenko I, Fedorenko A, Mazarji M, Sushkova S, Mandzhieva S, Nazarenko A, Butova V, Wong MH, Rajput VD (2023) Fabrication of bio-

char derived from different types of feedstocks as an efficient adsorbent for soil heavy metal removal. Sci Rep 13(1). https://doi.org/10.1038/s41598-023-27638-9
Chauhan S, Shafi T, Dubey BK, Chowdhury S (2022) Biochar-mediated removal of pharmaceutical compounds from aqueous matrices via adsorption. Waste Dispos Sustain Energy. Springer. https://doi.org/10.1007/s42768-022-00118-y
Chen CK, Chen JJ, Nguyen NT, Le TT, Nguyen NC, Chang CT (2021) Specifically designed magnetic biochar from waste wood for arsenic removal. Sustain Environ Res 31(1). https://doi.org/10.1186/s42834-021-00100-z
El-Nemr, MA, Aigbe, UO, Hassaan, MA, Ukhurebor, KE, Ragab, S, Onyancha, RB, Osibote, OA, & El Nemr, A (2022) The use of biochar-NH2 produced from watermelon peels as a natural adsorbent for the removal of Cu(II) ion from water. Biomass Conversion and Biorefinery. https://doi.org/10.1007/s13399-022-02327-1
Escudero-Curiel S, Acevedo-García V, Sanromán MÁ, Pazos M (2021) Eco-approach for pharmaceutical removal: thermochemical waste valorization, biochar adsorption, and electro-assisted regeneration. Electrochim Acta 389. https://doi.org/10.1016/j.electacta.2021.138694
Gasim MF, Choong ZY, Koo PL, Low SC, Abdurahman MH, Ho YC, Mohamad M, Suryawan IWK, Lim JW, Oh WD (2022) Application of biochar as functional material for remediation of organic pollutants in water: an overview. Catalysts 12(2). MDPI. https://doi.org/10.3390/catal12020210
Golding S (2021) The application of biochar can significantly improve the health and quality of soil, meaning it can play a crucial part in mitigating the environmental challenges of climate change, global warming, sustainable farming, pollution, resulting habitat destruction, disruption of the natural balance of ecosystems, and ecological imbalance. https://www.azolifesciences
Gopinath A, Divyapriya G, Srivastava V, Laiju AR, Nidheesh PV, Kumar MS (2021) Conversion of sewage sludge into biochar: a potential resource in water and wastewater treatment. Environ Res 194. Academic Press Inc. https://doi.org/10.1016/j.envres.2020.110656
Guy Laurent Zanli, BL, Tang, W, & Chen, J (2022) N-doped and activated porous biochar derived from cocoa shell for removing norfloxacin from aqueous solution: Performance assessment and mechanism insight. Environmental Research, 214. https://doi.org/10.1016/j.envres.2022.113951
Issaka E, Fapohunda FO, Amu-Darko JNO, Yeboah L, Yakubu S, Varjani S, Ali N, Bilal M (2022) Biochar-based composites for remediation of polluted wastewater and soil environments: challenges and prospects. Chemosphere 297. https://doi.org/10.1016/j.chemosphere.2022.134163
Kang Z, Jia X, Zhang Y, Kang X, Ge M, Liu D, Wang C, He Z (2022) A review on application of biochar in the removal of pharmaceutical pollutants through adsorption and persulfate-based AOPs. Sustainability 14(16). MDPI. https://doi.org/10.3390/su141610128
Katiyar R, Chen CW, Singhania RR, Tsai ML, Saratale GD, Pandey A, Di Dong C, Patel AK (2022) Efficient remediation of antibiotic pollutants from the environment by innovative biochar: current updates and prospects. Bioengineered 13(6):14730–14748. Taylor and Francis Ltd. https://doi.org/10.1080/21655979.2022.2108564
Keerthana, KB, Wu, SW, Wu, ME, & Kokulnathan, T. (2023) The United States Energy Consumption and Carbon Dioxide Emissions: A Comprehensive Forecast Using a Regression Model. Sustainability, 15(10). https://doi.org/10.3390/su15107932
Li R, Wu Y, Lou X, Li H, Cheng J, Shen B, Qin L (2023) Porous biochar materials for sustainable water treatment: synthesis, modification, and application. Water 15(3). MDPI. https://doi.org/10.3390/w15030395
Mazarji M, Bayero MT, Minkina T, Sushkova S, Mandzhieva S, Tereshchenko A, Timofeeva A, Bauer T, Burachevskaya M, Kızılkaya R, Gülser C, Keswani C (2021) Realizing United Nations sustainable development goals for greener remediation of heavy metals-contaminated soils by biochar: emerging trends and future directions. Sustainability 13(24). MDPI. https://doi.org/10.3390/su132413825
Niedziński T, Łabętowicz J, Stępień W, Pęczek T (2023) Analysis of the use of biochar from organic waste pyrolysis in agriculture and environmental protection. J Ecol Eng 24(4):85–98. https://doi.org/10.12911/22998993/159347

Osman AI, Fawzy S, Farghali M, El-Azazy M, Elgarahy AM, Fahim RA, Maksoud MIAA, Ajlan AA, Yousry M, Saleem Y, Rooney DW (2022) Biochar for agronomy, animal farming, anaerobic digestion, composting, water treatment, soil remediation, construction, energy storage, and carbon sequestration: a review. Environ Chem Lett 20(4):2385–2485. Springer Science and Business Media Deutschland GmbH. https://doi.org/10.1007/s10311-022-01424-x

Palansooriya KN, Li J, Dissanayake PD, Suvarna M, Li L, Yuan X, Sarkar B, Tsang DCW, Rinklebe J, Wang X, Ok YS (2022) Prediction of soil heavy metal immobilization by biochar using machine learning. Environ Sci Technol 56(7):4187–4198. https://doi.org/10.1021/acs.est.1c08302

Papageorgiou A, Azzi ES, Enell A, Sundberg C (2021) Biochar produced from wood waste for soil remediation in Sweden: carbon sequestration and other environmental impacts. Sci Total Environ 776. https://doi.org/10.1016/j.scitotenv.2021.145953

Paula M, Luque A, López JE, Saldarriaga JF (2022) Remediation of Pb-contaminated soil using biochar-based slow-release P fertilizer and biomonitoring employing bioindicators. https://doi.org/10.21203/rs.3.Rs-2289314/v1

Praveen S, Jegan J, Bhagavathi Pushpa T, Gokulan R, Bulgariu L (2022) Biochar for removal of dyes in contaminated water: an overview. Biochar 4(1). Springer. https://doi.org/10.1007/s42773-022-00131-8

Puga A, Moreira MM, Pazos M, Figueiredo SA, Sanromán MÁ, Delerue-Matos C, Rosales E (2022) Continuous adsorption studies of pharmaceuticals in multicomponent mixtures by agroforestry biochar. J Environ Chem Eng 10(1). https://doi.org/10.1016/j.jece.2021.106977

Qiu M, Liu L, Ling Q, Cai Y, Yu S, Wang S, Fu D, Hu B, Wang X (2022) Biochar for the removal of contaminants from soil and water: a review. Biochar 4(1). Springer. https://doi.org/10.1007/s42773-022-00146-1

Stylianou M, Christou A, Michael C, Agapiou A, Papanastasiou P, Fatta-Kassinos D (2021) Adsorption and removal of seven antibiotic compounds present in water with the use of biochar derived from the pyrolysis of organic waste feedstocks. J Environ Chem Eng 9(5):105868. https://doi.org/10.1016/j.jece.2021.105868

Tang W, Zanli BLGL, Jing F, Hu T, Chen J (2023) Low-temperature pyrolytic biochar is a preferred choice for sulfonamide-cu(II) contaminated soil remediation in tropical climate regions. Sci Total Environ 876:162792. https://doi.org/10.1016/j.scitotenv.2023.162792

Tisserant, A, & Cherubini, F (2019) Potentials, limitations, co-benefits, and trade-offs of biochar applications to soils for climate change mitigation. In Land 8(12) https://doi.org/10.3390/LAND8120179

Tong Z, Liu F, Tian Y, Zhang J, Liu H, Duan J, Bi W, Qin J, Xu S (2022) Effect of biochar on antibiotics and antibiotic resistance genes variations during co-composting of pig manure and corn straw. Front Bioeng Biotechnol 10. https://doi.org/10.3389/fbioe.2022.960476

Visiy EB, Djousse BMK, Sangodoyin A, Gbadegesin AS, Fonkou T (2022) Impact of the application of biochar previously used in domestic wastewater treatment on the growth of lettuce (Lactuca sativa). J Appl Nat Sci 14(3):963–970. https://doi.org/10.31018/jans.v14i3.3731

Wang Y, Gu X, Huang Y, Ding Z, Chen Y, Hu X (2022) Insight into biomass feedstock on formation of biochar-bound environmentally persistent free radicals under different pyrolysis temperatures. RSC Adv 12(30):19318–19326. https://doi.org/10.1039/d2ra03052g

Xiang L, Harindintwali JD, Wang F, Redmile-Gordon M, Chang SX, Fu Y, He C, Muhoza B, Brahushi F, Bolan N, Jiang X, Ok YS, Rinklebe J, Schaeffer A, Zhu YG, Tiedje JM, Xing B (2022) Integrating biochar, bacteria, and plants for sustainable remediation of soils contaminated with organic pollutants. Environ Sci Technol 56(23):16546–16566. American Chemical Society. https://doi.org/10.1021/acs.est.2c02976

Zahed MA, Salehi S, Madadi R, Hejabi F (2021) Biochar is a sustainable product for the remediation of petroleum-contaminated soil. Curr Res Green Sustain Chem 4. Elsevier B.V. https://doi.org/10.1016/j.crgsc.2021.100055

Zhao J, Gao F, Sun Y, Fang W, Li X, Dai Y (2021) New use for biochar derived from bovine manure for tetracycline removal. J Environ Chem Eng 9(4). https://doi.org/10.1016/j.jece.2021.105585

Chapter 10
Effect of Algae-Based Nanoparticles and Implications in Remediation

Njabulo S. Mdluli, Kgomotso G. Mabena, Mxolisi J. Kiwanuka, and Nomvano Mketo

Abstract Algae-based nanoparticles fall under the green synthesized biogenic nanoparticles. These nanomaterials are regarded as the best as they make use of green chemistry materials for their synthesis. However, the supermagnetic nanoparticles derived from algae have shown several limitations as they have adverse effects on microbes, plants, and animal life. The improper disposal of industrial waste and pollutants, including algae nanoparticles, can contaminate soil and groundwater. In this process, iron nanomaterials can migrate between different environments. Notably, biogenic nanoparticles, such as nano zero-valent iron, are highly reactive, posing potential environmental risks. This therefore poses a major toxic impact on microorganisms and directly and indirectly affects the environment. These nanoparticles especially the iron oxide derived from algae have the potential to disturb the growth of aquatic plants, animals, and algae species like *Chlorella vulgaris* which are key for aquatic life. It is therefore paramount to study the implications of algal nanoparticles in remediation and come up with solutions on how these nanoparticles can be made to have minimal side effects on the environment.

Keywords Algae · Nanoparticles · Remediation · Environment · Biogenic

10.1 Introduction

The Protista kingdom contains a group of primarily aquatic animals known as algae. Algae exhibit diverse life cycles, spanning from microscopic Micromonas to gigantic kelps reaching 60 m in length. Unlike plants and animals, algae boast a broader range of photosynthetic pigments (Yang et al. 2022). Beyond their ecological

N. S. Mdluli · K. G. Mabena · M. J. Kiwanuka · N. Mketo (✉)
Department of Chemistry, College of Science and Engineering and Technology, Florida Science Campus, University of South Africa, Roodepoort, Johannesburg, South Africa
e-mail: mketon@unisa.ac.za

A. Guldhe et al. (eds.), *Biomaterials for Environmental Applications: Detection and Remediation*, Environmental Challenges and Solutions,
https://doi.org/10.1007/978-3-032-12701-3_10

significance as oxygen producers and aquatic life sustenance, algae contribute substantially to human economies as sources of crude oil, food, and various medical and industrial products (Castillo-Henríquez et al. 2020). Algae have drawn a lot of attention since it has recently been employed in the environmentally friendly synthesis of several metal nanoparticles, including ZnO, Fe_3O_4, and Ag, which are used in the removal of both organic and inorganic pollutants in the environment (Satapathy et al. 2017). Biological synthesis by microorganisms holds enormous potential because it is eco-friendly and cost-effective, avoids toxicants and harsh chemicals, and requires less energy than physicochemical NPs production (Ahmed et al. 2017). Scientists have recently become interested in promoting the interaction of physical molecules with biological systems, and they have discovered that many bacteria create NPs via either intracellular or extracellular pathways when they interact with inorganics (Bolade et al. 2020). Recently, it has been found that bacteria (like actinomycetes), fungi, and yeasts may produce metal nanoparticles both extracellularly and intracellularly (Saif et al. 2016).

The production of green nanoparticles has emerged as a promising method for creating nanomaterials with diverse properties. Traditional physical and chemical approaches have limitations due to hazardous byproducts, expensive equipment, and lengthy procedures (Bolade et al. 2020). Biogenic synthesis offers a more eco-friendly, cost-effective, and energy-efficient alternative. This method utilizes living organisms like bacteria, fungi, and algae to produce nanoparticles (Ju-Nam and Lead 2008). Microalgae and cyanobacteria are particularly promising due to their rapid growth, high CO_2 sequestration rates, and ability to hyperaccumulate heavy metals without producing toxic byproducts (Sodhi et al. 2021). Algal extracts serve as natural reducing and capping agents, enabling the green synthesis of nanoparticles (Shiny et al. 2014). Furthermore, algal cells can absorb heavy metals and utilize them as nutrients, making them a valuable tool for removing hazardous contaminants (Adeleye et al. 2016). Despite their potential, microalgae-based nanoparticle production remains largely unexplored (Ezeuko et al. 2021).

10.2 Synthesis of Algae-Based Nanoparticles

Algae are diverse, aquatic organisms that undergo photosynthesis and belong to the kingdom Plantae. They have the unique ability to absorb and transform heavy metal ions (Rai et al. 2018). Algal extracts are rich in various valuable compounds, including carbohydrates, vitamins, nutrients, oils, and fatty acids. Additionally, they contain bioactive substances like antioxidants, as well as pigments such as chlorophyll, which act as natural reducing and stabilizing agents (Su et al. 2019). This has led to a new generation of NPs which are synthesized from these microorganisms. Algae-mediated synthesis of NPs has proven to be efficient and healthy as compared to the traditional modes (chemical and physical) of NPs synthesis (Vimercati et al. 2020).

The intracellular biosynthesis of nanoparticles (NPs) from bulk materials can be achieved through two methods. The first involves centrifuging cellular extracts during the logarithmic growth phase, rinsing to remove excess salts, and then combining the biomass with a precursor solution in an aqueous medium (Rai et al. 2018). The second approach entails culturing a specific microbe with an aqueous solution of bulk material under optimal conditions. The biosynthesis of gold nanoparticles (AuNPs) is indicated by colour changes to pink, ruby red, or purple, while the biosynthesis of silver nanoparticles (AgNPs) is indicated by a brown hue (Satapathy et al. 2017; Rautela et al. 2019; Maiti et al. 2014).

For instance, the synthesis of AgNPs from Desmodesmus sp. cells was accomplished within 24 h using a 5 mM $AgNO_3$ solution as a reducing agent. Additionally, Zhang et al. developed an eco-friendly and cost-effective method for synthesizing cadmium selenide quantum dots (CdSe QDs) from microalgae, proposing their application as a sensitive nanoprobe for imatinib detection (Ji et al. 2020). Scenedesmus sp. was utilized to manufacture silver nanohybrids (Ag/AgCl) exhibiting antibacterial activity against four bacterial species. Ana et al. employed Spirulina sp. LEB 18 to investigate AgNP production in conjunction with microalgal culture in silver-polluted industrial wastewater, suggesting the conversion of industrial effluents into nanoparticles via microalgal culture (Khan et al. 2022).

The biosynthesis process involves metal ions binding to the surface of algal cells and being processed by secreted metabolites, such as lipids, proteins, ions, pigments, antioxidants, non-protein DNA, RNA, and enzymes, which act as capping and reducing agents for NP formation. Following centrifugation to separate the biomass, the filtrate is treated with a precursor metal ion solution to produce NPs. Although extracellular production facilitates relatively easy NP purification, washing and blending of the biomass are essential preliminary steps (Yang et al. 2022).

The size, shape, and aggregation of NPs are influenced by their physicochemical properties, including pH, temperature, substrate concentration, and metal ions (Yang et al. 2022). A neutral pH has been shown to enhance NP capping and stability through interaction with the amine group of surface proteins, whereas a higher pH can augment the bio-reduction activity of functional groups, minimizing NP agglomeration (Satapathy et al. 2017). The bioactive substances present in the filtrate of the S. obliquus culture were utilized to create extracellularly generated AgNPs exhibiting antibacterial activity. Similarly, extracellularly generated AgNPs using C. vulgaris demonstrated photocatalytic activity against methylene blue dye. The extracellular synthesis of AgNPs was also achieved using the microalgae extract of Nannochloropsis sp. hexane (EAENH), with various biomolecules playing crucial roles in the reduction, capping, and stabilization of AgNPs during synthesis (Maiti et al. 2014). S. platensis has also demonstrated the extracellular synthesis of AuNPs in response to chloroauric acid at various concentrations (Durán-Álvarez et al. 2015).

10.3 Factors Affecting Algae-Based Synthesis

The branch of nanoscience that specializes in the algae-based synthesis of NPs is called Phyconanotechnology. Several characteristics, such as size, shape, and stability, influence the chemical and physical properties of NPs. These factors include pH, time, temperature, starting concentration, and metal type (Khanna et al. 2019).

Effect of Extract Concentration The concentration of the algal extract is reportedly directly proportional to the quantity and inversely proportional to the particle size of the NPs produced. However, increasing the concentration beyond a threshold concentration leads to agglomeration. Thus, it is paramount to standardize the concentration into an optimum along with other parameters (Mukherjee et al. 2021).

Effect of pH The pH level of the reaction environment has a profound impact on nanoparticle (NP) synthesis. Variations in hydrogen ion concentration significantly influence NP size and shape. This phenomenon occurs because the electrical charge of biomass and capping agents undergo substantial changes at different pH levels, affecting their ability to bind and reduce metal ions. Notably, research has shown that larger NPs tend to form at lower acidic pH values, whereas higher pH values yield smaller particles (Khan et al. 2022).

Effect of Temperature In general, as temperature rises, so does the pace of reaction and production. It has been noted that increasing the temperature can cause an increase in particle size. Additionally, alteration of the temperature also affects the shape of the NPs (Chugh et al. 2021). In an experiment, Prasad et al. discovered that the average size of nanoparticles synthesized by the brown algal Cystophora moniliformis at temperatures less than 65 °C was 75 nm (Prasad et al. 2013). He also found that nanoparticle aggregation occurs at higher temperatures (85–90 °C), and a cluster of particles with a size greater than 2 μm was observed at 95 °C. Thus, temperature optimization is critical, and it relies on the algal dye and technique of synthesis utilized.

Effect of Contact Time Generally speaking, the yield of nanoparticles produced increases with an increase in incubation time, but only temporarily. In the case of silver nanoparticles, for example, extending the duration causes agglomeration because of the NPs' instability. Furthermore, the organism being employed for synthesis determines how the incubation period affects things (Chugh et al. 2021).

Choice of Algae The synthesis of NPs is significantly impacted by the choice of algae. When choosing the type of algae for biosynthesis, it is important to consider the intrinsic properties of the microorganism such as growth rate, enzyme activity, biological pathways, and inoculum size (Uzair et al. 2020).

Bioactive Catalyst and Metallic Precursor The characteristics of the bioactive catalysts and additional metallic precursors affect the yield of nanoparticles that are generated. The reaction rate and the quality of the NPs are both highly impacted by the appropriate ratio of biocatalysts to metallic precursors (Li et al. 2021).

10.4 Characterization of Algae-Based Nanoparticles

Nanoparticles (NPs) biosynthesized from algae have been characterized by using a wide range of different analytical techniques such as TEM, SEM, UV-vis spectroscopy, FT-IR, XRD, XPS, AFM, and DLS. The formation of certain nanoparticles can be identified by observing colour changes in the reaction mixture at specific temperatures during synthesis. For example, the appearance of a brownish-violet or violet hue indicates the formation of silver nanoparticles, while a shift to purple or pink suggests the successful synthesis of gold nanoparticles (AuNPs) (Mukherjee et al. 2021).

The techniques above play different roles in the characterization of the nanoparticles. For instance, UV-Vis spectroscopy can be used to monitor the biosynthesis of metal NPs in aqueous solutions. This process is done by using absorption spectroscopy in the UV-Vis range (190–1100 nm) (Khan et al. 2019; Borah et al. 2020; Sharma et al. 2016). Generally, different materials have different absorption spectra, e.g., AgNPs have a spectral range between 400 and 450 nm (Verma et al. 2010), AuNPs range between 500 and 550 nm and ZnO-NPs range between 350 and 390 nm (Shukla and Iravani 2017). Metal nanoparticles (NPs) display distinct optical properties due to surface plasmon resonance (SPR), which arises from the interaction between free electrons in the metal NPs and light waves. This phenomenon produces a unique absorption band (lmax) that provides valuable information about the nanomaterial's characteristics, such as size, shape, aspect ratio, and dielectric constant. By analysing the bandwidth and amplitude of this absorption band, the particle size of metal NPs can be determined. This can be achieved using a UV-Vis spectrophotometer, which measures particle size in aqueous solutions by utilizing blue and green light rays within a specific spectrum (320–580 nm) (Gopu et al. 2021).

FT-IR spectroscopy plays a crucial role in identifying the functional groups present on nanoparticle surfaces. This analytical technique provides valuable insights into the surface chemistry and synthesis mechanisms involved. By examining the spectral range of 4000–400 cm^{-1}, researchers can pinpoint the specific biomolecules responsible for reducing, capping, and stabilizing nanoparticles (Dahoumane et al. 2016). A comparative analysis of the transmittance spectra of native extracts and reaction media helps identify the key biomolecules involved in nanoparticle synthesis. Typically, functional groups such as carbonyl, amine, and thiol are detected on nanoparticle surfaces. However, the complexity of biological systems can sometimes limit the effectiveness of FT-IR spectroscopy due to overlapping absorption peaks (Huq 2020).

To further investigate the interaction between the produced NPs and surrounding biomolecules, additional characterization techniques like XPS (X-ray photoelectron spectroscopy) can be used (Shah et al. 2015). Surface charge, hydrodynamic diameter, and particle stability in liquid form can be measured using DLS (dynamic light scattering) spectroscopy, while XRD (X-ray diffraction) data helps determine purity, crystalline size, geometry, orientation, and phases of the NPs (Mukherjee et al. 2021; Shah et al. 2015). XRD patterns are typically compared with a standard crystallographic database like JCPDS to obtain structural information and estimate particle size using the Debye-Scherer formula (Ullah et al. 2017). XRD is suitable for identifying both single and multiphase NPs. It should be noted that XRD diffractograms can be influenced by amorphous nanometals with varying inter-atomic lengths (Shah et al. 2015).

Advanced microscopic techniques such as TEM, SEM, and AFM can be used to study the morphology, size, distribution, and surface topography of the metal NPs (Abbasi et al. 2020; Oza et al. 2012). TEM and SEM are usually hyphenated with ADAX or EDS to further provide information about the metals present (Oza et al. 2012). In some studies, HR-TEM is used to determine the precise size, shape, and crystallinity of the biosynthesized NPs.

10.5 Implications of Algae Nanoparticles in the Environment

Iron nanoparticles have a wide range of environmental applications, but they also pose a threat when they come into contact with the environment (Dinesh et al. 2012). Improper industrial waste management, leaks, and remediation of pollution—particularly in groundwater and soil remediation where iron nanomaterial can migrate from one medium to another—can all have negative effects (Dinesh et al. 2012). Nano zero-valent iron (nZVI), one of the several forms of iron nanoparticles, is regarded as being particularly reactive. Zero-valent iron (Fe^0), which was once employed as permeable reactive barriers for the in-place treatment of groundwater, changes when contaminants and an exposed environment are present (Saif et al. 2016). Iron nanoparticles (NPs) have numerous hazardous effects on soil fauna and microorganisms that are both directly and indirectly harmful to the environment. Induced stress of nano-iron can result in significant toxicological effects of iron NPs on soil microorganisms and alterations in microbial biomass (Hou et al. 2018). Furthermore, nanosized zero-valent iron (nZVI) particles were found to induce transcriptional and proteomic stress responses in the soil bacterium Bacillus cereus (Fajardo et al. 2013).

The biogenic nanoparticles with magnetic character have also been studied on the ecological indicator of the health of an aquatic ecosystem is green algal growth (Ahmed et al. 2017). The effects of superparamagnetic iron oxide nanoparticles on the green alga *Chlorella vulgaris* have been studied. Three suspensions of varying chemical concentrations of iron oxide NPs were used to expose *C. vulgaris* cells to each suspension (Ahmed et al. 2017). Significant toxicity was present in superparamagnetic iron oxide nanoparticles, which also caused oxidative stress in algal cells, impaired their photochemical activity, and prevented cell division (Hou et al. 2018).

The growing presence of nanoparticles in the environment is partly attributed to their excessive use in consumer products. Silver nanoparticles (Ag NPs) dominate the market, accounting for 20% of all nanoparticle usage—a staggering threefold increase over carbon-based nanoparticles, the second most widely used type (Chen et al. 2019). As a result, key research priorities have emerged, focusing on the ecological and health impacts of accumulated Ag NPs, as well as their interactions with soil-plant systems (Satapathy et al. 2017).

Furthermore, nanoparticles (NPs) may exert influence on terrestrial plants and crops within agricultural systems, potentially compromising global food security and safety (Sharma et al. 2019). Research conducted by Wang et al. has revealed that accumulated silver nanoparticles (AgNPs) can translocate to the roots and shoots of two terrestrial crops, namely cowpea (*Vigna unguiculata*) and wheat (*Triticum aestivum*), raising concerns about their potential impact on crop health and productivity (Chen et al. 2009). It is worth noting that, some plants have the ability to phytoaccumulate CeO_2 NPs without modification, which could be dangerous for human health (Asadzadeh et al. 2021). Additionally, the buildup of these NPs triggers the production of antioxidant enzymes such as ascorbate peroxidase and catalase, which work to scavenge harmful reactive oxygen species. Similar to how NPs affect mycorrhizal associations, it's critical to comprehend how NPs affect the microbial relationships of plants in watery environments (Batterjee et al. 2022). Silver NPs were found to be damaging to the microorganisms connected to wetland plant life in microcosms. Interestingly, microorganisms (or microbial extracts) are environmentally friendly tools for the synthesis of NPs; but, NPs may inhibit microbial interaction, a crucial occurrence for the survival of the environment (Chen et al. 2019). In the context of an aquatic environment, interactions between plants and NPs can be particularly important. Algae are the foundation of the food chain, and certain types of algae and cyanobacteria cause an algal bloom that is harmful to the environment and drinking water. The Cu NPs can potentially reduce algal blooms and have good environmental effects in aquatic settings (Chen et al. 2019).

10.6 Remediation Strategies

Algae play a crucial role in environmental remediation through processes such as algal bioremediation, phytoremediation, and nutrient removal (Fig. 10.1) (Gomes et al. 2016). These microorganisms effectively absorb heavy metals and excess nutrients from contaminated water and soil, making them a sustainable solution for pollution management. The biosorption of heavy metals utilizes these nanoparticles to effectively absorb and eliminate contaminants from wastewater, demonstrating a high affinity for various heavy metal ions, including cadmium and lead (Priyadarshani et al. 2011). Additionally, nanoparticle-mediated bioremediation enhances the degradation of specific pollutants, making it a versatile approach for both soil and water remediation. In wastewater treatment, algae not only remove harmful pollutants but also produce valuable byproducts like biofuels and bioplastics, enhancing the

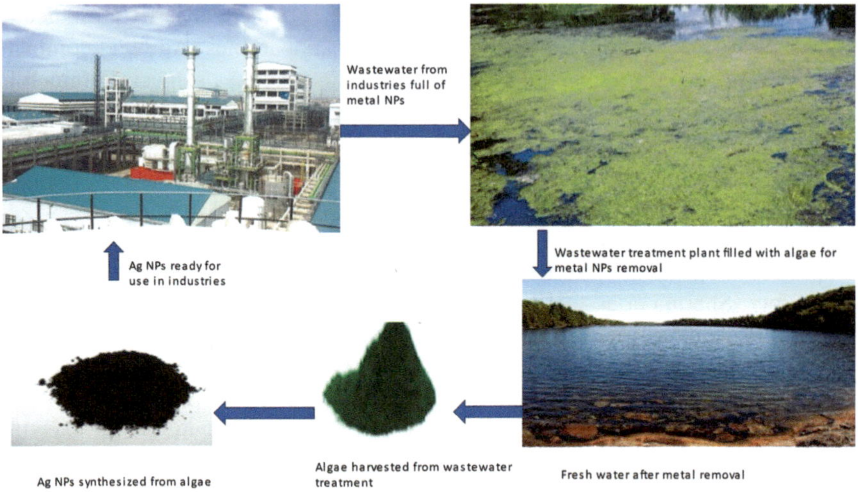

Fig. 10.1 Recycling of the biogenic NPs to prevent metal accumulation in the environment

economic viability of remediation efforts. Furthermore, phytoremediation leverages algae to extract pollutants, with nanoparticles increasing the surface area for absorption, thus improving efficiency. In the context of oil spills, algae-based nanoparticles can significantly enhance the degradation of oil pollutants, addressing the challenges of bioavailability (Macaulay and Rees 2014). Overall, the integration of algae in remediation strategies not only offers cost-effective solutions but also promotes eco-friendly practices in environmental management.

10.7 Conclusion

The effect and implication of algae-based nanoparticles in remediation have been successfully investigated. The synthesis of nanoparticles using algae is less toxic and environmentally friendly, but these nanoparticles when they are disposed into the environment pose to have negative effects. Iron and iron oxide nanoparticles are the most reported materials to have implications when they come in contact with the environment. Additionally, silver nanoparticles are also reported to have some implications, especially with the soil and plant systems. It can also be concluded that algae are effective in the removal of heavy metals due to their high removal efficiency for metals.

References

Abbasi BA et al (2020) Bioactivities of geranium wallichianum leaf extracts conjugated with zinc oxide nanoparticles. Biomolecules 10(1):1–19. https://doi.org/10.3390/biom10010038

Adeleye AS, Stevenson LM, Su Y, Nisbet RM, Zhang Y, Keller AA (2016) Influence of phytoplankton on fate and effects of modified Zerovalent iron nanoparticles. Environ Sci Technol 50(11):5597–5605. https://doi.org/10.1021/acs.est.5b06251

Ahmed S, Annu, Chaudhry SA, Ikram S (2017) A review on biogenic synthesis of ZnO nanoparticles using plant extracts and microbes: a prospect towards green chemistry. J Photochem Photobiol B 166:272–284. https://doi.org/10.1016/j.jphotobiol.2016.12.011

Asadzadeh SN, Malakootian M, Mehdipoor M, Neyestanaki DK (2021) The removal of tetracycline with biogenic CeO2 nanoparticles in combination with US/PMS process from aqueous solutions: kinetics and mechanism. Water Sci Technol 83(6):1470–1482. https://doi.org/10.2166/wst.2021.056

Batterjee MG, Nabi A, Kamli MR, Alzahrani KA, Danish EY, Malik MA (2022) Green hydrothermal synthesis of zinc oxide nanoparticles for UV-light-induced photocatalytic degradation of ciprofloxacin antibiotic in an aqueous environment. Catalysts 12(11):1–17. https://doi.org/10.3390/catal12111347

Bolade OP, Williams AB, Benson NU (2020) Green synthesis of iron-based nanomaterials for environmental remediation: a review. Environ Nanotechnol Monit Manag 13(2019):100279. https://doi.org/10.1016/j.enmm.2019.100279

Borah D et al (2020) Alga-mediated facile green synthesis of silver nanoparticles: photophysical, catalytic and antibacterial activity. Appl Organomet Chem 34(5):1–10. https://doi.org/10.1002/aoc.5597

Castillo-Henríquez L, Alfaro-Aguilar K, Ugalde-álvarez J, Vega-Fernández L, de Oca-Vásquez GM, Vega-Baudrit JR (2020) Green synthesis of gold and silver nanoparticles from plant extracts and their possible applications as antimicrobial agents in the agricultural area. Nanomaterials 10(9):1–24. https://doi.org/10.3390/nano10091763

Chen J, Wang YG, Li ZQ, Wang C, Li JF, Gu YJ (2009) Synthesis and characterization of magnetic nanocomposites with Fe3O4 core. J Phys Conf Ser 152:012041. https://doi.org/10.1088/1742-6596/152/1/012041

Chen F et al (2019) Algae response to engineered nanoparticles: current understanding, mechanisms and implications. Environ Sci Nano 6(4):1026–1042. https://doi.org/10.1039/c8en01368c

Chugh D, Viswamalya VS, Das B (2021) Green synthesis of silver nanoparticles with algae and the importance of capping agents in the process. J Genet Eng Biotechnol 19(1):126. https://doi.org/10.1186/s43141-021-00228-w

Dahoumane SA, Wujcik EK, Jeffryes C (2016) Noble metal, oxide and chalcogenide-based nanomaterials from scalable phototrophic culture systems. Enzym Microb Technol 95:13–27. https://doi.org/10.1016/j.enzmictec.2016.06.008

Dinesh R, Anandaraj M, Srinivasan V, Hamza S (2012) Engineered nanoparticles in the soil and their potential implications to microbial activity. Geoderma 173–174:19–27. https://doi.org/10.1016/j.geoderma.2011.12.018

Durán-Álvarez JC, Avella E, Ramírez-Zamora RM, Zanella R (2015) Photocatalytic degradation of ciprofloxacin using mono- (Au, Ag and Cu) and bi- (Au-Ag and Au-Cu) metallic nanoparticles supported on TiO2 under UV-C and simulated sunlight. Catal Today 266:175–187. https://doi.org/10.1016/j.cattod.2015.07.033

Ezeuko AS, Ojemaye MO, Okoh OO, Okoh AI (2021) Potentials of metallic nanoparticles for the removal of antibiotic resistant bacteria and antibiotic resistance genes from wastewater: a critical review. J Water Process Eng 41:102041. https://doi.org/10.1016/j.jwpe.2021.102041

Fajardo C, Saccà ML, Martinez-Gomariz M, Costa G, Nande M, Martin M (2013) Transcriptional and proteomic stress responses of a soil bacterium bacillus cereus to nanosized zero-valent iron (nZVI) particles. Chemosphere 93(6):1077–1083. https://doi.org/10.1016/j.chemosphere.2013.05.082

Gomes MADC, Hauser-Davis RA, de Souza AN, Vitória AP (2016) Metal phytoremediation: General strategies, genetically modified plants and applications in metal nanoparticle contamination. Ecotoxicol Environ Saf 134:133–147. https://doi.org/10.1016/j.ecoenv.2016.08.024

Gopu M et al (2021) Green biomimetic silver nanoparticles utilizing the red algae *Amphiroa rigida* and its potent antibacterial, cytotoxicity and larvicidal efficiency. Bioprocess Biosyst Eng 44(2):217–223. https://doi.org/10.1007/s00449-020-02426-1

Hou J, Wu Y, Li X, Wei B, Li S, Wang X (2018) Toxic effects of different types of zinc oxide nanoparticles on algae, plants, invertebrates, vertebrates and microorganisms. Chemosphere 193:852–860. https://doi.org/10.1016/j.chemosphere.2017.11.077

Huq MA (2020) Biogenic silver nanoparticles synthesized by *Lysinibacillus xylanilyticus* MAHUQ-40 to control antibiotic-resistant human pathogens *Vibrio parahaemolyticus* and *Salmonella typhimurium*. Front Bioeng Biotechnol 8(December):1–11. https://doi.org/10.3389/fbioe.2020.597502

Ji Y, Wen Y, Wang Z, Zhang S, Guo M (2020) Eco-friendly fabrication of a cost-effective cellulose nanofiber-based aerogel for multifunctional applications in cu(II) and organic pollutants removal. J Clean Prod 255:1–11. https://doi.org/10.1016/j.jclepro.2020.120276

Ju-Nam Y, Lead JR (2008) Manufactured nanoparticles: an overview of their chemistry, interactions and potential environmental implications. Sci Total Environ 400(1–3):396–414. https://doi.org/10.1016/j.scitotenv.2008.06.042

Khan I, Saeed K, Khan I (2019) Nanoparticles: properties, applications and toxicities. Arab J Chem 12(7):908–931. https://doi.org/10.1016/j.arabjc.2017.05.011

Khan F et al (2022) Prospects of algae-based green synthesis of nanoparticles for environmental applications. Chemosphere 293:1–14. https://doi.org/10.1016/j.chemosphere.2022.133571

Khanna P, Kaur A, Goyal D (2019) Algae-based metallic nanoparticles: synthesis, characterization and applications. J Microbiol Methods 163(June):105656. https://doi.org/10.1016/j.mimet.2019.105656

Li SN, Wang R, Ho SH (2021) Algae-mediated biosystems for metallic nanoparticle production: from synthetic mechanisms to aquatic environmental applications. J Hazard Mater 420(January):126625. https://doi.org/10.1016/j.jhazmat.2021.126625

Macaulay B, Rees D (2014) Bioremediation of oil spills: a review of challenges for research advancement. Ann Environ Sci 8:9–37. Available: www.aes.northeastern.edu.

Maiti S, Krishnan D, Barman G, Ghosh SK, Laha JK (2014) Antimicrobial activities of silver nanoparticles synthesized from *Lycopersicon esculentum* extract. J Anal Sci Technol 5(40):1–7. https://doi.org/10.1186/s40543-014-0040-3

Mukherjee A, Sarkar D, Sasmal S (2021) A review of green synthesis of metal nanoparticles using algae. Front Microbiol 12(August):1–7. https://doi.org/10.3389/fmicb.2021.693899

Oza G et al (2012) Facile biosynthesis of gold nanoparticles exploiting optimum pH and temperature of fresh water algae *Chlorella pyrenoidusa*. Pelagia Res Library 3(3):1405–1412

Prasad TNVKV, Kambala VSR, Naidu R (2013) Phyconanotechnology: synthesis of silver nanoparticles using brown marine algae *Cystophora moniliformis* and their characterisation. J Appl Phycol 25(1):177–182. https://doi.org/10.1007/s10811-012-9851-z

Priyadarshani I, Sahu D, Rath B (2011) Microalgal bioremediation: current practices and perspectives. J Biochem Technol 3(3):299–304

Rai PK et al (2018) Nanoparticle-plant interaction: implications in energy, environment, and agriculture. Environ Int 119:1–19. https://doi.org/10.1016/j.envint.2018.06.012

Rautela A, Rani J, Debnath Das M (2019) Green synthesis of silver nanoparticles from *Tectona grandis* seeds extract: characterization and mechanism of antimicrobial action on different microorganisms. J Anal Sci Technol 10(5):1–10. https://doi.org/10.1186/s40543-018-0163-z

Saif S, Tahir A, Chen Y (2016) Green synthesis of iron nanoparticles and their environmental applications and implications. Nano 6(11):1–26. https://doi.org/10.3390/nano6110209

Satapathy S, Kumar S, Sukhdane KS, Shukla SP (2017) Biogenic synthesis and characterization of silver nanoparticles and their effects against bloom-forming algae and synergistic effect

with antibiotics against fish pathogenic bacteria. J Appl Phycol 29(4):1865–1875. https://doi.org/10.1007/s10811-017-1091-9
Shah M, Fawcett D, Sharma S, Tripathy SK, Poinern GEJ (2015) Green synthesis of metallic nanoparticles via biological entities. Materials 8(11):7278–7308. https://doi.org/10.3390/ma8115377
Sharma A et al (2016) Algae as crucial organisms in advancing nanotechnology: a systematic review. J Appl Phycol 28(3):1759–1774. https://doi.org/10.1007/s10811-015-0715-1
Sharma D, Kanchi S, Bisetty K (2019) Biogenic synthesis of nanoparticles: a review. Arab J Chem 12(8):3576–3600. https://doi.org/10.1016/j.arabjc.2015.11.002
Shiny PJ, Mukherjee A, Chandrasekaran N (2014) Haemocompatibility assessment of synthesised platinum nanoparticles and its implication in biology. Bioprocess Biosyst Eng 37(6):991–997. https://doi.org/10.1007/s00449-013-1069-1
Shukla AK, Iravani S (2017) Metallic nanoparticles: green synthesis and spectroscopic characterization. Environ Chem Lett 15(2):223–231. https://doi.org/10.1007/s10311-017-0618-2
Sodhi KK, Kumar M, Singh DK (2021) Insight into the amoxicillin resistance, ecotoxicity, and remediation strategies. J Water Process Eng 39:1–13. https://doi.org/10.1016/j.jwpe.2020.101858
Su Y et al (2019) Metallic nanoparticles induced antibiotic resistance genes attenuation of leachate culturable microbiota: the combined roles of growth inhibition, ion dissolution and oxidative stress. Environ Int 128:407–416. https://doi.org/10.1016/j.envint.2019.05.007
Ullah H, Khan I, Yamani ZH, Qurashi A (2017) Sonochemical-driven ultrafast facile synthesis of SnO2 nanoparticles: growth mechanism structural electrical and hydrogen gas sensing properties. Ultrason Sonochem 34:484–490. https://doi.org/10.1016/j.ultsonch.2016.06.025
Uzair B et al (2020) Green and cost-effective synthesis of metallic nanoparticles by algae: safe methods for translational medicine. Bioengineering 7(4):1–22. https://doi.org/10.3390/bioengineering7040129
Verma VC, Kharwar RN, Gange AC (2010) Biosynthesis of antimicrobial silver nanoparticles by the endophytic fungus aspergillus clavatus. Nanomedicine 5(1):33–40. https://doi.org/10.2217/nnm.09.77
Vimercati L et al (2020) Nanoparticles: an experimental study of zinc nanoparticles toxicity on marine crustaceans. General overview on the health implications in humans. Front Public Health 8(May):1–19. https://doi.org/10.3389/fpubh.2020.00192
Yang Y et al (2022) A critical review on the interaction of iron-based nanoparticles with blue-green algae and their metabolites: from mechanisms to applications. Algal Res 64:102670–102670. https://doi.org/10.1016/j.algal.2022.102670

Chapter 11
Role of Phytosynthesized Nanoparticles for Remediation of Contaminants

Sajmina Khatun, Chandra Lekha Putta, and Aravind Kumar Rengan

Abstract Phytosynthesized nanomaterials are expanding swiftly in the realm of nanotechnology for removing environmental contaminants. Both organic and inorganic environmental contaminants are detrimental to human health. The conventional methods for eliminating contaminants are expensive and time-consuming. Therefore, a workable, affordable, and ecologically friendly technique like nanotechnology is desirable. Phytosynthesized nanoparticles, synthesized using plant extracts or plant-derived compounds, have gained attention as potential tools for the remediation of environmental contaminants. These nanoparticles offer several advantages, including their green synthesis approach, an abundance of precursor materials, and unique properties. They can effectively remove heavy metals from contaminated environments through adsorption or precipitation and enhanced catalytic properties. The reliable physicochemical properties, ecological sustainability, and sufficiency of phytosynthesized nanomaterials widen their applications in the remediation of environmental contaminants. This chapter gives insight into the source of environmental contaminations, impacts of environmental degradation on human populations, conventional remediation technologies and their limitations, types of phytosynthesized nanomaterials, their different synthesis methods, and recent developments. The diverse applications of phytosynthesized nanoparticles for environmental pollutant remediation were also highlighted. This chapter also explored the recent developments, limitations, and potential future applications to reap the rewards of phytosynthesized nanoparticles in contaminant remediations.

Keywords Phytosynthesis · Nanomaterials · Environmental contaminants · Heavy metals · Hyperaccumulations · Inorganic contaminants

S. Khatun · C. L. Putta · A. K. Rengan (✉)
Department of Biomedical Engineering, Indian Institute of Technology Hyderabad, Kandi, Sangareddy, India
e-mail: Sajmina2018@gmail.com; bm22resch11001@iith.ac.in; aravind@bme.iith.ac.in

A. Guldhe et al. (eds.), *Biomaterials for Environmental Applications: Detection and Remediation*, Environmental Challenges and Solutions,
https://doi.org/10.1007/978-3-032-12701-3_11

11.1 Introduction

Environmental contamination is the instigation of toxic compounds or pollutants into the air, water, or soil that has an adverse impact on human health, ecosystems, and the general well-being of living organisms. The type of contaminant, its concentration, the period of exposure, and the susceptibility of the afflicted species are some variables that determine how dangerous environmental pollution is. Some common ecological contaminants include (a) air pollutants, including emissions from industrial processes, vehicle exhaust, and the burning of fossil fuels. Air pollution can cause respiratory illnesses, cardiovascular issues, and early mortality. It also leads to climate change and smog formation; (b) water impurities include chemicals, heavy metals, fertilizers, pesticides, and untreated sewage. Water contamination can result in the deterioration of aquatic habitats, reduce biodiversity, and contaminate drinking water sources. This causes a significant risk to human health, causing waterborne diseases, organ damage, and cancer; (c) soil pollutants include industrial waste, pesticides, herbicides, and heavy metals. Soil contamination affects agricultural productivity, contaminates food products, and can lead to the bioaccumulation of toxins in the food chain. It also poses risks to groundwater quality and can have long-lasting effects on soil fertility; (d) hazardous waste caused by the improper disposal of harmful materials, such as radioactive compounds, asbestos, or certain chemicals, poses severe risks to the ecosystem and human health (Sharma et al. 2019). Accidental spills and leaks can cause severe environmental contamination, harming soil, water bodies, and ecosystems. This leads to biodiversity loss, health hazards, and long-term ecological imbalance.

The persistence of contaminants necessitates costly and complex cleanup efforts, impacting both current and future generations. There are several conventional methods available to cure environmental contaminants: physical treatment (sedimentation, filtration), chemical treatment (coagulation and flocculation, precipitation, oxidation/reduction), biological treatment (composting, bioremediation), thermal treatment (controlled burning of waste/incineration, thermal desorption), etc. Although conventional treatment methods for environmental contaminants are often effective, they have some limitations. These limitations can vary based on the specific pollutant and the treatment method. Common limitations associated with conventional treatment techniques include incomplete removal of contaminants, secondary waste generation, cost and scalability, energy and resource requirements, long treatment duration, and limited applicability to certain pollutants (Crini et al. 2019). Current research focuses on advancing treatment technologies with cost-effective, eco-friendly solutions. Innovations like nanotechnology, advanced oxidation, and biotechnological methods aim to overcome the limitations of traditional treatments.

Nanomaterials (NMs) have been established as promising tools for environmental applications, particularly in detecting and remediating contaminants. Nanomaterials offer distinctive properties such as large surface-to-volume ratios and enhanced reactivity, making them well-suited for environmental applications.

Nanomaterials are employed as sensing platforms for rapidly and sensitively detecting contaminants in the detection field, including heavy metals, organic pollutants, and microbial pathogens. Functionalized nanoparticles and nanocomposites (NCs) enable selective recognition and quantification of target analytes in complex environmental matrices. They can be utilized for the adsorption, catalytic degradation, or immobilization of contaminants, thereby enabling their removal or transformation. Nanoparticles (NPs) functionalized with specific functional groups or coatings enhance their affinity towards target pollutants, resulting in efficient and selective remediation. However, challenges remain in the practical implementation of nanomaterials for environmental applications. Concerns about nanoparticle stability, potential toxicity, and environmental fate must be addressed through comprehensive risk assessment and regulatory frameworks. Additionally, large-scale remediation projects should consider scale-up, nanoparticle synthesis, and deployment cost-effectiveness (Keller et al. 2023).

The increasing concern over environmental pollution has driven the approach of green nanomaterial synthesis for sustainable environmental applications, particularly in detection and remediation. Green nanomaterials derived from natural sources or synthesized using eco-friendly methods offer several advantages over conventional nanomaterials. They possess unique properties such as tunable physicochemical characteristics and low toxicity, making them ideal candidates for environmentally friendly applications (Gour and Jain 2019). Overall, phytosynthesized nanoparticles offer a promising and sustainable approach to environmental remediation, combining the advantages of eco-friendliness, cost-effectiveness, selectivity, and biodegradability.

11.2 Conventional Nano-remediation Approaches for Environmental Pollution

The use of nanomaterials for ecological restoration is determined by the characteristics of the nanostructure. The two primary conventional approaches used for nanoparticle-based remediation are chemical reduction and catalysis. Additionally, owing to the greater surface area, randomized binding site distribution, and several modification possibilities in the coating, nanoparticles have also been employed in the adsorption-based elimination process (Guerra et al. 2018). Nanoremediation of various environmental contaminations is represented in Fig. 11.1. Furthermore, NM-based membranes are employed in water filtration as their pores have the potential to hold large molecules in water effluents (Abdelfattah and El-Shamy 2024). Also, contact with the membrane layer selectively removes the less essential molecules. Metal oxides, quantum dots, carbon nanotubes (CNTs), and biopolymers are nanoparticles that remediate water, soil, and air.

Air pollution remains a major global challenge, impacting public health and the climate, with key pollutants, including CO, PM, ozone, Pb, and SO_2, often formed

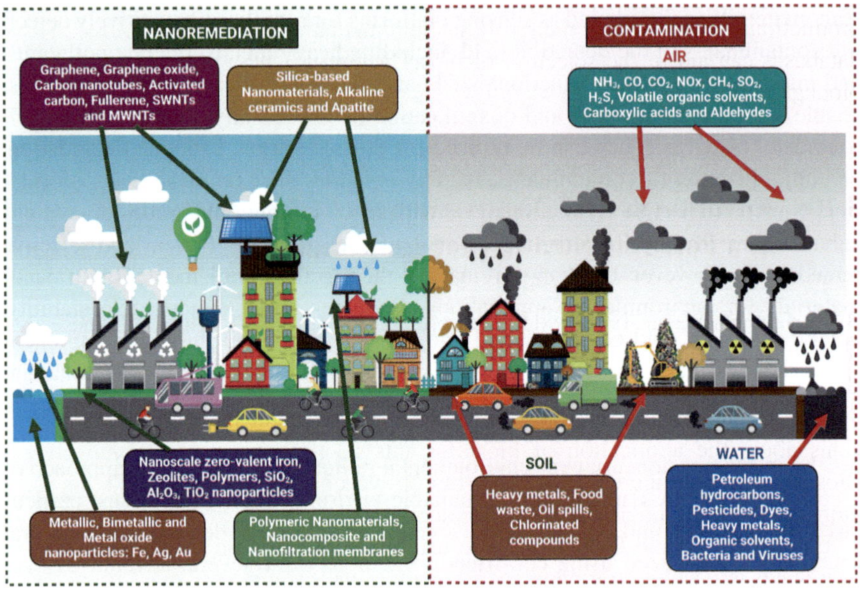

Fig. 11.1 Nanoremediation of ecological contamination

through chemical interactions of volatile organic compounds. To mitigate this issue, researchers have explored advanced nanomaterials like CNTs, graphite, graphene oxides (GO), and silica-based compounds, leveraging their high reactivity, porous structure, and thermal stability for effective pollution control (Guerra et al. 2018). Carbon-based and silica-based nanomaterials offer significant potential for air pollution management through remediation, pollution monitoring, and contaminant detection. Graphene oxide (GO), a key carbon-based nanomaterial, has been used in NH_3 gas sensors, SO_2 adsorption, and as a photocatalyst for degrading volatile compounds under UV radiation. Silica-based nanomaterials, with their large surface area and tunable properties, effectively remove CO_2, ketones, aldehydes, and other pollutants from the air. Additionally, amine-modified silica nanoparticles have shown promise in eliminating toxic metals like Pb, highlighting their role in mitigating industrial air pollution (Parashar et al. 2020; Mpouras et al. 2021).

Advanced restoration solutions are essential as the rapid development of industries and urbanization has further exacerbated soil degradation. Due to their high reactivity, surface modification capabilities, and tunable physical properties, nanomaterials have gained prominence in soil remediation. Their in-situ application allows for large-scale soil purification while reducing time and costs. Carbon-based nanomaterials, including CNTs, fullerenes, graphene, offer exceptional adsorption properties and eco-friendly advantages. CNTs, in particular, exhibit superior adsorption capacities for heavy metals like Pb^{2+}, Cu^{2+}, Ni^{2+}, and Zn^{2+}, which depend on pH, organic material content, and soil composition. Additionally, metal-based nanomaterials like nanoscale zero-valent iron (nZVI) are widely used for removing chlorinated compounds, heavy metals, and other pollutants through redox transformations.

Surface modifications using noble metals, biopolymers, and composite materials further enhance their functionality, ensuring efficient and sustainable soil remediation (Rajput et al. 2022).

Water pollution also poses a serious hazard to ecosystems and human health, making it a critical concern for environmental researchers. Various pollutants, including heavy metals, hydrocarbons, pesticides, and pharmaceutical residues, can degrade water quality, necessitating effective remediation techniques. The substantial surface area, adsorption capacity, and sensitivity of nanoparticles have made nanotechnology a promising option. Carbon-based nanomaterials, like activated carbon and CNTs, effectively remove contaminants like fluorides, heavy metals, and pharmaceuticals. Metal-based nanoparticles, including FeO, ZnO, and TiO_2, are widely used for dye adsorption, chromium neutralization, and photocatalytic degradation of pollutants. Polymer-based nanomaterials, such as chitosan and polyamide membranes, enhance water filtration by trapping contaminants and improving pollutant rejection. These nanoparticle-driven technologies offer cost-effective and efficient approaches to addressing water pollution, highlighting their potential for large-scale environmental applications (Peydayesh et al. 2020; Mpouras et al. 2021). Some conventional nanoremediation methods, their remediation process, advantages, and limitations are listed in Table 11.1.

Table 11.1 Conventional nanoremediation, remediation process, remediation media, and their advantages & disadvantages

Nanomaterials	Remediation process	Media type	Advantages	Limitations	References
Polymer-based NMs	Membrane filtration/ nanofiltration	Water	Using waste-derived polymers Compliance with other therapies	pH dependence Degrades under elevated temperatures	Peydayesh et al. (2020)
Carbon-based NMs	Adsorption	Air, soil, water	Compliance with other therapies Large surface area Excellent adsorption properties Microporosity Ecologically safe	ROS generation DNA damage Lysosomal damage Costly Rapid saturation	Mpouras et al. (2021)
Metal-based NMs	Adsorption, photocatalysis, photodegradation, oxidation, reduction	Soil, water	Removal of organic pollutants Large specific surface area Compliance with other therapies	Adverse effects Harmful to human health Detrimental to the environment	Aragaw et al. (2021)
Silica-based NMs	Adsorption, catalysis	Air, water	Adjustable pore size Compliance with other therapies Flexibility in surface modifications	Irregular particle size distribution	Parashar et al. (2020)

11.3 The Impact of Conventional Nanoremediation on Environmental Pollutants and the Importance of Phytosynthesized Nanoremediation

Conventional nanomaterials have proven effective in nanoremediation, aiding in the elimination of various inorganic and organic contaminants from water and soil. Their ability to eliminate toxic metals, synthetic dyes, and biological pollutants highlights their potential in water purification. As demand for nano-based remediation rises, biologically synthesized 1D hollow nanomaterials are emerging as multifunctional structural components. However, concerns over their potential medical hazards have gained attention due to increased environmental exposure. Iron (Fe) NPs have been investigated for industrial and ecological remediation applications. Yet, releasing Fe NPs into the environment is associated with potential hazards. The toxicity evaluation for Fe NPs was conducted in an early stage of the zebrafish model to study embryonic development. The findings demonstrated that particles more significant than 10 mg L^{-1} caused developmental abnormalities in the embryos, leading to death, delayed hatching, and deformity (Zhu et al. 2012). Additionally, the effects of Fe NPs on human and animal cells resulting in cytotoxicity have been examined. However, there aren't many reports of genotoxicity caused by Fe NPs in human cells. Recent research by Ahamed et al. examined the effects of Fe NPs on two human lung epithelial (A549) and skin epithelial (A431) cell lines (Ahamed et al. 2008). The results revealed that Fe NPs might cause genotoxicity in both cell lines. This genotoxicity is probably caused by generating reactive oxygen species (ROS), leading to cellular oxidative stress.

The biological consequences of TiO_2 NPs are currently being examined because of their extensive use, and concerns about possible hazards to the population and workers exposed to them are escalating. According to investigations, TiO_2 NPs can trigger apoptosis, nuclear degradation, oxidative stress, pro-inflammatory reactions in human endothelial cells, and chronic pulmonary inflammation in rats. Cerium oxide (CeO_2) nanoparticles have demonstrated excellent adsorption capabilities, making them effective in water remediation. However, research on their potential ecological and social impacts remains limited. Some studies suggest CeO_2 NPs may induce oxidative stress and even cause cell death in human lung epithelial cells, raising concerns about their safety. Despite these drawbacks, CeO_2 NPs exhibit remarkable catalytic properties, mimicking the activities of catalase and superoxide dismutase. This unique ability allows them to act as reactive oxygen species (ROS) scavengers in various biological contexts. As a result, they have shown potential in neuroprotection and reducing apoptosis in cells exposed to oxidative stress (Yang et al. 2016).

The assessment of the in vivo toxicity of copper (Cu) NPs differs based on the size of the nanoparticles. High Cu consumption can result in hemolysis, jaundice, liver cirrhosis, lipid profile alteration, oxidative damage, renal failure, and even mortality. Cu NPs cause nephrotoxicity, hepatotoxicity, and severe damage to the spleens of experimental animal models, as well as causing gill toxicity in zebrafish, based on animal studies. Cu NPs are cytotoxic and induce DNA damage, according to in vitro studies. The rapid advancement of silver (Ag) and gold (Au) NPs holds considerable potential for applications in numerous fields. Despite their enormous benefits for the environment usage, they impose short-term and long-term consequences on the environment's and organisms' well-being. The size of gold (Au) NPs affects the toxic nature of the particles. Human keratinocyte cells (HaCaT) were harmful to rod-shaped Au NPs, while human breast adenocarcinoma cells' mitochondrial integrity was compromised. Additionally, Ag NPs exhibited cytotoxic and genotoxic effects on human cells, resulting in mitochondrial failure and ROS production, which in turn induced DNA damage and chromosomal abnormalities (Babonaitė et al. 2024).

Phytosynthesized nanomaterials represent a paradigm shift in pollution control, offering a more sustainable and green alternative to conventional pollutant management strategies. These nanomaterials, synthesized using plant extracts, offer several advantages over traditional approaches, including their biocompatibility, low toxicity, and sustainability. Unlike chemical or physical methods, which often require harmful reagents or high energy consumption, phytosynthesized nanoparticles (NPs) are produced using natural resources, reducing the environmental impact. The properties that interact with various pollutants make them highly effective for water and soil remediation applications. Plant-derived nanoparticles can be functionalized to target specific pollutants, providing a more selective and efficient removal process. The eco-friendly nature of these nanoparticles, combined with their ability to degrade or immobilize pollutants, positions them as a promising tool for environmental remediation, particularly in areas where traditional methods are less effective or too costly (Ahmed et al. 2022).

11.4 Types and Methodologies for Phytosynthesized Nanomaterials

The eco-friendly and sustainable approach has garnered considerable attention within the realm of nanotechnology, owing to its inherent simplicity, cost-effectiveness, and diverse potential applications across multiple disciplines. Different types of nanoparticles, for example, gold, silver, copper-based, platinum, titanium dioxide, iron, zinc oxide, and bimetallic (Zhan et al. 2011; Suresh et al. 2015; Kumar et al. 2015; Nayak et al. 2015; Stan et al. 2015; Ganaie et al. 2016; Supraja et al. 2016; Yeasmin et al. 2017; Rao et al. 2017; Chaudhuri et al. 2019; Larayetan et al. 2019) can be phytosynthesized from different plant parts using various plant extracts. Some of the commonly studied types of phytosynthesized nanoparticles are included in Table 11.2.

Table 11.2 Types of nanomaterials derived from plant sources

Plant source	Part of plant	Nanomaterials	Size (nm)	Shape	References
Acorus calamus, *Cucurbita maxima*, and *Moringa oleifera*	Petals	Silver NPs	21–57	Spherical and rectangular	Nayak et al. (2015)
Acacia nilotica	Gum	Silver NPs	10–78	Spherical	Yeasmin et al. (2017)
Allium cepa, *Allium sativum*, and *Petroselinum crispum*	Leaves	Zinc oxide NPs	14–70	Hexagonal	Stan et al. (2015)
Aloe vera	Leaves	Copper oxide NPs	20	Spherical	Kumar et al. (2015)
Antigonon leptopus	Leaves	Au–Ag NPs	10–60	Spherical	Ganaie et al. (2016)
Azadirachta indica and *Ocimum tenuiflorum*	Leaf and flower	Gold NPs	20	Spherical	Rao et al. (2017)
Boswellia ovalifoliolata	Bark	Zinc oxide NPs	20	Spherical	Supraja et al. (2016)
Bursera serrata	Fruit	Gold NPs	10	Spherical, triangular, and tetragonal	Chaudhuri et al. (2019)
Cacumen platycladi	Leaves	Au–Pd NPs	7.4 ± 0.8	Spherical	Zhan et al. (2011)
Callistemon citrinus	Aerial parts	Silver NPs	29	Spherical	Larayetan et al. (2019)
Cassia fistula	Plant	Zinc oxide NPs	5–15	Crystalline	Suresh et al. (2015)

Based on the precursors, the techniques used to develop phytosynthesized nanoparticles are categorized into top-down and bottom-up methodologies, depicted in Fig. 11.2.

11.4.1 Top-Down Approach

In top-down techniques, the starting compound is used in large size to shrink to the nanoscale by applying chemical and physical methods. These procedures comprise electrochemical modifications and photochemical changes employing a wide range of techniques that produce stabilized nanoparticles, such as mechanical grinding, thermal and laser treatment, and ablation. These are less complicated to execute but can change the synthesized nanoparticles' characteristics and surface chemistry. Highly efficient lasers are used in ablation therapy to damage metallic plates, generating ablated nanoparticles in the liquid medium. The exfoliation time, laser intensity, wavelength, and the medium's absorption rate impact the effectiveness of this technique in producing nanoparticles. Cu NPs, Au NPs, and Ag NPs have been

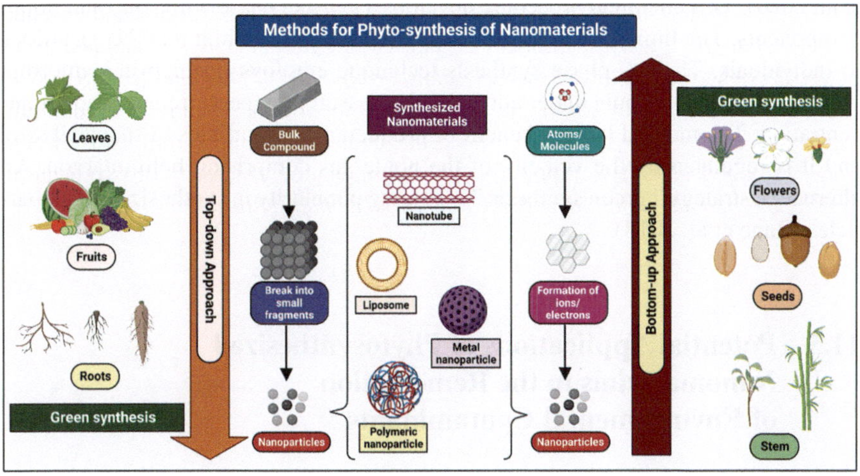

Fig. 11.2 Methodologies for synthesizing plant-derived nanomaterials

developed by this approach. In mechanical grinding, large particles are reduced to the appropriate smaller sizes. According to the requirements, an agitator operates at 75–500 rpm speed. The container size, milling rate, temperature, habitat, and use of inert gases affect the nanostructure size and uniformity (El-Eskandarany et al. 2021).

11.4.2 Bottom-Up Approach

Contrarily, bottom-up methods construct atoms/molecules to develop nanostructured blocks that ultimately culminate in the desired nanoparticles. It could be performed through a combination of biological and chemical approaches. Findings have unequivocally demonstrated that natural nanoparticle synthesis methods have minor disadvantages compared to alternative technologies. Mechanical procedures have low rates of manufacture along with substantial time, energy consumption, and economic expenses. Hazardous substances and solvents are used in chemical synthesis, resulting in undesirable derivatives and residues. The solid-state method to the synthesis of nanoparticles is practical and reasonable compared to numerous alternatives. Milling, calcination, and sintering are the stages involved. The particles produced by this technique are exceedingly stable and exceptionally rare to agglomerate. Second, removing solvents during preparation makes it easier to manipulate and synthesize nanoparticles on large amounts of Ag NPs. Metal-oxide nanoparticles can be produced effectively using the sol-gel method of synthesis (liquid-state technique), which has several other benefits. This strategy is suitable for synthesizing highly complicated nanocomposites with minimal production temperatures, better concentration, and efficient methodology. This technique has produced many metal-oxide nanoparticles, including ZnO NPs, TiO_2 NPs, and SnO_2 NPs (Parashar et al. 2020). This chemical procedure develops a gel-like phase with solid and liquid components. The limitation is that the organic solvents used in it may be hazardous to individuals. The gas-phase synthesis technique employs magnetron sputtering, which yields high outputs under controllable pressure, temperature, and vapor concentration. The method has the benefit of producing nanoparticles of the ideal size, and it is regulated by the velocity of the noble gas comprising helium/argon. An alternative strategy, "green synthesis," is gaining popularity in synthesizing nanoparticles (Rane et al. 2018).

11.5 Potential Applications of Phytosynthesized Nanomaterials in the Remediation of Environmental Contaminants

Nanomaterials produced through phytosynthesis can act as sensors or probes to find and monitor environmental pollutants. These nanoparticles can be functionalized to interact only with specific pollutants and offer sensitivity in detecting and remedying pollutants. They can be employed for various ecological monitoring purposes, such as soil pollution detection, water quality evaluation, air quality monitoring, heavy metal removal, and organic pollutant degradation. Some phytosynthesized nanomaterials and their potential applications are included in Table 11.3.

11.5.1 Water Purification

The concept of conscious water use, often referred to as the "water footprint," has gained prominence, especially in urban settings where municipalities strive to reconcile the escalating demand for public water supply with effective water quality and wastewater management strategies. Contaminated water causes severe public health issues like vomiting and diarrhea caused by microbes or long-term disorders like heavy metal bioaccumulation. Over the past few decades, a diverse array of wastewater purification methods has been developed, encompassing techniques such as solvent extraction, micro and ultrafiltration, adsorption, sedimentation and gravity separation, precipitation, coagulation, flotation, evaporation, oxidation, distillation, reverse osmosis, ion exchange, electrodialysis, and electrolysis. Among all the techniques, adsorption is one of the most important approaches for wastewater management because of its ease of use, low cost, and availability of a diverse spectrum of adsorbents. However, absorbents have some limitations, as different contaminants are present in water, and a single adsorbent cannot remove all pollutants (Ali et al. 2011).

A wide range of nanomaterials has been developed and utilized for the remediation of aquatic contaminants. As more environmentally friendly and less polluting

Table 11.3 Phytosynthesized nanoparticles, their source, essential experimental parameters for synthesis, and their potential application for removal of environmental contaminants

Nanoparticles type	Source/name of the plants	Experimental condition	Applications	References
Iron nanoparticles (KP-Fe NPs)	*Pyrus sinkiangensis Yu*	pH 5 Temperature: 55 °C Time: 120 min	Removal of Cr (VI) from aqueous media	Rong et al. (2020)
Iron oxide nanoparticles (FeO NPs)	*Pleurotus ostreatus*	pH 2–9 Temperature: 25 °C Time: 90 min	Solid-phase extraction has been employed for the efficient separation of Ni (II) and Pb (II) ions from water and food samples.	Özdemir et al. (2021)
Iron oxide nanoparticles (Fe_xO_y NPs)	*Luma apiculata* (*LAL*); *Eucalyptus globulus* (*EGL*): *Phragmites australis* (*PAL*)	pH 2–4 Temperature: 20–40 °C Time: 40 min	Removal of toxic chemical compounds like Cr (VI) from aqueous media	Salgado et al. (2019)
Copper oxide nanoparticles (CuO NPs)	*Punica granatum*	pH 4–10 Temperature: 25 °C Time: 24 h	Removal of toxic dyes	Peternela et al. (2018)
Silver nanoparticles (Ag NPs)	Extracts of tomato, onion, and acacia catechu	Temperature: 25 °C Time: 30 min	Removal of organic dyes	Chand et al. (2020)
Zinc oxide nanoparticles (ZnO NPs)	*Shorea robusta*	pH 6 Temperature: 60 °C Time: 60 min	Removed Pb^{2+} and Cd^{2+} ions from wastewater	Özdemir et al. (2021)
Silver nanoparticles (Ag NPs)	*Acalypha indica*	Temperature: 60 °C Time: 30 min	Inhibition against waterborne pathogens like *Escherichia coli* and *Vibrio cholerae*	Krishnaraj et al. (2010)
Magnetic iron nanoparticles (Fe_3O_4 NPs)	Olive oil	pH 6.44 Temperature: 80–100 °C Time: 30 min	Solid-liquid phase microextraction for determination of nickel ions in soil, potato, red tea, white tea, mushroom, lettuce, cabbage, apple, urban water, purified drinking water	Es'haghi et al. (2016)

(continued)

Table 11.3 (continued)

Nanoparticles type	Source/name of the plants	Experimental condition	Applications	References
Bentonite-supported "green" nanoscale zero-valent iron nanoparticles (B-nZVI NPs)	Green tea extracts	pH 2–6 Temperature: 75 °C Time: 24 h	Cr (VI) adsorption from soil	Soliemanzadeh and Fekri (2017)
Membranes containing reactive nanoparticles (Fe/Pd NPs)	Green tea extracts	pH 5.5 Temperature: 90 °C Time: 30 min	Removal of trichloroethylene from groundwater	Soliemanzadeh and Fekri (2017)

nanoparticle synthesis methods emerge, the need for green syntheses is rising for purifying contaminated water by adsorbing, degrading, or transforming pollutants in the water. Their high reactivity and large surface area enable efficient elimination of contaminants, including organic compounds and heavy metals. The phytosynthesized nanoparticles contribute to the remediation of contaminated water via adsorption and immobilization, catalytic degradation, antibacterial activity, enhanced filtration, and regeneration of water treatment materials. The ability to be functionalized to have particular binding sites characterizes phytosynthesized nanoparticles. They can effectively adsorb and immobilize various contaminants present in water. To stop the spread of these contaminants and lower their concentration in water, the nanoparticles can operate as a medium to trap and capture them. Certain phytosynthesized nanoparticles exhibit catalytic properties that enable the degradation of contaminants. For example, Bibi et al. synthesized Fe_2O_3 NPs via the green synthesis method using pomegranate (*P. granatum*) seed extract. Upon exposure to UV light irradiation, the Fe_2O_3 nanoparticles demonstrated notable photocatalytic activity in degrading reactive blue dye, achieving a degradation efficiency of up to 95.08% within 56 min. This finding suggests that *P. granatum* seed extract holds potential as a viable agent for dye degradation in wastewater treatment applications (Bibi et al. 2019).

Phytosynthesized nanoparticles can also enhance the absorption and filtration efficiency of contaminated water. When incorporated into membranes or filters, the nanoparticles can improve the removal of contaminants, including bacteria, viruses, etc. The nanoparticles can provide additional adsorption sites and create a more selective filtration medium. Research has highlighted the presence of numerous bioactive chemicals in various unconventional plants that play pivotal roles in the synthesis of nanoparticles. For example, Cusioli et al. applied *Moringa oleifera* seed extract functionalized with Fe NPs to synthesize an adsorbent for removing metformin from contaminated water. The adsorptive capacity of Moringa oleifera seeds is primarily attributed to the activity of lectin, a water-soluble protein that functions as a natural cationic polyelectrolyte. This protein facilitates the adherence of contaminants in aqueous media and enhances coagulation (Cusioli et al. 2020). Vidovix et al. employed a green synthesis approach using pomegranate leaves to produce CuO nanoparticles, which were subsequently applied in the adsorption of MB. The equilibrium study demonstrated a maximum adsorption capacity of 166.02 mg/g at a temperature of 318 °K, highlighting the advantages and potential applications of phytosynthesized nanoparticles in the remediation of contaminated water (Vidovix et al. 2019).

11.5.2 Soil Remediation

Soil supports biodiversity, carbon storage and climate regulation, erosion control, nutrient cycling, water filtration, and storage. Soil quality directly affects agricultural productivity and influences the nutritional composition of the crops grown. Recently, soil contamination has become a significant issue, similar to other contamination, due to enhanced environmental pollution. The challenges are in monitoring (in situ), obtaining effective mixing, and successful interfaces between the remedial agents and the objective pollutant in contaminated soil. Traditionally, ex situ remediation strategies *like surfactant flushing, thermal desorption, chemical oxidation, soil vapor extraction*, chemical reduction, immobilization, sequestration, and bioremediation have been developed to remove soil contaminants. But some limitations like time and energy consumption, economic ineptitude, or environmental unfriendliness hinder them from effective application in the remediation of soil contaminants (Qian et al. 2020).

Phytosynthesized nanocompounds offer a promising approach for remediating contaminated soil by enhancing pollutant uptake, degradation, stabilization, and immobilization. Their small particle size and large surface-to-volume ratio allow them to interact with contaminants, making them more accessible for plant absorption and facilitating more effective soil cleanup. Some of these nanoparticles exhibit catalytic properties, accelerating the breakdown of organic pollutants or converting toxic metals into less harmful forms. By promoting decomposition processes, they help reduce the concentration of pollutants in the soil. Additionally, they contribute to pollutant immobilization by forming complexes or adsorbing contaminants, preventing their uptake by plants and reducing groundwater contamination risks. Certain plant-derived nanoparticles also act as soil amendments, improving fertility by enhancing water retention, mineral absorption, and organic matter content, thereby supporting plant growth. Their ability to improve soil conditions indirectly aids in remediation by fostering the establishment of pollutant-degrading plant species. Moreover, phytosynthesized nanoparticles can be integrated with other remediation techniques, such as electrokinetic remediation or soil washing, to enhance contaminant removal efficiency (Romeh 2018).

11.5.3 Air Pollution Control

Air pollution poses substantial threats to human health and ecosystems, while airborne pathogens contribute to infectious diseases. Addressing these challenges requires reducing emissions, adopting cleaner technologies, and implementing sustainable practices. In this context, nanomaterials, particularly green-synthesized and phytosynthesized nanoparticles, have emerged as promising eco-friendly solutions for air purification. These nanoparticles leverage adsorption, catalysis, and plant-microbial interactions to enhance contaminant removal and remediation efficiency. A critical aspect of air pollution control is managing bioaerosols, which may carry bacteria, microbial components, or toxic byproducts. Plant-extract-based nanomaterials offer a safer alternative to conventional inorganic nanoparticles, like Ag NPs or CNTs, for bioaerosol mitigation. Hwang et al. developed antimicrobial air filters using nanoparticles derived from natural plant (*Euscaphis japonica*) extracts. These filters were evaluated for pressure drop, antibacterial effect, filtration effectiveness, key chemical factors, and cytotoxicity. The results demonstrated enhanced antibacterial activity against airborne bacteria like *Staphylococcus epidermidis* and *Micrococcus luteus*, suggesting effective bacterial aerosol inactivation. The active compounds in the plant (*Euscaphis japonica*) extract, including 1(β)-O-galloyl pedunculagin, kaempferol-3-O-glucoside, and quercetin-3-O-glucuronide are believed to play a crucial role in bioaerosol suppression, further highlighting the potential of plant-derived nanomaterials for air purification applications (Hwang et al. 2015).

Sim et al. synthesized antimicrobial nanoparticles extracted from the *Sophora flavescens* plant coated onto activated carbon filters. Activated carbon filters, which have a large surface area, a high adsorption capacity, and specific surface reactivity, have a variety of uses, including dehumidification, air filtration, and water purification. However, because carbon materials have a strong affinity for microorganisms, bacteria may accumulate on activated carbon filters during refinement and can reduce gas adsorption effectiveness. The activated carbon filter was coated with *Sophora flavescens* NPs to combat this microbial contamination. The coating of *Sophora flavescens* NPs in the activated carbon filter boosted the efficacy of *Staphylococcus epidermidis* inactivation. Enhanced absorption of toxic volatile compounds was also observed, suggesting that an activated carbon filter coated with plant extract nanoparticles could be a promising approach to dealing with bioaerosol and gaseous pollution (Sim et al. 2014).

11.5.4 Heavy Metal Removal

Heavy metal pollution has emerged as a pressing global issue, exacerbated by industrial activities and disturbances in natural biogeochemical cycles, necessitating urgent attention and effective management strategies. Unlike organic pollutants, heavy metals are non-decomposable, accumulating in soils and water bodies, posing serious risks to ecosystems and human health. Their presence in soil, originating from industries, mining, and agriculture, can impair plant growth, reduce soil fertility, and disrupt microbial communities. In aquatic systems, heavy metals bioaccumulate in the food chain, affecting fish, invertebrates, and algae, leading to biodiversity loss and ecological imbalances. Moreover, their entry into the food chain through contaminated plants and water further amplifies health risks, making heavy metal pollution a critical environmental challenge (Mitra et al. 2022). Therefore, to mitigate the adverse effects of heavy metals, it is essential to implement proper waste management practices, control industrial emissions, promote sustainable agriculture, and engage in remediation efforts to clean up contaminated sites. Regular monitoring and control measures are essential to minimizing heavy metal pollution and protecting the ecosystem and human health.

The phytosynthesized nanoparticles have emerged as an eco-friendly and sustainable approach to removing heavy metals from the environment. These nanoparticles offer unique properties and interactions that can effectively mitigate the negative impacts of heavy metal pollution and contribute to environmental remediation efforts. The phytosynthesized nanoparticles remove heavy metals by precipitation, redox reaction, chelation, and enhanced uptake. The phytosynthesized nanomaterials can prompt the precipitation of heavy metal ions, leading to their removal from the solution. The nanoparticles can act as nucleation sites for forming metal precipitates, thereby facilitating the exclusion of heavy metals from the aqueous phase. These nanoparticles can undergo redox reactions with heavy metal ions, transforming them into less toxic forms or their complete removal. The nanoparticles can act as electron donors or acceptors, facilitating the conversion of heavy metals to their respective oxidation states. The phytosynthesized nanoparticles often contain biomolecules, such as proteins, enzymes, and phytochemicals, which possess chelating properties. These biomolecules can intricate with heavy metal ions, establishing stable complexes that are less toxic and more easily removable from the environment. For example, Tagesse et al. synthesized Ag NPs with the *Moringa stenopetal* extract to remove Cr from an aqueous solution. These green synthesized Ag NPs were observed to remove 97% of Cr from the aqueous solution (Tagesse 2021). These findings demonstrated that green synthesized nanoparticles could be promising nanosorbents for the remediation of heavy metal contamination.

11.5.5 Organic Pollutant Degradation

Organic pollutants refer to pharmaceuticals, personal care goods, and chemicals that contain carbon atoms. These pollutants can have various detrimental effects on the environment. Some of the significant harmful effects of organic pollutants include toxicity, bioaccumulation, biomagnification, and disruption of ecosystems, as well as water, air, and soil contamination. Many organic pollutants are toxic to living organisms, including humans, animals, and plants. Depending on exposure level and duration, they can cause acute or chronic poisoning. These pollutants can disrupt cellular processes and damage organs, leading to death. These pollutants tend to pile up in the tissues of organisms through the process of bioaccumulation. They often resist degradation, persisting in the environment for a long time. As organisms consume contaminated food or water, the pollutants accumulate in their bodies, leading to higher concentrations in developed trophic levels of the food chain. These pollutants can also undergo biomagnification, increasing their concentrations as they move up the food chain. Predatory organisms that consume tainted prey can accumulate higher levels of contaminants in their bodies. Organic pollutants can also disrupt ecological balance and harm ecosystems. They can affect the growth of plants, impair the reproductive success of animals, and alter the composition of communities. This disruption can have far-reaching consequences, including reduced biodiversity and ecosystem instability (Wang et al. 2023).

There are numerous conventional ways to degrade organic pollutants in environments, such as advanced oxidation processes, adsorption, membrane treatment, and combination processes, but most are expensive and time-consuming. Recently, phytosynthesized nanoparticles have emerged as a viable approach to enhancing the deterioration of organic pollution. The efficacy of phytosynthesized nanomaterials in organic pollutant degradation can vary based on aspects like nanoparticle properties, pollutant characteristics, reaction conditions, and the unique application scenario. Phytosynthesized nanoparticles can exhibit catalytic properties that facilitate the degradation of organic pollutants. These nanoparticles can act as catalysts, accelerating chemical reactions involved in the breakdown of contaminants (Choudhary et al. 2018). For example, Choudhary et al. developed Ag NPs using *Cuminum cyminum* seed extract. These phytosynthesized Ag NPs were observed to catalyze the deterioration of various organic dyes like methyl blue (MB), methyl red (MR), and rhodamine B (RhB) dyes in 2 min (Choudhary et al. 2018). Bogireddy et al. synthesized the biogenic Ag NPs using *Sterculia acuminata* fruit extract. The synthesized biogenic Ag NPs depicted excellent catalytic activity in the reduction of various organic dyes like 4-NP, MO, phenol red (PR), direct blue 24 (DB24), and MB (Bogireddy et al. 2016). It has been concluded that this exceptional dye degradation efficiency of biogenic Ag NPs may pave the way for new industrial and environmental applications.

Phytosynthesized nanoparticles can participate in redox chemical reactions, facilitating the deprivation of organic pollutants. These nanoparticles can transfer electrons to or accept electrons from contaminants, leading to their transformation or breakdown into less harmful compounds. For example, Fe NPs derived from plant extracts have shown the potential to degrade various organic pollutants through redox reactions. Some phytosynthesized nanoparticles, such as TiO_2 NPs derived from plant extracts, also exhibit photocatalytic activity. These nanoparticles can absorb light energy and generate ROS, such as hydroxyl radicals (•OH), which are highly reactive and capable of destroying organic pollutants. Plant-derived nanoparticles can also exhibit synergistic effects when combined with other degradation techniques. For example, they can be used with advanced oxidation or biological treatments to enhance the overall degradation efficiency. The nanoparticles can act as catalysts or provide ROS, complementing the degradation mechanisms of other treatment approaches (Abdelfattah and El-Shamy 2024).

11.6 Limitations and Future Prospects

Phytosynthesized nanoparticles, derived from plant extracts or plant-based compounds, offer several advantages, like eco-friendliness, cost-effectiveness, and ease of synthesis. However, they also have certain limitations that need to be considered. Here are some limitations: (a) standardization and reproducibility of phytosynthesized nanoparticles can vary based on plant species, extraction methods, and environmental considerations. This lack of standardization can result in variations in nanoparticle properties, hindering consistent and reproducible synthesis. (b) Control over the shape, size, and surface properties of phytosynthesized nanoparticles can be challenging. These factors greatly influence their physicochemical properties and performance in specific applications. (c) The limited stability and shelf-life of phytosynthesized nanoparticles due to aggregation, oxidation, or degradation over time. Ensuring their stability under different storage and application conditions is essential for their practical use. (d) Scale-up and production cost of synthesizing phytosynthesized nanoparticles is typically performed on a laboratory scale. Scaling up production to meet industrial demands while maintaining quality and cost-effectiveness remains challenging. (e) Although the eco-toxicity and environmental impact of phytosynthesized nanoparticles are generally considered environmentally friendly, their potential eco-toxicity and long-term environmental impacts need further investigation. Assessing their behavior, persistence, and possible effects on ecosystems is essential.

Additionally, several prospects can enhance their potential application and efficiency in the remediation of environmental contaminants, like (a) optimization of synthesis protocols, further research can focus on optimizing and standardizing synthesis protocols for phytosynthesized nanoparticles. This includes identifying key factors influencing NPs properties and developing reproducible methods for consistent synthesis. (b) Tailoring properties through plant selection: Different plant species possess unique phytochemical compositions, which can be harnessed to tailor the properties of phytosynthesized nanoparticles. Exploring various plant sources and their extract variations can enable the synthesis of nanoparticles with desired characteristics. (c) Functionalization and hybridization of nanoparticles with specific ligands, polymers, or metals to enhance their stability, reactivity, and selectivity. Also, hybridizing these nanoparticles with other nanomaterials can improve properties and multi-functionality. (d) Advanced characterization techniques can provide detailed insights into the physical and chemical properties of phytosynthesized nanoparticles. This can aid in understanding their properties, optimizing synthesis parameters, and correlating them with performance in remediation and other applications. (e) Regulatory considerations: as the field progresses, it is necessary to establish guidelines for the safe and responsible use of phytosynthesized nanoparticles. Assessing their potential risks, conducting toxicity studies, and implementing proper disposal strategies are essential for their commercialization and widespread adoption.

11.7 Conclusion

In conclusion, phytosynthesized nanoparticles have emerged as an eco-friendly approach offering promising potential for the remediation of environmental contaminants. These nanoparticles, synthesized using plant extracts or plant-derived compounds, have shown applications in various remediation processes. They can be utilized to remove heavy metals, degrade organic pollutants, and purify water resources. Moreover, essential phytochemicals of plant-extracted nanoparticles contribute to the bioaerosol inactivation and mitigation of airborne toxins. Phytosynthesized nanoparticles can also aid in immobilizing and stabilizing contaminants in soil and removing heavy metals from the aqueous phase. The phytosynthesized Ag NPs have been demonstrated to catalyze the degradation of various organic dyes, including MB, MR, and RhB colors. However, standardization, scalability, stability, and potential eco-toxicity of phytosynthesized nanoparticles need to be attended.

Further investigations are required to optimize synthesis techniques, ensure reproducibility, improve scalability, and assess these nanoparticles' long-term effects and environmental impacts. The future prospects of phytosynthesized nanoparticles in remediation are promising. Advancements in synthesis techniques, including the control of nanoparticle properties, will enhance their remediation capabilities. Integrating phytosynthesized nanoparticles into hybrid materials and developing targeted delivery systems can improve efficiency and selectivity. Additionally, integrating these nanoparticles with other technologies and considering regulatory aspects will contribute to their safe and responsible use. Overall, phytosynthesized nanoparticles play a significant role in the remediation of environmental contaminants. With continued research, standardization, and their synthesis and application advancements, these nanoparticles can contribute to sustainable and effective remediation strategies, improving environmental quality and human well-being.

References

Abdelfattah I, El-Shamy AM (2024) A comparative study for optimizing photocatalytic activity of TiO_2-based composites with ZrO_2, ZnO, Ta_2O_5, SnO, Fe_2O_3, and CuO additives. Sci Rep 14:1–26. https://doi.org/10.1038/s41598-024-77752-5

Ahamed M, Karns M, Goodson M et al (2008) DNA damage response to different surface chemistry of silver nanoparticles in mammalian cells. Toxicol Appl Pharmacol 233:404–410. https://doi.org/10.1016/j.taap.2008.09.015

Ahmed SF, Mofijur M, Rafa N et al (2022) Green approaches in synthesising nanomaterials for environmental nanobioremediation: technological advancements, applications, benefits and challenges. Environ Res 204:111967. https://doi.org/10.1016/j.envres.2021.111967

Ali I, Khan TA, Asim M (2011) Removal of arsenic from water by electrocoagulation and electrodialysis techniques. Sep Purif Rev 40:25–42. https://doi.org/10.1080/15422119.2011.542738

Aragaw TA, Bogale FM, Aragaw BA (2021) Iron-based nanoparticles in wastewater treatment: a review on synthesis methods, applications, and removal mechanisms. J Saudi Chem Soc 25:101280. https://doi.org/10.1016/j.jscs.2021.101280

Babonaitė M, Striogaitė E, Grigorianaitė G, Lazutka JR (2024) In vitro evaluation of DNA damage induction by silver (Ag), gold (Au), silica (SiO_2), and aluminum oxide (Al_2O_3) nanoparticles in human peripheral blood mononuclear cells. Curr Issues Mol Biol 46:6986–7000. https://doi.org/10.3390/cimb46070417

Bibi I, Nazar N, Ata S et al (2019) Green synthesis of iron oxide nanoparticles using pomegranate seeds extract and photocatalytic activity evaluation for the degradation of textile dye. J Mater Res Technol 8:6115–6124. https://doi.org/10.1016/j.jmrt.2019.10.006

Bogireddy NKR, Kiran Kumar HA, Mandal BK (2016) Biofabricated silver nanoparticles as green catalyst in the degradation of different textile dyes. J Environ Chem Eng 4:56–64. https://doi.org/10.1016/j.jece.2015.11.004

Chand K, Cao D, Eldin Fouad D et al (2020) Green synthesis, characterization and photocatalytic application of silver nanoparticles synthesized by various plant extracts. Arab J Chem 13:8248–8261. https://doi.org/10.1016/j.arabjc.2020.01.009

Chaudhuri K, Hasan SN, Barai AC et al (2019) Green synthesis and characterization of gold nanoparticles using *Bursera serrata* fruit extract. Asian J Pharm Pharmacol 5:332–343. https://doi.org/10.31024/ajpp.2019.5.2.17

Choudhary MK, Kataria J, Sharma S (2018) Evaluation of the kinetic and catalytic properties of biogenically synthesized silver nanoparticles. J Clean Prod 198:882–890. https://doi.org/10.1016/j.jclepro.2018.09.015

Crini G, Lichtfouse E, Wilson LD, Morin-Crini N (2019) Conventional and non-conventional adsorbents for wastewater treatment. Environ Chem Lett 17:195–213. https://doi.org/10.1007/s10311-018-0786-8

Cusioli LF, Quesada HB, de Brito Portela Castro AL et al (2020) Development of a new low-cost adsorbent functionalized with iron nanoparticles for removal of metformin from contaminated water. Chemosphere 247:125852. https://doi.org/10.1016/j.chemosphere.2020.125852

El-Eskandarany MS, Al-Hazza A, Al-Hajji LA et al (2021) Mechanical milling: a superior nanotechnological tool for fabrication of nanocrystalline and nanocomposite materials. Nanomaterials 11:2484. https://doi.org/10.3390/nano11102484

Es'haghi Z, Vafaeinezhad F, Hooshmand S (2016) Green synthesis of magnetic iron nanoparticles coated by olive oil and verifying its efficiency in extraction of nickel from environmental samples via UV-vis spectrophotometry. Process Saf Environ Prot 102:403–409. https://doi.org/10.1016/j.psep.2016.04.011

Ganaie SU, Abbasi T, Abbasi SA (2016) Rapid and green synthesis of bimetallic Au–Ag nanoparticles using an otherwise worthless weed Antigonon leptopus. J Exp Nanosci 11:395–417. https://doi.org/10.1080/17458080.2015.1070311

Gour A, Jain NK (2019) Advances in green synthesis of nanoparticles. Artif Cells Nanomed Biotechnol 47:844–851. https://doi.org/10.1080/21691401.2019.1577878

Guerra F, Attia M, Whitehead D, Alexis F (2018) Nanotechnology for environmental remediation: materials and applications. Molecules 23:1760. https://doi.org/10.3390/molecules23071760

Hwang GB, Heo KJ, Yun JH et al (2015) Antimicrobial air filters using natural *Euscaphis japonica* nanoparticles. PLoS One 10:1–14. https://doi.org/10.1371/journal.pone.0126481

Keller AA, Ehrens A, Zheng Y, Nowack B (2023) Developing trends in nanomaterials and their environmental implications. Nat Nanotechnol 18:834–837. https://doi.org/10.1038/s41565-023-01409-z

Krishnaraj C, Jagan EG, Rajasekar S et al (2010) Synthesis of silver nanoparticles using *Acalypha indica* leaf extracts and its antibacterial activity against water borne pathogens. Colloids Surf B Biointerfaces 76:50–56. https://doi.org/10.1016/j.colsurfb.2009.10.008

Kumar PPNV, Shameem U, Kollu P et al (2015) Green synthesis of copper oxide nanoparticles using *Aloe vera* leaf extract and its antibacterial activity against fish bacterial pathogens. BioNanoScience 5:135–139. https://doi.org/10.1007/s12668-015-0171-z

Larayetan R, Ojemaye MO, Okoh OO, Okoh AI (2019) Silver nanoparticles mediated by *Callistemon citrinus* extracts and their antimalaria, antitrypanosoma and antibacterial efficacy. J Mol Liq 273:615–625. https://doi.org/10.1016/j.molliq.2018.10.020
Mitra S, Chakraborty AJ, Tareq AM et al (2022) Impact of heavy metals on the environment and human health: novel therapeutic insights to counter the toxicity. J King Saud Univ Sci 34:101865. https://doi.org/10.1016/j.jksus.2022.101865
Mpouras T, Polydera A, Dermatas D et al (2021) Multi wall carbon nanotubes application for treatment of Cr(VI)-contaminated groundwater; modeling of batch & column experiments. Chemosphere 269:128749. https://doi.org/10.1016/j.chemosphere.2020.128749
Nayak D, Pradhan S, Ashe S et al (2015) Biologically synthesised silver nanoparticles from three diverse family of plant extracts and their anticancer activity against epidermoid A431 carcinoma. J Colloid Interface Sci 457:329–338. https://doi.org/10.1016/j.jcis.2015.07.012
Özdemir S, Serkan Yalçın M, Kılınç E (2021) Preconcentrations of Ni(II) and Pb(II) from water and food samples by solid-phase extraction using *Pleurotus ostreatus* immobilized iron oxide nanoparticles. Food Chem 336:127675. https://doi.org/10.1016/j.foodchem.2020.127675
Parashar M, Shukla VK, Singh R (2020) Metal oxides nanoparticles via sol–gel method: a review on synthesis, characterization and applications. J Mater Sci Mater Electron 31:3729–3749. https://doi.org/10.1007/s10854-020-02994-8
Peternela J, Silva MF, Vieira MF et al (2018) Synthesis and impregnation of copper oxide nanoparticles on activated carbon through green synthesis for water pollutant removal. Mater Res 21:1–11. https://doi.org/10.1590/1980-5373-mr-2016-0460
Peydayesh M, Mohammadi T, Nikouzad SK (2020) A positively charged composite loose nanofiltration membrane for water purification from heavy metals. J Membr Sci 611:118205. https://doi.org/10.1016/j.memsci.2020.118205
Qian Y, Qin C, Chen M, Lin S (2020) Nanotechnology in soil remediation—applications vs. implications. Ecotoxicol Environ Saf 201:110815. https://doi.org/10.1016/j.ecoenv.2020.110815
Rajput VD, Minkina T, Upadhyay SK et al (2022) Nanotechnology in the restoration of polluted soil. Nanomaterials 12:1–16. https://doi.org/10.3390/nano12050769
Rane AV, Kanny K, Abitha VK, Thomas S (2018) Methods for synthesis of nanoparticles and fabrication of nanocomposites. In: Synthesis of inorganic nanomaterials. Elsevier, pp 121–139
Rao Y, Inwati GK, Singh M (2017) Green synthesis of capped gold nanoparticles and their effect on Gram-positive and Gram-negative bacteria. Futur Sci OA 3:FSO239. https://doi.org/10.4155/fsoa-2017-0062
Romeh AAA (2018) Green silver nanoparticles for enhancing the phytoremediation of soil and water contaminated by fipronil and degradation products. Water Air Soil Pollut 229:147. https://doi.org/10.1007/s11270-018-3792-3
Rong K, Wang J, Zhang Z, Zhang J (2020) Green synthesis of iron nanoparticles using Korla fragrant pear peel extracts for the removal of aqueous Cr(VI). Ecol Eng 149:105793. https://doi.org/10.1016/j.ecoleng.2020.105793
Salgado P, Márquez K, Rubilar O et al (2019) The effect of phenolic compounds on the green synthesis of iron nanoparticles (FexOy-NPs) with photocatalytic activity. Appl Nanosci 9:371–385. https://doi.org/10.1007/s13204-018-0931-5
Sharma BM, Bečanová J, Scheringer M et al (2019) Health and ecological risk assessment of emerging contaminants (pharmaceuticals, personal care products, and artificial sweeteners) in surface and groundwater (drinking water) in the Ganges River basin, India. Sci Total Environ 646:1459–1467. https://doi.org/10.1016/j.scitotenv.2018.07.235
Sim KM, Kim KH, Hwang GB et al (2014) Development and evaluation of antimicrobial activated carbon fiber filters using Sophora flavescens nanoparticles. Sci Total Environ 493:291–297. https://doi.org/10.1016/j.scitotenv.2014.06.002
Soliemanzadeh A, Fekri M (2017) The application of green tea extract to prepare bentonite-supported nanoscale zero-valent iron and its performance on removal of Cr(VI): effect of relative parameters and soil experiments. Microporous Mesoporous Mater 239:60–69. https://doi.org/10.1016/j.micromeso.2016.09.050

Stan M, Popa A, Toloman D et al (2015) Enhanced photocatalytic degradation properties of zinc oxide nanoparticles synthesized by using plant extracts. Mater Sci Semicond Process 39:23–29. https://doi.org/10.1016/j.mssp.2015.04.038

Supraja N, Prasad TNVKV, Krishna TG, David E (2016) Synthesis, characterization, and evaluation of the antimicrobial efficacy of *Boswellia ovalifoliolata* stem bark-extract-mediated zinc oxide nanoparticles. Appl Nanosci 6:581–590. https://doi.org/10.1007/s13204-015-0472-0

Suresh D, Nethravathi PC, Udayabhanu et al (2015) Green synthesis of multifunctional zinc oxide (ZnO) nanoparticles using *Cassia fistula* plant extract and their photodegradative, antioxidant and antibacterial activities. Mater Sci Semicond Process 31:446–454. https://doi.org/10.1016/j.mssp.2014.12.023

Tagesse W (2021) Adsorptive removal of chromium (VI) using silver nanoparticles synthesized via green approach with the extract of moringastenopetala. Orient J Chem 37:380–387. https://doi.org/10.13005/ojc/370217

Vidovix TB, Quesada HB, Januário EFD et al (2019) Green synthesis of copper oxide nanoparticles using *Punica granatum* leaf extract applied to the removal of methylene blue. Mater Lett 257:8–11. https://doi.org/10.1016/j.matlet.2019.126685

Wang S, Zheng N, Sun S et al (2023) Bioaccumulation of organophosphorus flame retardants in marine organisms in Liaodong Bay and their potential ecological risks based on species sensitivity distribution. Environ Pollut 317:120812. https://doi.org/10.1016/j.envpol.2022.120812

Yang Y, Mao Z, Huang W et al (2016) Redox enzyme-mimicking activities of CeO_2 nanostructures: intrinsic influence of exposed facets. Sci Rep 6:1–7. https://doi.org/10.1038/srep35344

Yeasmin S, Datta HK, Chaudhuri S et al (2017) In-vitro anti-cancer activity of shape controlled silver nanoparticles (AgNPs) in various organ specific cell lines. J Mol Liq 242:757–766. https://doi.org/10.1016/j.molliq.2017.06.047

Zhan G, Huang J, Du M et al (2011) Green synthesis of Au–Pd bimetallic nanoparticles: single-step bioreduction method with plant extract. Mater Lett 65:2989–2991. https://doi.org/10.1016/j.matlet.2011.06.079

Zhu X, Tian S, Cai Z (2012) Toxicity assessment of iron oxide nanoparticles in zebrafish (*Danio rerio*) early life stages. PLoS One 7:e46286. https://doi.org/10.1371/journal.pone.0046286

Chapter 12
Exploring the Potential of Fungus-Mediated Synthesis of Nanoparticles and Its Application in Remediation of Contaminated Environment

Shoaib Mukhtar and Imran Uddin

Abstract An environmentally responsible and long-lasting method of reducing environmental contamination is fungus bioremediation. This chapter uses a natural synthesis strategy to investigate the synergistic coupling of new nanomaterial utilization with fungus-based remediation solutions. The main objective is to reduce the ecological impact of conventional remediation techniques while addressing environmental damage. Because of their diverse enzymatic machinery, fungi are essential for the breakdown of both organic and inorganic contaminants. This chapter concludes the significance of the remediation potential of fungi by using nanomaterials, utilizing their natural capabilities. Because of their special physicochemical characteristics, nanoparticles provide a viable way to increase the effectiveness and selectivity of bioremediation procedures. Environmentally friendly processes and sustainable resources are utilized in the eco-friendly synthesis of nanomaterials, which is in line with the ideas of ecologically responsible behavior. This method encourages the development of scalable and affordable solutions while reducing the usage of dangerous chemicals. By combining environmentally acceptable nanomaterial synthesis with fungal-mediated processes, this chapter aims to further the subject of bioremediation. The study's findings have great potential for creating sustainable and environmentally friendly solutions to a range of environmental problems, promoting a cleaner and healthier world.

S. Mukhtar
Research Group of Environmental and Inorganic Photochemistry, Center for Natural Sciences, Faculty of Engineering, University of Pannonia,
Veszprém P.O BOX 158, H-8201, Hungary
e-mail: mukhtar.shoaib@phd.mk.uni-pannon.hu

I. Uddin (✉)
Department of Conservative Dentistry and Endodontics, Saveetha Dental College and Hospital, Saveetha Institute of Medical and Technical Sciences, Saveetha University, Chennai, Tamil Nadu, India

A. Guldhe et al. (eds.), *Biomaterials for Environmental Applications: Detection and Remediation*, Environmental Challenges and Solutions,
https://doi.org/10.1007/978-3-032-12701-3_12

Keywords Fungus · Bioremediation · Ecofriendly · Nanomaterials · Green synthesis

12.1 Introduction

The natural world has developed a multitude of highly efficient arrangements of proteins, nucleic acids, and other macromolecules, which are capable of carrying out complex activities that remain challenging for humans to replicate in a laboratory setting (Lutz et al. 2013). One issue that has garnered significant attention from scientists of materials is the advancement of extremely efficient tiny functional materials by natural means, using sophisticated and innovative methods. The current increase in interest in nanoscience and nanotechnology may be attributed to the persistent efforts to develop functioning tiny structures that mimic those found in nature (Kumar et al. 2024). Nanomaterials are used in many consumer goods, including electronics, because of their special mechanical, chemical, and physical characteristics. These nanoparticles can be synthesized using green approach which can be seen in (Fig. 12.1) (Solomon and D'Souza 2011; Abdussalam-Mohammed 2019).

In biology, there are many magnificent examples of artificial structures that are put together in an organized way. Biological systems can make artificial materials through processes that happen at temperatures, pressures, and pH levels that are close to those found in nature (Suganya and Mahalingam 2017). Nature has evolved many methods for the production of inorganic substances at the micro- and

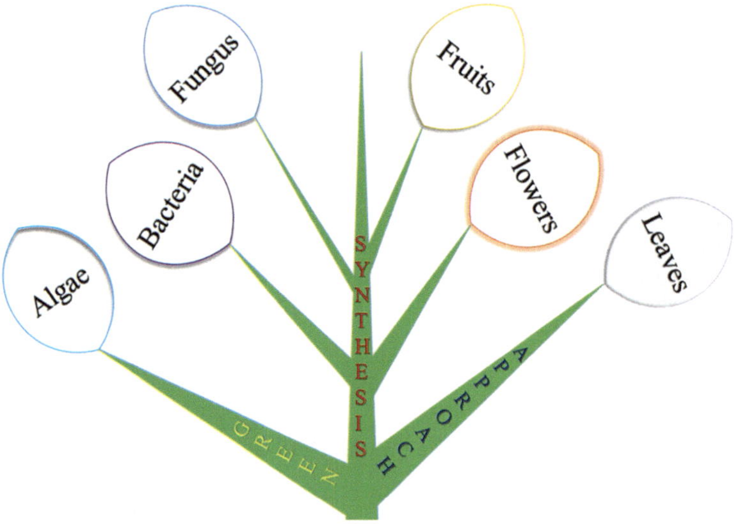

Fig. 12.1 Green approach for synthesis of nanoparticles

nanoscales. A crucial aspect of nanotechnology is developing reliable and ecologically friendly techniques for manufacturing materials on a nanoscale (Reyes et al. 2009). Recently, the possibility of using microorganisms as bioreactors has been investigated for synthesizing nanoparticles. These approaches use biomolecules such as DNA, proteins, enzymes, and microorganisms like fungi and bacteria to reduce metal ions and generate nanoparticles. These microorganisms provide potential for environmentally friendly and energy-efficient operations. Nanoparticles can be synthesized in circumstances of room temperature and neutral to mild pH, which closely resemble ambient settings, without the use of hazardous chemicals, solvents, or expensive organometallic precursors. Obtaining these substances through physical or chemical means is not feasible. This technique has brought about a significant change in the creation of nanoparticles, representing a shift in the prevailing paradigm. Due to the detrimental effects of heavy metal ions on biological systems, bacteria, yeast, and fungi use a detoxification process whereby they convert these ions into a non-toxic metallic form. Proteins and other biomolecules spontaneously functionalize the surfaces of the synthesized nanomaterials. This not only facilitates the attachment of additional ligands to the surfaces, but also helps to prevent agglomeration, promote stable suspension in water, and provide necessary surface passivation (Sandoval-Cárdenas et al. 2017). These advantages are particularly valuable in wet-chemical methods. The deliberate use of microbes for the synthesis of nanomaterials offers a strategic approach to produce a diverse array of nanomaterials, including basic metals as well as complex systems like metal oxides and sulfides. In addition to metal nanoparticles, several studies have also addressed production of semiconductor nanoparticles, including ZnS and CdS, and other similar materials (Sonawane et al. 2023). Fungi are an excellent option for establishing extracellular processes due to their ease of cultivation and handling, as well as their ability to produce several enzymes (Rizzo et al. 2020). Fungi have a high capacity for protein synthesis and enzyme production. Moreover, organic nanoparticles derived from a specific fungal enzyme possess several desirable attributes. Extracellular biosynthesis of nanoparticles is superior to intracellular biosynthesis for extensive production due to the ease of separating and cleaning nanoparticles from the extracellular reaction solution (Baskar and Selvam 2021). Fungal species like *Phaenerochaete chrysosporium*, *Trichothecium* sp., *Fusarium oxysporum*, *Colletotrichum* sp., *A. fumigatus*, *F. semitectum*, *T. viride*, *Aspergillus niger*, *Coriolus versicolor*, *Penicillium brevicompactum*, *Phoma glomerata*, *Cladosporium cladosporioides*, *Penicillium fellutanum*, and *Volvariella volvaceae* have all been investigated in recent years for the synthesis of nanoparticles. In comparison with other microbes, fungi provide many advantages. In bioreactors or other chambers, fungus mycelial mesh is more resilient than plant materials or bacteria under various circumstances, including flow pressure and agitation. They are simple to handle, meticulous to grow, and simple to manufacture. Reductive proteins secrete more extracellularly, and these secretions are more manageable in downstream processing. Furthermore, since the nanoparticles that precipitate outside of cells have no extraneous biological components, they may be used directly in a variety of applications. This conjugation of functions has ushered in a new age in bioremediation and is expected to yield multifunctional nanoparticles. We shall attempt to shed some light on the production of nanoparticles by fungi and the growing bioremediation uses of this process here.

12.2 Fungus-Mediated Synthesis of Metal, Metal Oxide, Metal Sulfides, and Bimetallic Nanoparticles

The remarkable optoelectronic properties of metallic nanoparticles make their manufacturing process improvements very desirable. Anticipating such unique properties and potential uses, numerous forms, such as nanorods/nanowires, nanoplates, nano cubes, nano dumbbells, nano discs, nano tapes/nanobelts, nanotetrapods/arrows, and countless others, have been successfully synthesized using a range of chemical and physical techniques. Fungus-mediated metal, metal oxide, and metal sulfide synthesis and bimetallic nanoparticles have been discussed below.

12.2.1 Fungus-Mediated Synthesis of Metal Nanoparticles

In their 2021 study, Wang and colleagues used liquid extracts from *Aspergillus sydowii* fungi to create silver nanoparticles outside of living cells. The researchers first noticed that silver nanoparticles were forming when the solution changed color during the process. They confirmed this particle formation by measuring light absorption at 420 nm using ultraviolet spectroscopy. X-ray diffraction analysis provided additional verification of the nanoparticles structure. The research revealed that these silver nanoparticles were spherical in shape with varied sizes between 1 and 24 nm. The particles displayed a cubic crystalline structure, meaning their atoms were arranged in a regular, cube-like pattern.

The researcher tested different conditions to find the best way to make these nanoparticles. They adjusted three important factors: temperature, acidity level (pH), and the amount of silver salt used. Through this optimization process, they discovered the ideal conditions were using 1.5 mM silver concentration, maintaining a pH of 8.0, and keeping the temperature at 50 °C.

When the researchers tested these silver nanoparticles for medical applications, they found impressive results. The particles showed strong antifungal properties, effectively fighting against several types of disease-causing fungi that affect humans. Additionally, the nanoparticles demonstrated anticancer effects when tested on two specific cancer cell lines in laboratory conditions—HeLa cells (cervical cancer) and MCF-7 cells (breast cancer)—by preventing these harmful cells from multiplying (Wang et al. 2021).

Gajbhiye's 2009 study showed that the everyday fungus Alternaria alternata can be used to make silver nanoparticles (Ag-NPs) outside the fungal cells. The researchers then tested whether adding these Ag-NPs could boost the antifungal power of the drug fluconazole against four different fungi: *Fusarium semitectum, Phoma glomerata, Phoma herbarum,* and *Candida albicans.* When the silver nanoparticles were mixed with fluconazole, the drug worked better against all test organisms. The strongest extra effect was seen with Candida albicans, where the area of fungal growth inhibition expanded the most. There was a moderate improvement against *Phoma glomerata* and *Trichoderma* species, but no noticeable boost in activity against *Phoma herbarum or Fusarium semitectum* (Gajbhiye et al. 2009). In 2021, Boruah reported that chitosan nanoparticles were produced using a single commercial supply of chitosan as well as four distinct fungal sources: *Trichoderma viride, Beauveria bassiana, Metarhizium anisopliae,* and *Fusarium oxysporum.* Research using UV-vis spectroscopy on synthesized nanoparticles revealed absorption peaks in the 310.02–342.00 nm range. The produced nanoparticles' surfaces were found to be stable and positively charged in the natural world. Synthesized nanoparticles are almost spherical in form, according to research using electron microscopy. The nanoparticles of chitosan are made from *T. viride, B. bassiana, M. anisopliae,* and *F. oxysporum,* and commercial goods had average sizes of 273.20, 172.50, 78.36, 89.03, and 300.10 nm, according to DLS analysis. The results of the compatibility research between *T. asperellum* and chitosan nanoparticles revealed a very compatible response at the concentration that was examined. In vitro tests using *T. asperellum* and chitosan nanoparticles against *Fusarium oxysporum, Rhizoctonia solani,* and *Sclerotium rolfsii* showed that the combination suppressed the pathogens' mycelial development more effectively than Trichoderma alone and carbendazim at 0.1% (Boruah and Dutta 2021).

Actinobacterial spp. were used in the biological production of anisotropic metal nanoparticles, as described by Bharde et al. (2007). The resulting biogenic gold nanoparticles have a morphology of triangles and hexagons, with edge lengths ranging from 0.05 to 2 μm. The triangular gold particles' surface topography revealed that their thickness is around 9 nm. In the electromagnetic spectrum, these flat particles exhibit significant absorption in the near infrared (NIR) region. It was discovered that each nanotriangles were single crystals that were strongly aligned in the (111) crystallographic orientation. The cytochrome oxidase in bacteria is in charge of mediating the reduction of gold ions, according to a biochemical examination of the process. Along the bacterial cell wall, capping proteins connected to the triangle particle surface function as shape-directing agents. *Actinobacter* spp. produce hexagonal and triangular gold particles with a high affinity for gold ions. *Actinobacter* spp. are also used to synthesize silver nanoparticles, which have flat, plate-like shapes. Proteins known as silver-binding proteins may bind and stabilize silver nanoparticles, hence mediating the silver ion biological reduction (Bharde et al. 2007). Pritish et al. used the fungus Fusarium acuminatum to produce gold nanoparticles (AuNPs) extracellularly. Au-NPs' mycosynthesis was performed by subjecting the filtrate of fungal cells to a 1 mM solution of HAuCl4, as fungi produce extracellular enzymes that synthesize nanoparticles. The absorbance peak we saw fell between 520 and 550 nm, which is the gold nanoparticles' surface plasmon absorbance. Proteins were discovered to cap the nanoparticles that were synthesized for this study. The unique development of the solution's crystalline gold nanoparticles was shown by the XRD data. TEM investigation revealed the spherical and polydispersed AuNPs, with a 17 nm average size, in the 8–28 nm range. Additionally, standardized the variables that impact pH and temperature on the biogenesis of gold nanoparticles, and salt content. It was discovered that the ideal conditions for the synthesis of Au-NP were an acidic pH, a temperature of 37 °C, and a 1 mM salt concentration (Tidke et al. 2014).

The 2012 study by Mishra and colleagues showed that both the clear broth (supernatant) and the live-cell filtrate from the fungus *Penicillium rugulosum*, grown on a standard potato dextrose medium, can be harnessed to make gold nanoparticles. Researchers confirmed the particles' formation and measured their size and shape using several techniques such as optical absorption measurements in the UV–visible range, infrared vibrational analysis (FTIR), surface chemical profiling via X-ray photoelectron spectroscopy, and crystalline structure assessment using X-ray diffraction. Because the nanoparticles made from the broth fraction had the nicest, most uniform shapes, the team then tweaked two key factors—solution acidity (pH) and the gold-salt concentration—to fine-tune particle size and form. Finally, they demonstrated a practical application by attaching the optimized gold nanoparticles to purified DNA from *Staphylococcus aureus* and *Escherichia coli*, paving the way for bio-conjugation uses (Mishra et al. 2012).

In their 2004 research, Bansal and colleagues discovered a straightforward way to make zirconia nanocrystals at room temperature using proteins released by fungi. The process works by exposing the fungus *Fusarium oxysporum* to water solutions containing zirconia-fluoride compounds (ZrF_6^{2-} anions). The fungus releases certain proteins into the surrounding liquid, these proteins weigh between 24 and 28 kDa and are quite similar to a protein called silicatein. These proteins break down the metal compounds through a process called hydrolysis, which creates zirconia nanoparticles outside the fungal cells. This discovery opens up exciting possibilities for using biological methods to mass-produce important oxide materials that have technological applications (Bansal et al. 2004). In 2005, Bansal claimed that biological systems synthesize inorganic compounds at conditions that are similar to ambient temperature, pressure, and pH neutrality. This is shown by the process of biosilicification seen in marine creatures like diatoms, but the production of silica in laboratories requires very high pH and temperature levels. Here, we demonstrate that *Fusarium oxysporum*, a fungus, may be challenged with the corresponding aqueous anionic complexes SiF_6^{2-} and TiF_6^{2-}, to form particles of silica and titania. However, silica crystallization requires calcination at 300 °C, extracellular proteins facilitate the easy formation of crystalline titania particles at ambient temperature by hydrolyzing the anionic complexes (Bansal et al. 2005).

12.2.2 Fungus-Mediated Synthesis of Metal Oxide Nanoparticles

In 2012, Rajakumar and his research team developed an innovative, environmentally friendly approach to create titanium dioxide nanoparticles (TiO_2 NPs) using the fungus Aspergillus flavus. The fungus acted as both a reducing agent (helping to form the particles) and a capping agent (keeping them stable). The researchers used several analytical techniques to study these nanoparticles: infrared spectroscopy (FTIR), X-ray diffraction (XRD), atomic force microscopy (AFM), scanning electron microscopy (SEM), and transmission electron microscopy (TEM). The X-ray diffraction results showed that more TiO_2 nanoparticles were successfully formed, revealing two different crystal structures: anatase and rutile at specific angles.

When examined under a scanning electron microscope, the TiO_2 nanoparticles appeared as oval and spherical shapes, mostly existing as individual particles with some clusters. Their sizes ranged from 62 to 74 nm. The atomic force microscopy revealed rough surfaces on the TiO_2 particles, which indicated the temperature effects during crystal formation. Testing these nanoparticles against harmful bacteria showed promising results. The synthesized TiO_2 nanoparticles could effectively stop the growth of *E. coli* bacteria at a concentration as low as 40 μg/mL, which matched the results from standard antibacterial testing methods (Rajakumar et al. 2012).

In 2013, Venkatesh and colleagues demonstrated an eco-friendly way to grow zinc oxide (ZnO) nanorods using the common soil fungus Fusarium solani. They collected the dried fungal product, then heated it to 350 °C for 1 h in an oxygen-rich furnace (a process called calcination) to turn it into pure ZnO nanopowder.

To understand how the material behaved under heat, they ran a thermogravimetric/differential thermogravimetric (TGA/DTG) analysis, which tracks weight changes as a sample is heated.

Next, they probed the crystal structure with X-ray diffraction (XRD). A sharp peak at $2\theta = 36.2°$ revealed the characteristic (101) plane of hexagonal zinc oxide, confirming that the powder was indeed ZnO. Finally, scanning electron microscopy (SEM) images showed that these particles took the form of slender rods, each about 60–95 nm in diameter, true nanorods formed by this straightforward, fungus-driven method (Venkatesh et al. 2013). In 2008, Uddin and colleagues reported a production of Bi_2O_3 nanoparticles at room temperature, using a 5–8 nm size range. This was achieved by extracellularly utilizing bismuth nitrate as a precursor, the plant pathogenic fungus *Fusarium oxysporum* is being challenged. The strong coating of proteins that naturally functionalizes the as-synthesized particle surfaces gives them excellent stability in the aqueous medium. Selective area electron diffraction structural analysis, powder XRD, and high-resolution transmission electron microscopy show that the particles are mostly single crystalline and primarily crystallize in alpha-phase with monoclinic structure (Uddin et al. 2008).

12.2.3 Fungus-Mediated Synthesis of Metal Sulfides Nanoparticles

Suganya and her team (2017) showed that you can turn common oyster mushrooms (*Pleurotus florida*) into tiny zinc sulfide (ZnS) crystals using a gentle, "green" method. They soaked mushroom extracts in water, mixed in zinc and sulfur sources, and watched the solution change color, an indication that nanoparticles were forming, later confirmed by UV–visible spectroscopy. To purify the particles, they used a water-ethanol precipitation step, then collected the solid ZnS nanocrystals.

Once isolated, these mushroom-made ZnS particles were examined inside and out. Infrared spectroscopy (FTIR) revealed which chemical groups were attached, X-ray diffraction (XRD) showed the crystals' lattice structure, and powerful microscopes SEM and TEM let the researchers see each particle's shape and size. Energy-dispersive X-ray analysis (EDAX) confirmed the elemental makeup.

Finally, the team tested whether these ZnS nanoparticles could help plants grow. They treated green gram (*Vigna radiata*) seeds with different nanoparticle concentrations and found that seeds exposed to ZnS germinated faster and more vigorously than untreated seeds. In other words, ZnS nanoparticles made by mushrooms not only form beautifully ordered crystals but also give a real boost to seed sprouting (Suganya and Mahalingam 2017).

The 2009 publication by Reyes et al. noted that the optical and electrical features of CdS nanoparticles have garnered interest. Traditional approaches to its synthesis have some drawbacks, including problematic high-pressure and particle size needs, and the use of harmful substances. Inorganic elements may be supplied intracellularly or extracellularly by several microbes; this much is known. This study details the effects of exposing *Fusarium* sp. biomass to a $CdSO_4$ solution prior to its use in a sustainable synthesis process for CdS nanoparticles. Radiation spectroscopy, UV-visible light, and atomic force microscopy were employed to characterize the nanoparticles (Reyes et al. 2009).

In their 2017 study, Sandoval-Cárdenas and colleagues showed how the fungus *Fusarium oxysporum f.* sp. *lycopersici* can "grow" tiny cadmium sulfide (CdS) quantum dots just outside its cells. First, they soaked the fungal threads (mycelia) in a cadmium sulfate solution at 30 °C for 12 days. As the reaction proceeded, the mixture gradually turned yellow, a visual clue that CdS nanoparticles were forming. They then filtered out the fungal biomass with simple paper filters, leaving the newly made quantum dots suspended in the liquid.

To confirm their creation, the team used UV–visible spectroscopy and fluorescence measurements. The filtered solution showed a strong absorbance between 300 and 500 nm and bright green emission at 520 nm signals that never appeared in control samples without cadmium sulfate. Under a transmission electron microscope, these biologically produced dots looked like perfect little spheres, each about 2–6 nm across. Energy-dispersive X-ray analysis (EDX) further verified that they contained cadmium and sulfur.

Importantly, Sandoval-Cárdenas et al. emphasized the value of running both positive and negative controls when using UV–Vis and fluorescence spectroscopy. Controls helped them be confident that the optical signals they saw truly came from fungal biosynthesis of CdS quantum dots and not from unrelated background signals (Sandoval-Cárdenas et al. 2017).

In 2007, Uddin and colleagues showed that simply mixing live cultures of the fungus Fusarium oxysporum with bismuth nitrate pentahydrate ($Bi(NO_3)_3{\cdot}5H_2O$) and sodium sulfite (Na_2SO_3) at room temperature leads to the formation of bright, water-dispersible bismuth sulfide (Bi_2S_3) nanoparticles. Under the electron microscope, these biosynthesized particles appear as nearly spherical dots about 15 nm across. Powder X-ray diffraction and selected-area electron diffraction both confirm they're highly crystalline with an orthorhombic lattice.

These Bi_2S_3 nanoparticles are naturally capped by fungal proteins, giving them a stable band gap of 3.04 eV and excellent shelf life in water. Researchers further characterized them using infrared spectroscopy (FTIR), ultraviolet spectroscopy (UV), photoluminescence (PL), X-ray photoelectron spectroscopy (XPS), and thermogravimetric analysis (TGA).

Finally, Uddin et al. demonstrated a biomedical application: they labeled the nanoparticles as a SPECT–CT imaging probe, injected them into rats, and successfully tracked their distribution in the body over time, calculating both biodistribution patterns and blood-clearance rates (Uddin et al. 2022).

12.2.4 Synthesis of Bimetallic Nanoparticles Mediated by Fungi

Senapati et al. (2005) showed that the fungus *Fusarium oxysporum* can be used to "grow" tiny gold–silver alloy particles outside its cells. By mixing equal parts of gold chloride ($HAuCl_4$) and silver nitrate ($AgNO_3$), each at 1 mM, with different amounts of fungal biomass, they produced round alloy nanoparticles that measured between 8 and 14 nm across. The exact gold-to-silver ratio in each particle depended on how much the natural cofactor NADH the fungus released: more NADH shifted the mix toward one metal or the other. This simple, room-temperature method yielded remarkably stable bimetallic nanostructures just by harnessing the fungus's own chemistry (Senapati et al. 2005).

Longoria and colleagues (2011) screened the filamentous fungus *Neurospora crassa* and found it can "brew" both silver and gold nanoparticles on its own. They exposed the fungus to dilute metal solutions (1 mM $AgNO_3$ for silver, 1 mM $HAuCl_4$ for gold) and then used scanning and transmission electron microscopy (SEM and TEM) plus energy-dispersive X-ray spectroscopy (EDS) to confirm what appeared under the microscope. Silver particles formed by the fungus averaged about 11 nm across, while gold particles were around 32 nm and most were nicely round. By mixing the two metal salts in different proportions, they even managed to produce alloyed Au/Ag nanoparticles, as verified by EDS. Thin slices of the fungal cells examined under TEM showed that *Neurospora crassa* not only makes these nanoparticles but can also accumulate them inside its hyphae. Because *N. crassa* grows quickly, isn't harmful, and naturally reduces and stabilizes metal ions, it's a promising "nanofactory" that's easy and inexpensive to cultivate (Castro-Longoria et al. 2011).

According to Kumar et al. (2007), there is a lot of interest in the development of synthetic processes for semiconductor nanocomposites. Although there are many chemical processes accessible and widely utilized, they often demand greater temperatures, involve dangerous chemicals, and use a lot of energy. Conversely, biological systems synthesis of inorganic materials is typified via procedures that occur at atmospheric pressure and at ambient temperature. For the very first time, they describe how the fungus *Fusarium oxysporum*, when cultivated with a $CdCl_2/SeCl_4$ mixture, produces very bright CdSe quantum dots at room temperature (Kumar et al. 2007). Similarly, many more fungus-mediated nanoparticles have been reported and listed in Table 12.1:

Table 12.1 List of fungus-mediated synthesis of nanoparticles

SN	Fungi	Types of NPs	References
1	*Aspergillus fumigatus*	Ag	Bhainsa and D'Souza (2006)
2	*Aspergillus niger*	Ag	Gade et al. (2008)
3	*Coriolus versicolor*	Ag	Sanghi and Verma (2009)
4	*Colletotrichum* sp.	Au	Shankar et al. (2003)
5	*F. oxysporum*	Au	Gericke and Pinches (2006)
6	*F. oxysporum*	Au	Mukherjee et al. (2002)
7	*Volvariella volvacea*	Au–Ag	Philip (2009)
8	*Aspergillus oryzae*	Fe	Tarafdar and Raliya (2013)
9	*F. oxysporum*	Magnetite	Bharde et al. (2006)
10	*F. oxysporum*	Pt	Riddin et al. (2006)
11	*F. oxysporum*	$SrCO_3$	Rautaray et al. (2004)
12	*Rhizopus nigricans*	Cr	Bai and Abraham (2001)
13	*Aspergillus niger*	Pb	Kapoor et al. (1999)
14	*Aspergillus niger*	Cd	Kapoor et al. (1999)
15	*Aspergillus niger*	Cu	Kapoor et al. (1999)
16	*Trichoderma atroviride*	Cu	Errasquín and Vázquez (2003)
17	*Aspergillus niger*	Ni	Kapoor et al. (1999)
18	*Penicillium spinulosum*	Cd	Townsley et al. (1986)
19	*Trichoderma atroviride*	Co	Kacprzak and Malina (2005)
20	*A. nidulans*	As	Maheswari and Murugesan (2009)

12.3 Applications of Fungus-Mediated Nanoparticles in Bioremediation

Bioremediation is the use of natural resources to transform contaminants into chemicals that are innocuous for the purpose of waste management. Nanotechnology plays a significant role in detecting heavy metals using fungal nanoparticles, which are smaller than a micron and even individual atoms. Even though there are a number of physical and chemical methods for creating nanoparticles, they present numerous drawbacks. Hence, it is essential to devise an environmentally sustainable approach for the production of nanoparticles. Fungi have an extensive range of uses in the process of bioremediation for heavy metals. Fungi are a superior option for bioremediating heavy metals because of their rapid development and exceptional ability to bind to these metals. Due to their rapid multiplication cycle and high morphological and genetic adaptability, they may be efficiently and cost-effectively cultured in large numbers. Fungi use many mechanisms, similar to other microbes, to detoxify heavy metals. In general, fungi use two principal mechanisms for metal detoxification. The first technique is called biosorption, which involves metals adhering to the fungus's surface. The second approach is known as bioaccumulation, which involves the absorption of metals into the fungal cells via cellular metabolism. One important stage in the biological remediation of heavy metals is biosorption. The following is a list of the numerous studies that have been conducted.

In 2013, Bennet and colleagues explored how naturally occurring microorganisms could help clean up hexavalent chromium (Cr^{6+}) pollution by collecting six different species from contaminated soil and water. They measured how well these organisms reduced chromium contamination using a chemical test called the 1,5-diphenylcarbazide method, which involved checking light absorption at a specific wavelength (540 nm).

The research revealed two different cleanup strategies among microorganisms. Three bacterial species: *Klebsiella pneumoniae*, *Bacillus firmus*, and *Mycobacterium* species were particularly good at soaking up the toxic Cr^{6+} directly into their cell bodies like tiny sponges. Meanwhile, three fungal species: *Aspergillus flavus*, another *Aspergillus species*, and *A. niger* took a different approach by chemically converting the dangerous Cr^{6+} into the much less harmful Cr^{3+} form through interactions with their cell walls. These results show real promise for using these microorganisms as natural cleanup crews for environments contaminated with cancer-causing and gene-damaging hexavalent chromium, offering a biological solution to a serious environmental health problem (Bennett et al. 2013).

In 2018, Acosta-Rodríguez et al. successfully isolated a fungus, specifically identified as *A. niger*, that has the ability to thrive in low molecular weight media (LMM) containing several heavy metals. This fungus likely has resistance to these metals. The fungal biomass effectively eliminated several heavy metals (chromium (VI) and zinc (II) with 100% efficiency, mercury (II) with 83.2% efficiency, and fluorine (I) with 83% efficiency) at diverse pH settings (with 4.0 for lead (II), 5.0 for zinc (II) and cobalt (II), and 5.5 for mercury (II), 28 °C, using 1 g of fungal biomass). These results ultimately suggest that this fungus may be employed to remove heavy metals from polluted soils and water sources (Acosta-Rodríguez et al. 2018).

In 2013, Mukherjee and colleagues worked with a specific fungus called *Aspergillus foetidus* MTCC 8876, which they obtained from the Microbial Type Culture Collection and Gene Bank (MTCC). This particular strain was originally found at a wastewater treatment plant in Kalyani and had been stored in the culture collection for research purposes.

The researchers first discovered that this fungus was special because it could survive in environments with unusually high levels of lead contamination. To test just how much lead it could handle, they grew the fungus in laboratory dishes containing a standard growth medium called Czapek-Dox, but with different amounts of lead added to it.

The observation was quite remarkable: the fungus actually thrived when exposed to lead concentrations up to 200 mg/L, levels that would kill most other organisms. However, when they pushed the lead levels even higher, the fungus finally started struggling and its growth began to slow down. These results suggest that *A. foetidus* has developed natural defense systems that allow it to not only survive toxic lead exposure but absorb large amounts of the metal from its surroundings. This makes it an excellent candidate for use in bioremediation essentially using living organisms to clean up lead pollution from contaminated water sources (Chakraborty et al. 2013).

Alothman et al. (2020) developed simple, eco-friendly fungal biosorbents from two common molds: *Penicillium chrysogenum* and *Aspergillus ustus* to clean up toxic metals in water. They grew and processed the fungal biomass, then tested how well it grabbed onto cadmium (Cd^{2+}), copper (Cu^{2+}), and lead (Pb^{2+}) under different conditions, including changes in pH, temperature, and how long the biomass and metal solution were in contact.

When they applied these new biosorbents to real wastewater samples collected from the western, eastern, and northern regions of Saudi Arabia, they found metal concentrations ranging from 2.6 to 234 μg/L. In particular, copper levels in every sample exceeded the World Health Organization's safe limit for drinking water. The fungal materials proved highly effective at pulling out these heavy metals, demonstrating that biosorption with fungal biomass is not only powerful but also a low-cost, green alternative for treating contaminated effluents (Alothman et al. 2020).

In 2014, Kurniati and co-workers set out to see if fungi from forest soil could clean up mercury contamination. They started by growing six fungal isolates on potato dextrose agar plates spiked with 25 mg/L of Hg^{2+}, then measured how well each one grew. One strain, later named KRP1, stood out with the highest growth ratio (0.831), so the team identified it by sequencing its 18S rRNA and examining its colony and cell shapes. The results showed a 98% match to *Aspergillus flavus*.

To test mercury removal, they ran each experiment three times under both shaking and still conditions. Impressively, A. flavus KRP1 removed about 97.5% of the mercury in stirred cultures and 98.7% in stationary ones. These findings suggest that this particular A. flavus strain could serve as an effective, biosorption-based solution for cleaning up mercury(II) from contaminated liquids (Kurniati et al. 2014).

In their 2017 study, Fawzy et al. identified *Rhizopus stolonifier* as the fungus species with the highest resistance to the hazardous metals tested (Pb, Cd, Cu, and Zn), followed by *Macrophomina phaseolina*. The study revealed that the levels of contamination in the soils were correlated with the metal's toxicity, as well as the physico-chemical characteristics such as pH, conductivity, organic matter, and carbonate content. This research validated the efficacy of both biological fungal substrates (*M. phaseolina* and *R. stolonifier*) and chemicals (MO, Zeolite, $CaCO_3$, and phosphate) in the process of bioremediation for contaminated soils. These substances were shown to effectively decrease the levels of various heavy metals when compared to the control group, with fungus exhibiting particularly promising results. The addition of *M. phaseolina* was more effective in decreasing the concentration of chemically accessible heavy metals in the soils under investigation (Fawzy et al. 2017).

In March 2015, Mehran Mohammadian Fazli and colleagues collected fungi from various sites in Zanjan province, Iran, looking for strains that could survive and clean up cadmium contamination. They tested seven species: *Aspergillus versicolor*, *Aspergillus fumigatus*, two *Paecilomyces* strains (sp.9 and sp.G), *Trichoderma* sp., *Microsporum* sp., and *Cladosporium* sp. for both their ability to tolerate cadmium and to accumulate it.

Among these, *A. versicolor* stood out: when grown on agar containing 100 mg/L cadmium, it maintained a tolerance index of 0.8, indicating strong resistance. Even more impressively, it absorbed nearly 7 mg of cadmium/g of its dry mycelial weight more than any other tested fungus. By comparison, *Paecilomyces* sp.9 took up about 5.88 mg/g, *A. fumigatus* about 5.24 mg/g, *Microsporum* sp. about 5.08 mg/g, and *Trichoderma* sp. roughly 4.56 mg/g of cadmium. Interestingly, a fungus's ability to survive high cadmium levels didn't always match its capacity to store the metal. In the end, the researchers highlighted *A. versicolor* as the most promising candidate for future cadmium bioremediation efforts (Fazli et al. 2015).

In 2009, Ezzouhri and colleagues collected 36 yeast and filamentous fungal strains from sites in Tangier, Morocco, that were heavily contaminated with metals. These fungi fell into five main groups: *Aspergillus*, *Penicillium*, *Fusarium*, *Alternaria*, and *Geotrichum* and were tested for their ability to grow in the presence of lead (Pb^{2+}), chromium (Cr^{6+}), copper (Cu^{2+}), zinc (Zn^{2+}), and cadmium (Cd^{2+}).

Most isolates showed strong tolerance to lead, chromium, copper, and zinc. Only one strain; a Penicillium species was able to survive exposure to cadmium. The exact level of resistance depended on both the fungal species and the site where it was collected. For lead, the minimum inhibitory concentrations (MICs) ranged from 20 to 25 mM; for copper and zinc, from 15 to 20 mM; and for chromium, from 10 to 15 mM.

Among all the fungi tested, the Aspergillus and Penicillium strains displayed the highest overall metal tolerance. They not only withstood these harsh conditions but often grew better than control cultures that contained no heavy metals. Because of their remarkable resilience, these fungi are promising candidates for developing biological treatments to remove toxic metals from contaminated wastewater (Ezzouhri et al. 2009).

According to Bibi et al. (2018), removing all of the Cr from contaminated soils completely is quite difficult. Nevertheless, this problem may be reduced by using plant growth-boosting bacteria as a bioremediation technique. Cr-resistant endophytes were isolated from wholesome flora close to the Mardan University campus. From the designated plants, 114 different species of endophytic fungi were identified. Four of the 114 strains that have been found have made *L. sativa* resistant to Cr. Based on their ITS region (18 S rDNA) homology, the strains were categorized as *Fusarium proliferatum*, *Penicillium radicum*, *Rhizopus* sp., and *Aspergillus fumigatus*. The isolates have effectively removed chromium from the culture medium and soil, while also converting it from the highly hazardous hexavalent form to the less toxic trivalent form. As a result, they have aided in the recovery of normal growth in chromium-stressed *L. sativa*. The *Rhizopus* sp. CUC23 mostly accumulated Cr and detoxified it inside its cells, whereas *Outside of their cells, A. fumigatus ML43* and *P. radicum PL17 detoxified up to 95% of Cr.* The findings indicate that the chosen endophytic strains have the potential to serve as biofertilizers for cultivating crops in Cr polluted soils, promoting both health and safety (Bibi et al. 2018). Within the field of bioremediation, this fungus-mediated nanoparticles have many applications, and Fig. 12.2 illustrates how they are used in different industries. Because they effectively remove organic pollutants, microbiological contaminants, and heavy metals, they are essential to the treatment of water.

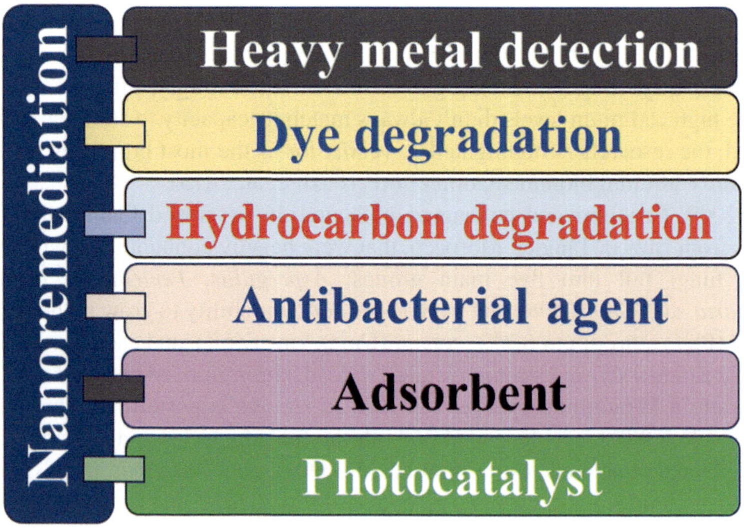

Fig. 12.2 Applications of fungus-mediated nanoparticles in nanoremediation

12.4 Ecofriendly Prospectives of Fungus-Mediated Nanoparticles in Bioremediation

In bioremediation, fungal-mediated nanoparticles provide a plethora of environmentally beneficial opportunities, revolutionizing environmental cleaning via effective and sustainable methods. This novel approach uses the special powers of fungus to create nanoparticles, offering a green substitute for conventional remediation techniques as well as a number of advantages for the environment. Fungi have the capacity to precisely target and eliminate certain pollutants, in contrast to some traditional techniques that could unintentionally damage non-target creatures or disturb ecosystems. This focused strategy reduces unintended environmental harm, guaranteeing that the rehabilitation procedure is efficient and ecologically conscious. Such as nanoparticles of lead and cadmium carbonate crystals were produced by a fungus *Fusarium oxysporum* as a heavy-metal cleanup method where Pb^{2+} and Cd^{2+} react with CO_2 to form $PbCO_3$ and $CdCO_3$ crystals, according to Sanyal et al. (2005) and the use of *Penicillium aurantiogriseum*, *Penicillium citrinum*, and *Penicillium waksmanii* in the green production of copper oxide nanoparticles as documented by Honary et al. (2012).

12.5 Economical Aspects of Bioremediation of Nanoparticles Using Fungal Synthesis of Nanoparticles

In order to address environmental contamination brought on by a variety of pollutants, including organic pollutants and heavy metals, bioremediation—particularly the production of nanoparticles by fungal synthesis-has shown promise and sustainability. This novel approach has important economic ramifications in addition to providing an environmentally friendly answer. The fact that fungus can grow in a variety of habitats and settings is one of the main economic benefits of using them to synthesize nanoparticles. It has been shown that fungi including *Aspergillus*, *Fusarium*, and *Penicillium* may efficiently produce nanoparticles by reducing metal ions (Acosta-Rodríguez et al. 2018; Alothman et al. 2020). Comparing these microorganisms to more conventional chemical synthesis techniques, their cultivation is generally less costly, and they are comparatively simple to produce. This results in reduced manufacturing costs, which adds to the economic appeal of fungal-based nanoparticle synthesis. Fungal nanoparticle synthesis is economically viable in part because of its scalability. Large-scale fungal cultivation in bioreactors enables the mass synthesis of nanoparticles. Fungal synthesis can produce nanoparticles in large batches that are readily scaled up to suit the needs of various remediation projects. This approach offers an affordable way to clean polluted surroundings. Moreover, using industrial and agricultural waste as a source of nutrients is one benefit of using fungus in the creation of nanoparticles (Haruma et al. 2019). By recycling waste resources, this not only lowers the total cost of agriculture but also advances the idea of the circular economy. The economic burden associated with conventional disposal techniques is lessened by turning trash into beneficial materials for fungal development, which promotes a more economical and sustainable approach.

12.6 Future Aspects

Fungal synthesis of nanoparticles for bioremediation is a promising field that might revolutionize environmental cleaning techniques and lead to major breakthroughs in the future. More investigation into the genetic and metabolic processes behind the process is probably in store, opening the door to the engineering of fungi to produce nanoparticles with optimal efficiency. This may result in the emergence of fungal strains that are more effective in reducing metal and producing more nanoparticles. The future success of fungal nanoparticle production also depends on its scalability. Developing large-scale manufacturing techniques, such as bioreactors designed for the growth of production of nanoparticles and fungus, may be the focus of efforts. Real-time tracking of remediation activities may be possible with the integration of sophisticated monitoring and evaluation tools, such as nano sensors and imaging technology.

12.7 Conclusion

Overall, a new approach to environmentally benign bioremediation has emerged with the combination of fungal-mediated nanoparticle production and their use in remediation of polluted environments. A sustainable solution to contamination problems may be found in the inherent powers of fungus and green synthesis technologies. The produced nanoparticles have multipurpose properties, such as the ability to trap heavy metals and break down contaminants. This study demonstrates the promise of a more sustainable future by demonstrating how nature-based methods and cutting-edge technology may work together in harmony to improve remediation strategies.

References

Abdussalam-Mohammed W (2019) Review of therapeutic applications of nanotechnology in medicine field and its side effects. J Chem Rev 1(3):243–251. https://doi.org/10.33945/SAMI/JCR.2019.3.5

Acosta-Rodríguez I, Cardenás-González JF, Pérez ASR, Oviedo JT, Martínez-Juárez VM (2018) Bioremoval of different heavy metals by the resistant fungal strain *Aspergillus niger*. Bioinorg Chem Appl 2018:3457196. https://doi.org/10.1155/2018/3457196

Alothman ZA et al (2020) Low cost biosorbents from fungi for heavy metals removal from wastewater. Sep Sci Technol 55(10):1766–1775. https://doi.org/10.1080/01496395.2019.1608242

Bai RS, Abraham TE (2001) Biosorption of Cr (VI) from aqueous solution by *Rhizopus nigricans*. Bioresour Technol 79(1):73–81. https://doi.org/10.1016/S0960-8524(00)00107-3

Bansal V, Rautaray D, Ahmad A, Sastry M (2004) Biosynthesis of zirconia nanoparticles using the fungus *Fusarium oxysporum*. J Mater Chem 14(22):3303–3305. https://doi.org/10.1039/b407904c

Bansal V et al (2005) Fungus-mediated biosynthesis of silica and titania particles. J Mater Chem 15(26):2583–2589. https://doi.org/10.1039/b503008k

Baskar D, Selvam AK (2021) Biosynthesis of nanoparticles using microorganisms. In: Microbes for a sustainable environment and human welfare: advancements and opportunities. Nova Science Publishers

Bennett RM, Cordero PRF, Bautista GS, Dedeles GR (2013) Reduction of hexavalent chromium using fungi and bacteria isolated from contaminated soil and water samples. Chem Ecol 29(4):320–328. https://doi.org/10.1080/02757540.2013.770478

Bhainsa KC, D'Souza SF (2006) Extracellular biosynthesis of silver nanoparticles using the fungus *Aspergillus fumigatus*. Colloids Surf B Biointerfaces 47(2):160–164. https://doi.org/10.1016/j.colsurfb.2005.11.026

Bharde A et al (2006) Extracellular biosynthesis of magnetite using fungi. Small 2(1):135–141. https://doi.org/10.1002/smll.200500180

Bharde A, Kulkarni A, Rao M, Prabhune A, Sastry M (2007) Bacterial enzyme mediated biosynthesis of gold nanoparticles. J Nanosci Nanotechnol 7(12):4369–4377. https://doi.org/10.1166/jnn.2007.891

Bibi S et al (2018) Bioremediation of hexavalent chromium by endophytic fungi; safe and improved production of *Lactuca sativa* L. Chemosphere 211:653–663. https://doi.org/10.1016/j.chemosphere.2018.07.197

Boruah S, Dutta P (2021) Fungus mediated biogenic synthesis and characterization of chitosan nanoparticles and its combine effect with *Trichoderma asperellum* against *Fusarium oxysporum*, *Sclerotium rolfsii* and *Rhizoctonia solani*. Indian Phytopathol 74(1):81–93. https://doi.org/10.1007/s42360-020-00289-w

Castro-Longoria E, Vilchis-Nestor AR, Avalos-Borja M (2011) Biosynthesis of silver, gold and bimetallic nanoparticles using the filamentous fungus *Neurospora crassa*. Colloids Surf B Biointerfaces 83(1):517–520. https://doi.org/10.1016/j.colsurfb.2010.10.035

Chakraborty S, Mukherjee A, Das TK (2013) Biochemical characterization of a lead-tolerant strain of *Aspergillus foetidus*: an implication of bioremediation of lead from liquid media. Int Biodeterior Biodegrad 84:134–142. https://doi.org/10.1016/j.ibiod.2012.05.031

Errasquín EL, Vázquez C (2003) Tolerance and uptake of heavy metals by *Trichoderma atroviride* isolated from sludge. Chemosphere 50(1):137–143. https://doi.org/10.1016/S0045-6535(02)00485-X

Ezzouhri L et al (2009) Heavy metal tolerance of filamentous fungi isolated from polluted sites in Tangier, Morocco. Afr J Microbiol Res 3(2):35–48

Fawzy EM, Abdel Motaal FF, El Zayat SA (2017) Biosorption of heavy metals onto different eco-friendly substrates. J Bioremed Biodegr 08(03):1000394. https://doi.org/10.4172/2155-6199.1000394

Fazli MM, Soleimani N, Mehrasbi M, Darabian S, Mohammadi J, Ramazani A (2015) Highly cadmium tolerant fungi: their tolerance and removal potential. J Environ Health Sci Eng 13(1):19. https://doi.org/10.1186/s40201-015-0176-0

Gade AK, Bonde P, Ingle AP, Marcato PD, Durán N, Rai MK (2008) Exploitation of *Aspergillus niger* for synthesis of silver nanoparticles. J Biobased Mater Bioenergy 2(3):243–247. https://doi.org/10.1166/jbmb.2008.401

Gajbhiye M, Kesharwani J, Ingle A, Gade A, Rai M (2009) Fungus-mediated synthesis of silver nanoparticles and their activity against pathogenic fungi in combination with fluconazole. Nanomedicine 5(4):382–386. https://doi.org/10.1016/j.nano.2009.06.005

Gericke M, Pinches A (2006) Biological synthesis of metal nanoparticles. Hydrometallurgy 83(1–4):257–262. https://doi.org/10.1016/j.hydromet.2006.03.019

Haruma T, Yamaji K, Ogawa K, Masuya H, Sekine Y, Kozai N (2019) Root-endophytic *Chaetomium cupreum* chemically enhances aluminium tolerance in *Miscanthus sinensis* via increasing the aluminium detoxicants, chlorogenic acid and oosporein. PLoS One 14(2):e0212644. https://doi.org/10.1371/journal.pone.0212644

Honary S, Barabadi H, Gharaei-Fathabad E, Naghibi F (2012) Green synthesis of copper oxide nanoparticles using *Penicillium aurantiogriseum*, *Penicillium citrinum* and *Penicillium waksmanii*. Dig J Nanomater Biostruct 7(3):999–1005

Kacprzak M, Malina G (2005) The tolerance and Zn2+, Ba2+ and Fe3+ accumulation by *Trichoderma atroviride* and *Mortierella exigua* isolated from contaminated soil. Can J Soil Sci 85(2):283–290. https://doi.org/10.4141/S04-018

Kapoor A, Viraraghavan T, Cullimore DR (1999) Removal of heavy metals using the fungus *Aspergillus niger*. Bioresour Technol 70(1):95–104. https://doi.org/10.1016/S0960-8524(98)00192-8

Kumar SA, Ansary AA, Abroad A, Khan MI (2007) Extracellular biosynthesis of CdSe quantum dots by the fungus, *Fusarium oxysporum*. J Biomed Nanotechnol 3(2):190–194. https://doi.org/10.1166/jbn.2007.027

Kumar T et al (2024) Historical perspective of nanotechnology and functionalized nanomaterials. In: Materials horizons: from nature to nanomaterials, vol Part F1329. Springer. https://doi.org/10.1007/978-981-99-3021-0_1

Kurniati E et al (2014) Potential bioremediation of mercury-contaminated substrate using filamentous fungi isolated from forest soil. J Environ Sci (China) 26(6):1223–1231. https://doi.org/10.1016/S1001-0742(13)60592-6

Lutz J-F, Ouchi M, Liu DR, Sawamoto M (2013) Sequence-controlled polymers article outline sequence-controlled biological polymerization processes sequence-controlled polymerizations based on chemical approaches properties and promises of sequence-controlled polymers outlook related items in science. Science 341:1238149

Maheswari S, Murugesan AG (2009) Remediation of arsenic in soil by *Aspergillus nidulans* isolated from an arsenic-contaminated site. Environ Technol 30(9):921–926. https://doi.org/10.1080/09593330902971279

Mishra A, Tripathy SK, Il Yun S (2012) Fungus mediated synthesis of gold nanoparticles and their conjugation with genomic DNA isolated from *Escherichia coli* and *Staphylococcus aureus*. Process Biochem 47(5):701–711. https://doi.org/10.1016/j.procbio.2012.01.017

Mukherjee P et al (2002) Extracellular synthesis of gold nanoparticles by the fungus *Fusarium oxysporum*. ChemBioChem 3(5):461–463. https://doi.org/10.1002/1439-7633(20020503)3:5<461::AID-CBIC461>3.0.CO;2-X

Philip D (2009) Biosynthesis of Au, Ag and Au-Ag nanoparticles using edible mushroom extract. Spectrochim Acta A Mol Biomol Spectrosc 73(2):374–381. https://doi.org/10.1016/j.saa.2009.02.037

Rajakumar G et al (2012) Fungus-mediated biosynthesis and characterization of TiO_2 nanoparticles and their activity against pathogenic bacteria. Spectrochim Acta A Mol Biomol Spectrosc 91:23–29. https://doi.org/10.1016/J.SAA.2012.01.011

Rautaray D, Sanyal A, Adyanthaya SD, Ahmad A, Sastry M (2004) Biological synthesis of strontium carbonate crystals using the fungus *Fusarium oxysporum*. Langmuir 20(16):6827–6833. https://doi.org/10.1021/la049244d

Reyes LR, Gómez I, Garza MT (2009) Biosynthesis of cadmium sulfide nanoparticles by the fungi *Fusarium* sp. Int J Green Nanotechnol Biomed 1(1):B90–B95. https://doi.org/10.1080/19430850903149936

Riddin TL, Gericke M, Whiteley CG (2006) Analysis of the inter- and extracellular formation of platinum nanoparticles by *Fusarium oxysporum* f. sp. lycopersici using response surface methodology. Nanotechnology 17(14):3482. https://doi.org/10.1088/0957-4484/17/14/021

Rizzo J, Rodrigues ML, Janbon G (2020) Extracellular vesicles in fungi: past, present, and future perspectives. Front Cell Infect Microbiol 10:346. https://doi.org/10.3389/fcimb.2020.00346

Sandoval-Cárdenas DI, Gómez-Ramírez M, Rojas-Avelizapa NG, Vidales-Hurtado MA (2017) Synthesis of cadmium sulfide nanoparticles by biomass of *Fusarium oxysporum* f. sp. *Lycopersici*. Nano Res 46:179–191. https://doi.org/10.4028/www.scientific.net/JNanoR.46.179

Sanghi R, Verma P (2009) Biomimetic synthesis and characterisation of protein capped silver nanoparticles. Bioresour Technol 100(1):501–504. https://doi.org/10.1016/j.biortech.2008.05.048

Sanyal A, Rautaray D, Bansal V, Ahmad A, Sastry M (2005) Heavy-metal remediation by a fungus as a means of production of lead and cadmium carbonate crystals. Langmuir 21(16):7220–7224. https://doi.org/10.1021/la047132g

Senapati S, Ahmad A, Khan MI, Sastry M, Kumar R (2005) Extracellular biosynthesis of bimetallic Au-Ag alloy nanoparticles. Small 1(5):517–520. https://doi.org/10.1002/smll.200400053

Shankar SS, Ahmad A, Pasricha R, Sastry M (2003) Bioreduction of chloroaurate ions by geranium leaves and its endophytic fungus yields gold nanoparticles of different shapes. J Mater Chem 13(7):1822–1826. https://doi.org/10.1039/b303808b

Solomon M, D'Souza GGM (2011) Recent progress in the therapeutic applications of nanotechnology. Curr Opin Pediatr 23(2):215–220. https://doi.org/10.1097/MOP.0b013e32834456a5

Sonawane H, Deore J, Rajshri S, Chavan P (2023) Synthesis of ZnS nanomaterials and their applications via green approaches: an overview. BioNanoScience 13(2):879–890. https://doi.org/10.1007/s12668-023-01105-1

Suganya P, Mahalingam PU (2017) Green synthesis and characterization of zinc sulphide nanoparticles from macro fungi *Pleutrous florida*. IOSR J Appl Chem 10(7–3):37–42

Tarafdar JC, Raliya R (2013) Rapid, low-cost, and Ecofriendly approach for iron nanoparticle synthesis using *Aspergillus oryzae* TFR9. J Nanopart 2013:141274. https://doi.org/10.1155/2013/141274

Tidke PR, Gupta I, Gade AK, Rai M (2014) Fungus-mediated synthesis of gold nanoparticles and standardization of parameters for its biosynthesis. IEEE Trans Nanobioscience 13(4):397–402. https://doi.org/10.1109/TNB.2014.2347803

Townsley CC, Ross IS, Atkins AS (1986) Biorecovery of metallic residues from various industrial effluents using filamentous fungi. CABI Digital Library

Uddin I et al (2022) Starch-assisted synthesis of Bi_2S_3 nanoparticles for enhanced dielectric and antibacterial applications. ACS Omega 7(46):42438–42445. https://doi.org/10.1021/acsomega.2c05593

Uddin I, Adyanthaya S, Syed A, Selvaraj K, Ahmad A, & Poddar P (2008) Structure and microbial synthesis of sub-10 nm Bi2O3 nanocrystals. J Nanosci Nanotechnol 8(8):3909–3913. https://doi.org/10.1166/jnn.2008.179

Venkatesh KS, Palani NS, Krishnamoorthi SR, Thirumal V, Ilangovan R (2013) Fungus mediated biosynthesis and characterization of zinc oxide nanorods. In: AIP conference proceedings, vol 1536. American Institute of Physics. https://doi.org/10.1063/1.4810116

Wang D, Xue B, Wang L, Zhang Y, Liu L, Zhou Y (2021) Fungus-mediated green synthesis of nano-silver using *Aspergillus sydowii* and its antifungal/antiproliferative activities. Sci Rep 11(1):10356. https://doi.org/10.1038/s41598-021-89854-5

Chapter 13
Use of Bacterial Reduced Nanoparticle as a Tool for Environmental Remediation: An Overview

Nilesh A. Sonune

Abstract Because of rapid industrialization, contaminants such as heavy metals, different types of dyes, pharmaceutical waste, and chlorinated compounds are constantly dumped in nature in recent years. The elimination of these pollutants from the atmosphere is a big challenge in the world. If they are not treated may cause an adverse effect on the ecosystem and human health. In recent years, remediation by nanotechnology has received much attention due to the unique physical and chemical properties of nanoparticles. A combination of bioremediation and nanotechnology known as nanobiotechnology could overcome the limitations of traditional methods as nanobioremediation is eco-friendly and reduces the cost for removal of contaminants. Among microorganisms, bacteria are preferred for nanoparticle synthesis as of their high growth rate and more biomass protection ability. Several metals are used for the synthesis of bacteria nanoparticles including silver, platinum, gold, palladium, titanium, etc. The bacteria have the ability to produce nanoparticles with diverse size, shape, and composition with diverse physical and chemical properties and have more efficacy than bulk materials due to large surface-to-volume ratio, easy handling, economic viability, and high reactivity. This section focuses on the outline of synthesis of bacterial reduced nanoparticles, their mechanisms, factors affecting the nanoparticle synthesis, and their application in bioremediation.

Keywords Bacterial reduced nanoparticles · Green technology · Nanotechnology · Nanoparticle · Nanomaterials bioremediation

N. A. Sonune (✉)
Department of Microbiology, Adarsha Science, J.B. Arts and Birla Commerce Mahavidyalaya, Dhamangaon Railway, Maharashtra, India

A. Guldhe et al. (eds.), *Biomaterials for Environmental Applications: Detection and Remediation*, Environmental Challenges and Solutions,
https://doi.org/10.1007/978-3-032-12701-3_13

13.1 Introduction

In recent years, pollution of the environment has been one of the most important issues in the world due to anthropogenic activities. Water is the most crucial component needed for the existence of all life forms, agriculture as well as industrialization. The discharge water from these sectors contains sewage, fertilizers, organic compounds, inorganic compounds (heavy metals), insecticides, and industrial effluents (Kapoor et al. 2021). If the contaminants are not treated, they enter into the water stream and lead to water and soil pollution that adversely affect aquatic ecosystem and soil health respectively (Sharma et al. 2015).

There is a variety of methods for the elimination of contaminants from the atmosphere. However these methods have several drawbacks such as energy energy-intensive, cost-expensive, production of hazardous toxic chemicals as byproducts. In the last few years, technologies for bioremediation have been invented to improve the effectiveness for the elimination of contaminants from the environments. However, it is less effective when pollutants are at high concentrations. All these issues can be overcome by the use of nanoparticles in bioremediation termed as nanobioremediation or nanotechnology (Sonune 2021).

Over the last few years, Nanotechnology has attracted great interest because of its potential impact on remediation. Nanotechnology focuses on small structures and materials with sizes ranging from a few nanometers to less than 100 nanometers. When compared to bulk materials with the same chemical composition, nanoparticles show different and considerably altered physical, chemical, and biological properties because of their high surface-to-volume ratio (Elfeky et al. 2020; Salem and Fouda 2021). Coupling of the biological processes and nanoparticles has shown success in improving effectiveness of bioremediation, speed up the process, and increasing the use of biochemistry in environmental science. Nanoparticles generated for bioremediation have little environmental threat and can offer further features and functions for the biochemical cycle (Majlesi et al. 2018).

The term nanobioremediation is the relationship between living organisms and nanoparticles. An appropriate relationship between nanoparticles and living organisms is crucial for nanobioremediation, which uses NPs to clean up contaminants from the environment. Certain factors are critical in this process such as nanotoxicity, nanonutrition, and NP size, all of which may have an impact on the remediation process (Sekoai et al. 2019). Nanobioremediation also depends on physical and chemical interactions between living organisms, nanomaterials include size and shape of nanomaterial, surface coating, chemical nature of the nanomaterials and pollutant, media, type of organism, temperature, and pH. These phenomena become complex because several parameters influence the interactions. In these media, temperature and pH play a crucial role for suitable growth of living organisms. Similarly, these parameters also influence the stability of the nanomaterials and pollutants (Sekoai et al. 2019; Wu et al. 2019).

13.2 Biosynthesis of Bacterial Nanoparticles

The use of living organisms to synthesize nanoparticles has become a rapidly expanding field of study in green nanotechnology worldwide. The synthesis of nanoparticles can be achieved through a variety of biological entities such as microorganisms and plants which can constantly form an attractive option to traditional physical and chemical processes. In an effort to find more affordable techniques to synthesize nanoproducts, microorganisms were included in the process. The main reaction that takes place in the bottom-up process of nanoparticle manufacturing is a reduction/oxidation. Metal compounds in their nanoparticles are usually reduced by microbial enzymes possessing antioxidant or removing capabilities (Prathna et al. 2010).

However, because of their fast growth rate, metabolism, simplicity of handling, and high tolerance, bacteria are among the finest options for converting metal ions to nanoparticles. Compared to other microbes, bacteria are simpler to genetically shape and manage for biomineralization of metal ions (Wang et al. 2017a; Vaseghi et al. 2018). Because of exposure to the high concentration of metal ions, it leads to the harmful and harsh environmental conditions for the bacteria. They have developed a variety of natural defense mechanisms, including extracellular precipitation, intracellular sequestration, variations in metal ion concentration, and efflux pumps, to enable them to live in these harsh and stressful conditions. Bacteria have successfully used these methods to synthesize nanoparticles (Jadoun et al. 2022).

There are numerous methods for the synthesis of nanoparticles such as chemical synthesis, sol-gel method, etc.; however, synthesis of nanoparticles by biological method is most acceptable because of their clean, eco-friendly, less demand of energy and non-toxic effect (Fig. 13.1) (Mishra et al. 2014). Nanoparticle synthesis by bacteria is a bottom-up approach that involves reduction/oxidation reaction. These reactions are because of extracellular or intracellular secondary metabolites of bacteria such as enzymes. These enzymes reduce metal into their respective nanoparticles. Nanoparticles can be biosynthesized extracellularly or intracellularly; however, extracellular biosynthesis is the most used method because of its low cost and no requirement of downstream processes. Following the addition of a precursor molecule, extracellular components such as enzymes and other secondary metabolites present in the cell-free extract perform the oxidation and reduction reaction necessary for the production of particles (Fig. 13.2) (Ahmad and Khan 2013).

After synthesis of bacterial nanoparticle, they are further characterized by several methods like scanning electron microscope, transmission electron microscopy, UV-visible spectroscopy, Fourier transform infrared spectroscopy, and X-ray diffraction (Fig. 13.3). Nanoparticles show variation in function based on their shape and size. The biologically produced nanoparticles have higher specific surface area and are higher catalytically reactive. Because of a capping agent released by particular bacteria, biosynthesized nanoparticles do not aggregate (Jeevan et al. 2012).

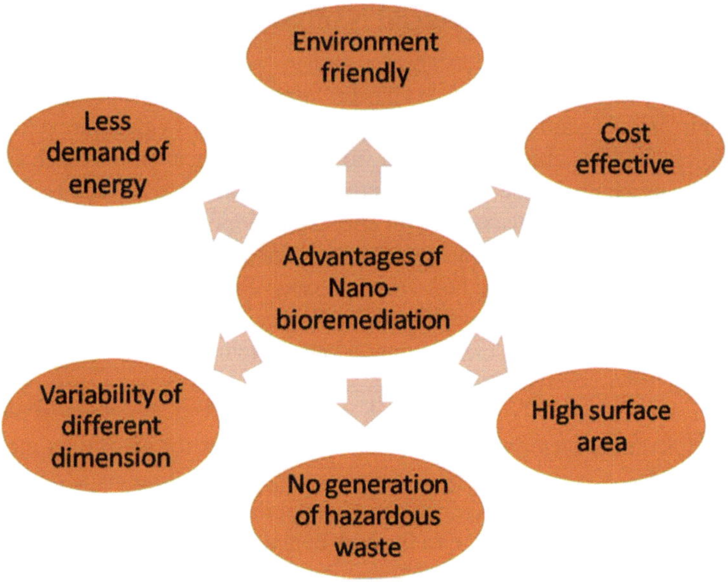

Fig. 13.1 Advantages of nanobioremediation

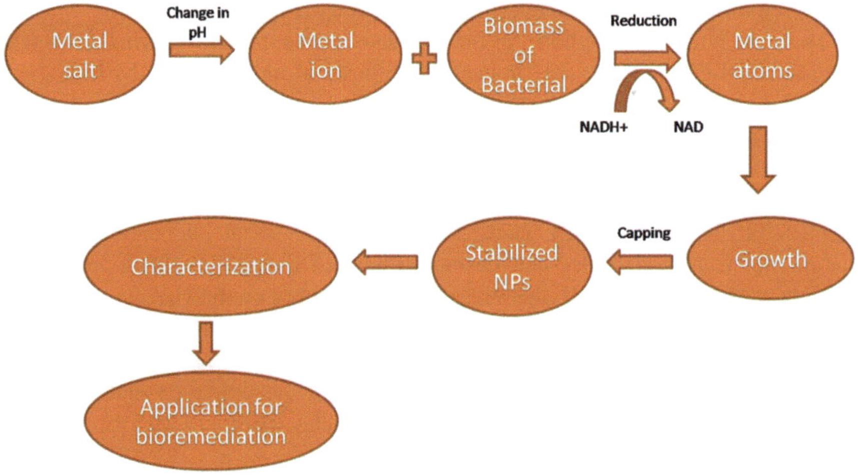

Fig. 13.2 Synthesis of bacterial reduced nanoparticles

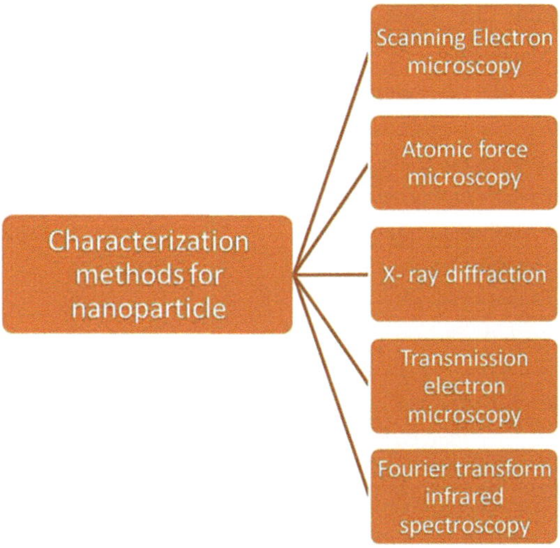

Fig. 13.3 Different methods of characterization of bacterial reduced nanoparticles

Numerous studies reported the bacterial nanoparticle synthesis for remediation of the environment, which is listed in Table 13.1.

13.3 Mechanisms of Bacterial Nanoparticles Synthesis

Mechanisms of nanoparticle production by several bacteria have been established. Biosynthesis of nanoparticles is done by culturing bacteria in a specific nutrient media containing the corresponding ions. Based on the localization site, nanoparticle synthesis by bacteria is classified into extracellular and intracellular. Metal ions can enter the bacterial cell, through active transport, through ion channels, endocytosis, or penetration through the lipid membrane (Grzelczak and Liz-Marzán 2014; Yin et al. 2016). The extracellular nanoparticle synthesis involves enzyme secretion, bioreduction, and particle capping. According to a number of studies, the simplified downstream and purification procedures associated with extracellular nanoparticle manufacturing make it the better approach over intracellular ones (Li et al. 2011; Singh and Singh 2019).

One of the most vital enzymes responsible for the nanoparticles synthesis in bacteria is nitrate reductase. During the bioreduction process, individual enzymes play vital role in the transport of electrons from donors to the positive metal ion (Banerjee and Ravishankar Rai 2018; Siddiqi et al. 2018). Functional groups of protein such as –NH2, –OH, –SH, and –COOH in some bacteria play a vital in synthesis and stabilization of nanoparticles. These groups provide binding sites for fixation of metal ions and localize them on the cell wall or in the periplasmic space (Gupta and Padmnabhan 2018).

Table 13.1 Synthesis of bacterial nanoparticles for environmental remediation

Name of bacteria	Nanoparticle	References
Pseudomonas aeruginosa	Cadmium sulfide	Raja et al. (2022)
Bacillus marisflavi	Silver	Ahmed et al. (2022)
Bacillus paralicheniformis	Silver	Martins et al. (2017)
Pseudochrobactrum sp.	Zinc oxide	Martins et al. (2017)
Pseudonomas putida	Manganese oxide	Furgal et al. (2015)
Desulfovibrio vulgaris	Platinum	Martins et al. (2017)
Escherichia coli	Palladium	Ahmed et al. (2022)
Bacillus pumilus	Silver	Martins et al. (2017)
Sphingomonas paucimobilis	Silver	Martins et al. (2017)
Escherichia coli	Silver	Ghorbani (2013)
B. amyloliquefaciens	Titanium oxide	Khan and Fulekar (2016)
Lactobacillus sp.	Titanium	Prasad et al. (2007)
Citrobacter freundii	Selenium	Wang et al. (2017a, b)
Pseudomonas putida	Silver	Thamilselvi and Radha (2013)
E. coli	Zinc oxide	Somu and Paul (2018)
Deinococcus radiodurans	Iron	Kim et al. (2019)
Enterococcus faecalis	Iron oxide	Samuel et al. (2021)
Shewanella loihica	Palladium	Misra et al. (2022)
Enterococcus faecalis	Palladium	Misra et al. (2022)
Pseudomonas aeruginosa	Gold	Husseiny et al. (2007)
Magnetospirillum magnetotacticum	Magnetite	Philipse and Maas (2002)
Pseudomonas putida NCIM 2650	Silver	Thamilselvi and Radha (2013)
Lactobacillus strains	Gold	Nair and Pradeep (2002)
Rhodobacter sphaeroides	Zinc sulfide	Bai et al. (2006)
Pseudomonas aeruginosa	Lanthanum	Korbekandi et al. (2012)
Pseudomonas stutzeri AG259	Silver	Haefeli et al. (1984)
Rhodopseudomonas capsulate	Gold	He et al. (2008)
Lactobacillus strains	Titanium	Prasad et al. (2007)
Escherichia coli MC4100	Gold	Deplanche and Macaskie (2008)
Escherichia coli	Cadmium sulfide	Sweeney et al. (2004)
Desulfovibrio magneticus strain RS-1	Magnetite	Pósfai et al. (2006)
Klebsiella pneumonia	Silver	Shahverdi et al. (2007)
Escherichia coli	Silver	Mahanty et al. (2013)
Enterobacter cloacae	Selenium	Kessi et al. (1999)
Klebsiella aerogenes	Cadmium sulfide	Holmes et al. (1995)
Desulfovibrio vulgaris	Uranium,	Kashefi et al. (2001)
Escherichia coli DH5α	Gold	Du et al. (2007)
Desulfobacteraceae	Zinc sulfide	Labrenz et al. (2000)
Geobacillus sp.	Gold	Correa-Llantén et al. (2013)

(continued)

Table 13.1 (continued)

Name of bacteria	Nanoparticle	References
Bacillus megatherium D01	Gold	Wen et al. (2009)
Clostridium thermoaceticum	Cadmium sulfide	Cunningham and Lundie (1993)
Bacillus cereus	Silver	Sunkar and Nachiyar (2012)
Bacillus subtilis	Silver	Saifuddin et al. (2009)
Aeromonas sp. SH10	Silver	Rai et al. (2006)
Corynebacterium sp. SH09	Silver	Zhang et al. (2005)
Bacillus subtilis 168	Gold	Southam and Beveridge (1994)
Serratia nematodiphila	Silver	Malarkodi et al. (2013)
Thermoanaerobacter ethanolicus TOR-39	Cobalt, Nickel & Chromium	Yeary et al. (2006)
Rhodopseudomonas palustrus	Cadmium sulfide	Bai et al. (2009)
Pseudomonas stutzeri AG259	Silver	Klaus et al. (1999)
Bacillus selenireducens	Selenite	Baesman et al. (2007)
P. stutzeri	Selenite	Narayanan and Sakthivel (2010)
P. aeruginosa SNT1	Selenium	Yadav et al. (2008)
Serratia nematodiphila	Silver	Malarkodi et al. (2013)
Bacillus strain CS11	Silver	Das et al. (2014)
Bacillus methylotrophicus	Silver	Wang et al. (2016)
Novosphingobium species	Silver	Du et al. (2017)
Brevibacillus formosus	Gold	Srinath et al. (2017)

The alternative ways for nanoparticle synthesis were also established. According to this mechanism, multiple biological components may be involved in the production of nanoparticles. In this, the principal reducing agent is nitrate reductase that converts metals into nanoparticles (Gupta and Padmnabhan 2018; Ali et al. 2019). Cofactors such as NADH are also required for enzymatic reaction. Certain studies showed competition for the negatively charged binding site between protons and metal ions is strongly correlated with rising pH (Sintubin et al. 2009; Kalishwaralal et al. 2010).

13.4 Factors Affecting Nanoparticles Synthesis

Several factors influence the synthesis, characterization, and application of nanoparticles (Fig. 13.4). The key parameters which affect the biosynthesis of nanoparticles are described below.

(a) *Methods*: Variety of methods available for nanoparticle synthesis that include physical method, chemical method, or biological method. Every method has

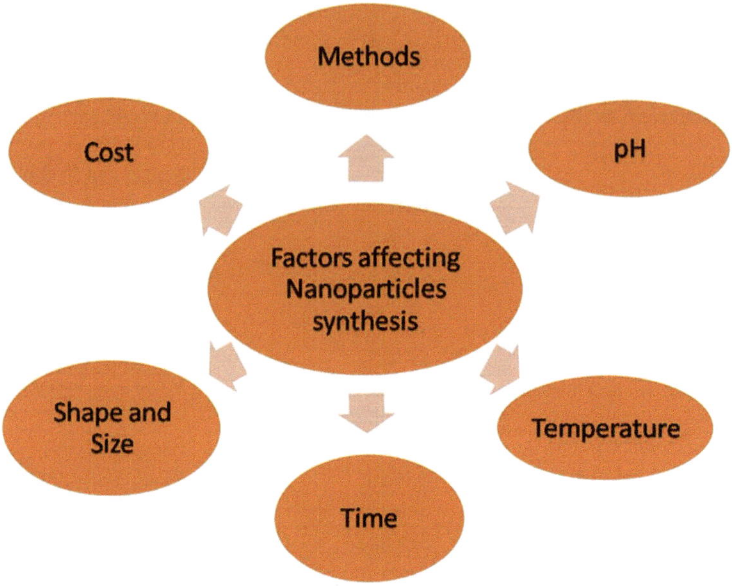

Fig. 13.4 Factors affecting nanoparticles synthesis

particular benefits and limitations. However, among them, the only biological method is eco-friendly and most satisfactory method because it uses nontoxic materials for the synthesis of nanoparticles (Vadlapudi and Kaladhar 2014).

(b) *pH*: pH of the solution or media is a key element in the bacterial nanoparticle synthesis. By changing the pH of the solution, nanoparticle size can be controlled because it affects the size and texture of nanoparticles (Patra and Baek 2014).

(c) *Temperature*: It is the most important factor which affects the nanoparticle synthesis and it determines nanoparticles nature. The report suggests that biological method needs ambient temperature or less than 100 °C that very low as compared to other methods (Rai et al. 2006).

(d) *Time*: Incubation time of reaction medium significantly affects the bacterial nanoparticle synthesis. The nanoparticles' characteristics can be changed with time. The storage condition and exposure to light of the reaction media also affect characteristics of nanoparticles (Darroudi et al. 2011).

(e) *Shape and Size*:

The properties of nanoparticles were determined by particle shape and size. It has been observed that when nanoparticles size decreases and it reaches to the nanometer scale, their melting point drops. The chemical characteristics are influenced by the appearance of nanoparticles. When one nanoparticle comes into touch with the surface of another nanoparticle, a modification in the chemical characteristics has been observed. The development of tunable nanoparticles is aided by this approach (Baer et al. 2013).

(f) *Cost*: Another crucial element is the cost involved in creating nanoparticles. The costs associated with producing nanoparticles must be kept under control in order to support future uses. The chemical method gives more yields in short time, although it is a more expensive process. Similarly, the physical method needs high-priced equipment. The synthesis of nanoparticles by biological processes is inexpensive and applied in at large scale (Gour and Jain 2019).

13.5 Application of Bacterial Nanoparticles in Remediation

Bacterial nanoparticles are one of the best strategies for the elimination of ecological pollutants from contaminated sites. It is potential technology and has many advantages such as environmentally friendly, cost-effective, no generation of hazardous wastes, high surface area, minimum energy requirement and variability of different dimensions as compared to traditional methods. Among the microorganisms, bacteria are used for remediation due to their ability to produce more biomass and high growth rate (Vanalakar 2018).

Soil and water are the most important natural resource that plays a major role in the economy. Because of man-made activities such as household discharges, heavy metals, chlorinated compounds, dyes, pharmaceutical and biomedical waste, etc. from the industries and mines are released into the soil and water that leads to environmental pollution. Indigenous bacteria from contaminant sites have good tolerance capacity and hence they play an important role in remediation (Sonune 2021). In this section, the application of bacterial nanoparticles for bioremediation of some major pollutants is discussed.

13.5.1 Degradation of Dyes

Several industries such as cosmetic, paper, textile, food and leather require different types of dyes. It leads to the generation of wastewater in huge quantity and contains several dyes and it causes harmful effects on humans, plants, and animals. Bacterial nanoparticles play an important role in achieving remediation of dyes from such contaminated water because of the large surface area with a significant number of active sites (Mahanty et al. 2020). Several studies reported the use of different bacterial nanoparticles in the bioremediation of dyes. The nanoparticle prepared from magnetic Fe_3O_4 and *Enterobacter cloacae* sp. A3 showed decolorization and biodegradation of Basic Red 46 azo dye. The percentage of declorization by this nanoparticle was found to be 91.60% at the optimum dosage of nanoparticle, 37 °C temperature and pH 7 after 24 h (Bekhit et al. 2020). Biosynthesized SnO_2-nanoparticles by *Erwinia herbicola* showed a dye degradation capacity and showed degradation of methyl orange (94%), methylene blue (93%), and erichrome black (98%) (Srivastava and Mukhopadhyay 2014). Several types of dyes include congo red, reactive black-5, malachite green and direct blue-1 were degraded by copper nanoparticles synthesized from Escherichia sp. SINT7 (Mandeep and Shukla 2020). Another report showed degradation of Napthol Green B dye by *Pseudoalteromonas* species under anaerobic conditions. The black colored iron-sulfur nanoparticle was synthesized by this species during degradation process that inhibits production of H_2S and metal complexes (Cheng et al. 2019). The other study reported potential application of *Bacillus pumillus* for bioremediation for Congo red and its removal percentage was 13% (Fulekar et al. 2015).

13.5.2 Removal of Heavy Metals

Because of the rapid industrialization, environmental contaminants such as heavy metals are becoming more prevalent and pose a major threat to public health. Prolonged exposure to certain heavy metals causes severe disorders in the humans, including immune weakness, osteoporosis, allergies, and respiratory failure. The accumulation of heavy metals in the environment is a major issue because of their inability to decompose. Thus, it is crucial to convert these hazardous heavy metals into a less harmful form. This can be achieved by using cost-effective bacterial nanoparticle (Alvarez et al. 2017; Yadav et al. 2017). Bacteria chelate the heavy metal or may convert them into nanoparticle of corresponding salt or ions in the soil. These biosynthesized nanoparticles in the soil can be extracted and can be used for industrial purpose or can enhance the soil and plant growth activity that leads to soil remediation. The report suggests that bacteria present in mines are tolerant to several metals including Cu, Al, Zn, Co, and Hg. Bacteria absorb them and convert them into the nanoparticles that can be easily harvested and used for industrial purposes (Salvadori et al. 2014). Cao et al. (2020) reported that polyvinylpyrrolidone-coated iron oxide nanoparticles with *Halomonas* sp. were efficient for bioremediation of Pb and Cd. The two toxic metals namely vanadium and chromium were efficiently removed from wastewater by *Shewanella loihica* PV-4. The removal percentage of vanadium and chromium after 27 days was found to be 71% and 91%, respectively (Wang et al. 2017a, b). Diba et al. (2021) reported the synthesis of lead nanoparticle with *Bacillus* sp. A21 and *Oceanobacillus* sp. A22 strains and lead and nickel nanoparticles with *Salinicoccus* sp. A43. *Micrococcus* sp. isolated from industrial wastewater showed bioremediation of nickel (80.9%) and lead (91.06%) (Benmalek and Fardeau 2016). Ha et al. (2016) reported application of palladium nanoparticles for the removal of hexavalent chromium from wastewater. Asksonthong et al. (2016) reported lead bioremediation efficiency of two bacteria namely *Halomonas elongate* and *Tetragenococcus halophilus.*

13.5.3 Catalytic Dehalogenation

Because of the low water solubility, resistance to oxidation and flame resistance, chlorinated aromatic compounds are extensively used in several industrial applications. However, over-application of these compounds leads to the pollution of soil, water, air, and sediments. Because these substances are resistant to degradation, eradication can be challenging. Chlorinated aromatic compounds have been broken down using a variety of reductive and oxidative techniques, but a simple approach is required to deal with these difficult xenobiotics (Kapoor et al. 2021). The synthesized Pd-based nanoparticles with *Desulfovibrio desulfuricans, Desulfovibrio vulgaris,* and *Desulfovibrio* sp. Oz-7 were found to be efficient for the dehalogenation of aromatic compounds and reported that the rate of dechlorination of these nanoparticles was higher as compared to chemical-Pd-nanoparticles (Fang et al. 2019). Biosynthesized palladium and gold nanoparticles with *Bacillus* sp. GP showed catalytic activity for reduction of 4-nitrophenol (Zhang and Hu 2018). The study of De Windt et al. (2005) reported the biosynthesis of palladium nanoparticle by *Shewanella oneidensis* MR-1, both intracellularly and extracellularly. Excellent dehalogenation performance was demonstrated by these generated Pd-NPs toward polychlorinated biphenyl congeners.

13.5.4 Removal of Pharmaceutical and Hospital Wastewater Pollutants

It was recently discovered that a large number of harmful xenobiotics and chemical compounds found in the pharmaceutical industry and medical waste effluents reach the aquatic system directly or through contaminated polluted areas. Traditional methods are employed to evaluate a few common pollutants from the wastewater. However, they are not applicable for the estimation of complex harmful contaminants and their byproducts (Malik et al. 2022). This can be overcome by the application of bacterial nanoparticle. Martins et al. (2017) prepared platinum and palladium nanoparticles with *Desulfovibrio vulgaris.* These nanoparticles showed removal of four pharmaceutical products including ciprofloxacin, sulfamethoxazole, ibuprofen, and 17β-estradiol. Bharti et al. (2021) reported that manganese oxide nanoparticle prepared by using *Pseudomonas putida* showed steroids estrone and 7 ά-ethinylestradiol removal efficiencies from the aqueous solution. Diclofenac is an anti-inflammatory halogenated drug. It is present in biomedical waste which causes contamination of the environment. The study of Azeez et al. (2019) reported that synthesis of Pd NPs by utilizing *Shewanella oneiedensis* MR-1 was efficient for the dehalogenation of aromatic compounds.

13.6 Conclusion

The conventional nanoparticle synthesis has many disadvantages including energy intensiveness, cost-expensive, and production of hazardous toxic chemicals. As environmental awareness increased, it has led to the development of green nanotechnology, which is a more reliable, eco-friendly technique for the removal of contaminants, and it also overcomes the disadvantages of conventional methods. Bacteria nanoparticle synthesis has received most attention compared to other microorganisms in green nanotechnology because of high metabolic activity, the growth rate of bacteria, and easy to control. The mechanism of bacterial nanoparticle synthesis is not well understood. However, many studies reported stable nanoparticle synthesis for remediation of toxic compounds by selecting proper bacteria and optimum conditions such as pH, temperature, incubation time, etc.

References

Ahmad T, Khan W (2013) Size variation of gold nanoparticles synthesized using tannic acid in response to higher chloroauric acid concentrations. World J Nano Sci Eng 3:62–68. https://doi.org/10.4236/wjnse.2013.33009

Ahmed SF, Mofijur M, Ahmed B, Mehnaz T, Mehejabin F, Maliat D, Hoang AT, Shafiullah GM (2022) Nanomaterials as a sustainable choice for treating wastewater. Environ Res 214:113807. https://doi.org/10.1016/j.envres.2022.113807

Ali J, Ali N, Wang L, Waseem H, Pan G (2019) Revisiting the mechanistic pathways for bacterial mediated synthesis of noble metal nanoparticles. J Microbiol Methods 159:18–25. https://doi.org/10.1016/j.mimet.2019.02.010

Alvarez A, Saez JM, Davila Costa JS, Colin VL, Fuentes MS, Cuozzo SA, Benimeli CS, Polti MA, Amoroso MJ (2017) Actinobacteria: current research and perspectives for bioremediation of pesticides and heavy metals. Chemosphere 166:41–62. https://doi.org/10.1016/j.chemosphere.2016.09.070

Asksonthong R, Siripongvutikorn S, Usawakesmanee W (2016) Evaluation of harmful heavy metal (Hg, Pb andCd) reduction using Halomonas elongata and Tetragenococcus halophilus for protein hydrolysate product.Funct Foods Health Dis 6:195. https://doi.org/10.31989/ffhd.v6i4.240

Azeez L, Adejumo AL, Lateef A, Adebisi SA, Adetoro RO, Adewuyi SO, Tijani KO, Olaoye S (2019) Zero-valent silver nanoparticles attenuate Cd and Pb toxicities on Moringa oleifera via immobilization and induction of phytochemicals. Plant Physiol Biochem 139:283–292. https://doi.org/10.1016/j.plaphy.2019.03.030

Baer DR, Engelhard MH, Johnson GE, Laskin J, Lai J, Mueller K, Munusamy P, Thevuthasan S, Wang H, Washton N, Elder A, Baisch BL, Karakoti A, Kuchibhatla SVNT, Moon D (2013) Surface characterization of nanomaterials and nanoparticles: important needs and challenging opportunities. J Vac Sci Technol A 31:50820. https://doi.org/10.1116/1.4818423

Baesman SM, Bullen TD, Dewald J, Zhang D, Curran S, Islam FS, Beveridge TJ, Oremland RS (2007) Formation of tellurium nanocrystals during anaerobic growth of bacteria that use Te oxyanions as respiratory electron acceptors. Appl Environ Microbiol 73:2135–2143. https://doi.org/10.1128/AEM.02558-06

Bai H-J, Zhang Z-M, Gong J (2006) Biological synthesis of semiconductor zinc sulfide nanoparticles by immobilized Rhodobacter sphaeroides. Biotechnol Lett 28:1135–1139. https://doi.org/10.1007/s10529-006-9063-1

Bai HJ, Zhang ZM, Guo Y, Yang GE (2009) Biosynthesis of cadmium sulfide nanoparticles by photosynthetic bacteria Rhodopseudomonas palustris. Colloids Surf B Biointerfaces 70:142–146. https://doi.org/10.1016/j.colsurfb.2008.12.025

Banerjee K, Ravishankar Rai V (2018) A review on mycosynthesis, mechanism, and characterization of silver and gold nanoparticles. BioNanoSci 8:17–31. https://doi.org/10.1007/s12668-017-0437-8

Bekhit F, Ibrahem S, Attia A (2020) Decolorization and degradation of the Azo dye by bacterial cells coated with magnetic iron oxide nanoparticles. Environ Nanotechnol Monit Manag 14:100376. https://doi.org/10.1016/j.enmm.2020.100376

Benmalek Y, Fardeau M-L (2016) Isolation and characterization of metal-resistant bacterial strain from wastewater and evaluation of its capacity in metal-ions removal using living and dry bacterial cells. Int J Environ Sci Technol 13. https://doi.org/10.1007/s13762-016-1048-6

Bharti B, Kumar R, Kumar H, Li H, Zha X, Ouyang F (2021) Advanced applications and current status of green nanotechnology in the environmental industry. In: Green functionalized nanomaterials for environmental applications, pp 303–331. https://doi.org/10.1016/B978-0-12-823137-1.00012-9

Cao X, Alabresm A, Chen YP, Decho AW, Lead J (2020) Improved metal remediation using a combined bacterial and nanoscience approach. Sci Total Environ 704:135378. https://doi.org/10.1016/j.scitotenv.2019.135378

Cheng S, Li N, Jiang L, Li Y, Xu B, Zhou W (2019) Biodegradation of metal complex Naphthol Green B and formation of iron-sulfur nanoparticles by marine bacterium Pseudoalteromonas sp CF10-13. Bioresour Technol 273:49–55. https://doi.org/10.1016/j.biortech.2018.10.082

Correa-Llantén DN, Muñoz-Ibacache SA, Castro ME, Muñoz PA, Blamey JM (2013) Gold nanoparticles synthesized by Geobacillus sp. strain ID17 a thermophilic bacterium isolated from Deception Island, Antarctica. Microb Cell Fact 12:75. https://doi.org/10.1186/1475-2859-12-75

Cunningham DP, Lundie LL (1993) Precipitation of cadmium by Clostridium thermoaceticum. Appl Environ Microbiol 59:7–14. https://doi.org/10.1128/aem.59.1.7-14.1993

Darroudi M, Ahmad MB, Zamiri R, Zak AK, Abdullah AH, Ibrahim NA (2011) Time-dependent effect in green synthesis of silver nanoparticles. Int J Nanomed 6:677–681. https://doi.org/10.2147/IJN.S17669

Das VL, Thomas R, Varghese RT, Soniya EV, Mathew J, Radhakrishnan EK (2014) Extracellular synthesis of silver nanoparticles by the Bacillus strain CS 11 isolated from industrialized area. 3 Biotech 4:121–126. https://doi.org/10.1007/s13205-013-0130-8

De Windt W, Aelterman P, Verstraete W (2005) Bioreductive deposition of palladium (0) nanoparticles on Shewanella oneidensis with catalytic activity towards reductive dechlorination of polychlorinated biphenyls. Environ Microbiol 7:314–325. https://doi.org/10.1111/j.1462-2920.2005.00696.x

Deplanche K, Macaskie LE (2008) Bio recovery of gold by Escherichia coli and Desulfovibriodesulfuricans. Biotechnol Bioeng 99:1055–1064. https://doi.org/10.1002/bit.21688

Diba H, Cohan RA, Salimian M, Mirjani R, Soleimani M, Khodabakhsh F (2021) Isolation and characterization of halophilic bacteria with the ability of heavy metal bioremediation and nanoparticle synthesis from Khara salt lake in Iran. Arch Microbiol 203:3893–3903. https://doi.org/10.1007/s00203-021-02380-w

Du L, Jiang H, Liu X, Wang E (2007) Biosynthesis of gold nanoparticles assisted by Escherichia coli DH5?? And its application on direct electrochemistry of hemoglobin. Electrochem Commun 9:1165–1170. https://doi.org/10.1016/j.elecom.2007.01.007

Du J, Singh H, Yi T-H (2017) Biosynthesis of silver nanoparticles by Novosphingobium sp. THG-C3 and their antimicrobial potential. Artif Cells Nanomed Biotechnol 45:211–217. https://doi.org/10.1080/21691401.2016.1178135

Elfeky AS, Salem SS, Elzaref AS, Owda ME, Eladawy HA, Saeed AM, Awad MA, Abou-Zeid RE, Fouda A (2020) Multifunctional cellulose nanocrystal/metal oxide hybrid, photo-degradation, antibacterial and larvicidal activities. Carbohydr Polym 230:115711. https://doi.org/10.1016/j.carbpol.2019.115711

Fang X, Wang Y, Wang Z, Jiang Z, Dong M (2019) Microorganism assisted synthesized nanoparticles for catalytic applications. Energies 12:190. https://doi.org/10.3390/en12010190

Fulekar M, Modi S, Pathak B (2015) Microbial synthesized silver nanoparticles for decolorization and biodegradation of Azo dye compound. https://doi.org/10.13074/jent.2015.06.152149

Furgal KM, Meyer RL, Bester K (2015) Removing selected steroid hormones, biocides and pharmaceuticals from water by means of biogenic manganese oxide nanoparticles in situ at ppb levels. Chemosphere 136:321–326. https://doi.org/10.1016/j.chemosphere.2014.11.059

Ghorbani HR (2013) Biosynthesis of silver nanoparticles by Escherichia coli. Asian J Chem 25:1247–1249. https://doi.org/10.14233/ajchem.2013.12805

Gour A, Jain NK (2019) Advances in green synthesis of nanoparticles. Artif Cells Nanomed Biotechnol 47:844–851. https://doi.org/10.1080/21691401.2019.1577878

Grzelczak M, Liz-Marzán LM (2014) The relevance of light in the formation of colloidal metal nanoparticles. Chem Soc Rev 43:2089–2097. https://doi.org/10.1039/C3CS60256G

Gupta R, Padmnabhan P (2018) Biogenic synthesis and characterization of gold nanoparticles by a novel marine bacteria Marinobacter algicola: progression from nanospheres to various geometrical shapes. J Microbiol Biotechnol Food Sci 8:732–737. https://doi.org/10.15414/jmbfs.2018.8.1.732-737

Ha C, Zhu N, Shang R, Shi C, Cui J, Sohoo I, Wu P, Cao Y (2016) Biorecovery of palladium as nanoparticles by enterococcus faecalis and its catalysis for chromate reduction. Chem Eng J 288:246–254. https://doi.org/10.1016/j.cej.2015.12.015

Haefeli C, Franklin C, Hardy K (1984) Plasmid-determined silver resistance in pseudomonas stutzeri isolated from a silver mine. J Bacteriol 158:389–392

He S, Zhang Y, Guo Z, Gu N (2008) Biological synthesis of gold nanowires using extract of Rhodopseudomonas capsulata. Biotechnol Prog 24:476–480. https://doi.org/10.1021/bp0703174

Holmes JD, Smith PR, Evans-Gowing R, Richardson DJ, Russell DA, Sodeau JR (1995) Energy-dispersive X-ray analysis of the extracellular cadmium sulfide crystallites of Klebsiella aerogenes. Arch Microbiol 163:143–147. https://doi.org/10.1007/BF00381789

Husseiny MI, El-Aziz MA, Badr Y, Mahmoud MA (2007) Biosynthesis of gold nanoparticles using Pseudomonas aeruginosa. Spectrochim Acta A Mol Biomol Spectrosc 67:1003–1006. https://doi.org/10.1016/j.saa.2006.09.028

Jadoun S, Chauhan NPS, Zarrintaj P, Barani M, Varma R, Chinnam S, Rahdar A (2022) Synthesis of nanoparticles using microorganisms and their applications: a review. Environ Chem Lett 20. https://doi.org/10.1007/s10311-022-01444-7

Jeevan P, Ramya K, Rena AE (2012) Extracellular biosynthesis of silver nanoparticles by culture supernatant of Pseudomonas aeruginosa. Indian J Biotechnol 11(1):72–76

Kalishwaralal K, Deepak V, Ram Kumar Pandian S, Kottaisamy M, BarathmaniKanth S, Kartikeyan B, Gurunathan S (2010) Biosynthesis of silver and gold nanoparticles using Brevibacterium casei. Colloids Surf B Biointerfaces 77:257–262. https://doi.org/10.1016/j.colsurfb.2010.02.007

Kapoor RT, Salvadori MR, Rafatullah M, Siddiqui MR, Khan MA, Alshareef SA (2021) Exploration of microbial factories for synthesis of nanoparticles – a sustainable approach for bioremediation of environmental contaminants. Front Microbiol 12:658294

Kashefi K, Tor JM, Nevin KP, Lovley DR (2001) Reductive precipitation of gold by dissimilatory Fe(III)-reducing bacteria and archaea. Appl Environ Microbiol 67:3275–3279. https://doi.org/10.1128/AEM.67.7.3275-3279.2001

Kessi J, Ramuz M, Wehrli E, Spycher M, Bachofen R (1999) Reduction of selenite and detoxification of elemental selenium by the phototrophic bacterium Rhodospirillum rubrum. Appl Environ Microbiol 65:4734–4740
Khan R, Fulekar MH (2016) Biosynthesis of titanium dioxide nanoparticles using Bacillus amyloliquefaciens culture and enhancement of its photocatalytic activity for the degradation of a sulfonated textile dye Reactive Red 31. J Colloid Interface Sci 475:184–191. https://doi.org/10.1016/j.jcis.2016.05.001
Kim HK, Jeong S-W, Yang JE, Choi YJ (2019) Highly efficient and stable removal of arsenic by live cell fabricated magnetic nanoparticles. Int J Mol Sci 20:3566. https://doi.org/10.3390/ijms20143566
Klaus T, Joerger R, Olsson E, Granqvist C-G (1999) Silver-based crystalline nanoparticles, microbially fabricated. Proc Natl Acad Sci U S A 96:13611–13614. https://doi.org/10.1073/pnas.96.24.13611
Korbekandi H, Iravani S, Abbasi S (2012) Optimization of biological synthesis of silver nanoparticles using lactobacillus casei subsp. casei. J Chem Technol Biotechnol. https://doi.org/10.1002/jctb.3702
Labrenz M, Druschel GK, Thomsen-Ebert T, Gilbert B, Welch SA, Kemner KM, Logan GA, Summons RE, De Stasio G, Bond PL, Lai B, Kelly SD, Banfield JF (2000) Formation of sphalerite (ZnS) deposits in natural biofilms of sulfate-reducing bacteria. Science 290:1744–1747. https://doi.org/10.1126/science.290.5497.1744
Li L, Hu Q, Zeng J, Qi H, Zhuang G (2011) Resistance and biosorption mechanism of silver ions by Bacillus cereus biomass. J Environ Sci 23:108–111. https://doi.org/10.1016/S1001-0742(10)60380-4
Mahanty A, Bosu R, Panda P, Netam S, Sarkar B (2013) Microwave assisted rapid combinatorial synthesis of silver nanoparticles using E coli culture supernatant. Int J Pharma Biosci 4:1030–1035
Mahanty S, Chatterjee S, Ghosh S, Tudu P, Gaine T, Bakshi M, Das S, Das P, Bhattacharyya S, Bandyopadhyay S, Chaudhuri P (2020) Synergistic approach towards the sustainable management of heavy metals in wastewater using mycosynthesized iron oxide nanoparticles: biofabrication, adsorptive dynamics and chemometric modeling study. J Water Process Eng 37:101426. https://doi.org/10.1016/j.jwpe.2020.101426
Majlesi Z, Ramezani M, Gerami M (2018) Investigation on some main glycosides content of Stevia rebaudian B. under different concentrations of commercial and synthesized silver nanoparticles. PBR. https://doi.org/10.18502/pbr.v4i1.140
Malarkodi C, Rajeshkumar S, Paulkumar K, Vanaja M, Jobitha GDG, Annadurai G (2013) Bactericidal activity of bio mediated silver nanoparticles synthesized by Serratia nematodiphila. Drug Invent Today 5:119–125. https://doi.org/10.1016/j.dit.2013.05.005
Malik S, Dhasmana A, Preetam S, Mishra YK, Chaudhary V, Bera SP, Ranjan A, Bora J, Kaushik A, Minkina T, Jatav HS, Singh RK, Rajpu VD (2022) Exploring microbial-based green nanobiotechnology for wastewater remediation: a sustainable strategy. Nanomaterials 12:4187. https://doi.org/10.3390/nano12234187
Mandeep, Shukla P (2020) Microbial nanotechnology for bioremediation of industrial wastewater. Front Microbiol 11:590631
Martins M, Mourato C, Sanches S, Noronha JP, Crespo MTB, Pereira IAC (2017) Biogenic platinum and palladium nanoparticles as new catalysts for the removal of pharmaceutical compounds. Water Res 108:160–168. https://doi.org/10.1016/j.watres.2016.10.071
Mishra A, Kumari M, Pandey S, Chaudhry V, Gupta KC, Nautiyal CS (2014) Biocatalytic and antimicrobial activities of gold nanoparticles synthesized by Trichoderma sp. Bioresour Technol 166:235–242. https://doi.org/10.1016/j.biortech.2014.04.085
Misra M, Chattopadhyay S, Sachan A, Ghosh Sachan S (2022) Microbially synthesized nanoparticles and their applications in environmental clean-up. Environ Technol Rev 11:18–32. https://doi.org/10.1080/21622515.2021.2011431

Nair B, Pradeep T (2002) Coalescence of nanoclusters and formation of submicron crystallites assisted by lactobacillus strains. Cryst Growth Des 2:293–298. https://doi.org/10.1021/cg0255164

Narayanan KB, Sakthivel N (2010) Biological synthesis of metal nanoparticles by microbes. Adv Colloid Interface Sci 156:1–13. https://doi.org/10.1016/j.cis.2010.02.001

Patra JK, Baek K-H (2014) Green nanobiotechnology: factors affecting synthesis and characterization techniques. J Nanomater 2014:1–12. https://doi.org/10.1155/2014/417305

Philipse AP, Maas D (2002) Magnetic colloids from magnetotactic bacteria: chain formation and colloidal stability. Langmuir 18:9977–9984. https://doi.org/10.1021/la0205811

Pósfai M, Moskowitz B, Arató B, Schüler D, Flies C, Bazylinski D, Frankel R (2006) Properties of intracellular magnetite crystals produced by Desulfovibrio magneticus strain RS-1. Earth Planet Sci Lett 249:444–455. https://doi.org/10.1016/j.epsl.2006.06.036

Prasad K, Jha A, Kulkarni A (2007) Lactobacillus assisted synthesis of titanium nanoparticles. Nanoscale Res Lett 2:248–250. https://doi.org/10.1007/s11671-007-9060-x

Prathna TC, Mathew L, Chandrasekaran N, Ashok M, Mukherjee A (2010) Biomimetic Synthesis ofNanoparticles: Science, Technology Applicability. BiomimS Learn Nat. InTech. https://doi.org/10.5772/8776

Rai A, Singh A, Ahmad A, Sastry M (2006) Role of halide ions and temperature on the morphology of biologically synthesized gold nanotriangles. Langmuir 22:736–741. https://doi.org/10.1021/la052055q

Raja RK, Hazir S, Balasubramani G, Sivaprakash G, Obeth ESJ, Boobalan T, Pugazhendhi A, Raj RHK, Arun A (2022) Green nanotechnology for the environment. In: Handbook of microbial nanotechnology. Elsevier, Amsterdam, pp 461–478

Saifuddin N, Wong CW, Yasumira AAN (2009) Rapid biosynthesis of silver nanoparticles using culture supernatant of bacteria with microwave irradiation. J Chem 6:61–70. https://doi.org/10.1155/2009/734264

Salem SS, Fouda A (2021) Green synthesis of metallic nanoparticles and their prospective biotechnological applications: an overview. Biol Trace Elem Res 199:344–370. https://doi.org/10.1007/s12011-020-02138-3

Salvadori MR, Ando RA, Oller do Nascimento CA, Correa B (2014) Intracellular Biosynthesis and Removal ofCopper Nanoparticles by Dead Biomass of Yeast Isolated from the Wastewater of a Mine in the BrazilianAmazonia. PLoS ONE 9(1): e87968. https://doi.org/10.1371/journal.pone.0087968

Samuel MS, Datta S, Chandrasekar N, Balaji R, Selvarajan E, Vuppala S (2021) Biogenic synthesis of iron oxide nanoparticles using enterococcus faecalis: adsorption of hexavalent chromium from aqueous solution and in vitro cytotoxicity analysis. Nanomaterials (Basel) 11:3290. https://doi.org/10.3390/nano11123290

Sekoai PT, Ouma CNM, Du Preez SP, Modisha P, Engelbrecht N, Bessarabov DG, Ghimire A (2019) Application of nanoparticles in biofuels: an overview. Fuel 237:380–397. https://doi.org/10.1016/j.fuel.2018.10.030

Shahverdi AR, Minaeian S, Shahverdi H, Jamalifar H, Nohi A-A (2007) Rapid synthesis of silver nanoparticles using culture supernatants of Enterobacteria: a novel biological approach. Process Biochem 42:919–923. https://doi.org/10.1016/j.procbio.2007.02.005

Sharma G, Jasuja ND, Kumar M, Ali MI (2015) Biological synthesis of silver nanoparticles by cell-free extract of *Spirulina platensis*. J Nanotechnol 2015:1–6. https://doi.org/10.1155/2015/132675

Siddiqi KS, Husen A, Rao RAK (2018) A review on biosynthesis of silver nanoparticles and their biocidal properties. J Nanobiotechnol 16:14. https://doi.org/10.1186/s12951-018-0334-5

Singh V, Singh A (2019) Role of microbially synthesized nanoparticles in sustainable agriculture and environmental management. In: Role of plant growth promoting microorganisms in sustainable agriculture and nanotechnology. Woodhead Publishing, pp 55–73. https://doi.org/10.1016/B978-0-12-817004-5.00004-X

Sintubin L, De Windt W, Dick J, Mast J, van der Ha D, Verstraete W, Boon N (2009) Lactic acid bacteria as reducing and capping agent for the fast and efficient production of silver nanoparticles. Appl Microbiol Biotechnol 84:741–749. https://doi.org/10.1007/s00253-009-2032-6

Somu P, Paul S (2018) Casein-based biogenic-synthesized zinc oxide nanoparticles simultaneously decontaminate heavy metals, dyes, and pathogenic microbes: a rational strategy for wastewater treatment. J Chem Technol Biotechnol 93:2962–2976

Sonune N (2021) Microbes: a potential tool for bioremediation. In: Rhizobiont in bioremediation of hazardous waste. Springer Singapore, pp 391–407. https://doi.org/10.1007/978-981-16-0602-1_17

Southam G, Beveridge TJ (1994) The in vitro formation of placer gold by bacteria. Geochim Cosmochim Acta 58:4527–4530. https://doi.org/10.1016/0016-7037(94)90355-7

Srinath BS, Namratha K, Byrappa K (2017) Eco-friendly synthesis of gold nanoparticles by gold mine bacteria Brevibacillus formosus and their antibacterial and biocompatible studies. IOSR J Pharm 7:53–60. wwwiosrphrorg

Srivastava N, Mukhopadhyay M (2014) Biosynthesis of SnO2 nanoparticles using bacterium Erwinia herbicola and their photocatalytic activity for degradation of dyes. Ind Eng Chem Res 53. https://doi.org/10.1021/ie5020052

Sunkar S, Nachiyar CV (2012) Biogenesis of antibacterial silver nanoparticles using the endophytic bacterium Bacillus cereus isolated from Garcinia xanthochymus. Asian Pac J Trop Biomed 2:953–959. https://doi.org/10.1016/S2221-1691(13)60006-4

Sweeney RY, Mao C, Gao X, Burt JL, Belcher AM, Georgiou G, Iverson BL (2004) Bacterial biosynthesis of cadmium sulfide nanocrystals. Chem Biol 11:1553–1559. https://doi.org/10.1016/j.chembiol.2004.08.022

Thamilselvi V, Radha KV (2013) Synthesis of silver nanoparticles from Pseudomonas putida NCIM 2650 in silver nitrate supplemented growth medium and optimization using response methodology. Dig J Nanomater Biostruct 8:1101

Vadlapudi V, Kaladhar DSVGK (2014) Review: green synthesis of silver and gold nanoparticles. Middle-East J Sci Res 19:834–842

Vanalakar SA (2018) Recent advances in synthesis of Cu2FeSnS4 materials for solar cell applications: a review. Sol Energy Mater Sol Cells 182. https://doi.org/10.1016/j.solmat.2018.03.021

Vaseghi Z, Nematollahzadeh A, Tavakoli O (2018) Green methods for the synthesis of metal nanoparticles using biogenic reducing agents: a review. Rev Chem Eng 34:529–559. https://doi.org/10.1515/revce-2017-0005

Wang C, Kim YJ, Singh P, Mathiyalagan R, Jin Y, Yang DC (2016) Green synthesis of silver nanoparticles by bacillus methylotrophicus, and their antimicrobial activity. Artif Cells Nanomed Biotechnol 44:1127–1132. https://doi.org/10.3109/21691401.2015.1011805

Wang G, Zhang B, Li S, Yang M, Yin C (2017a) Simultaneous microbial reduction of vanadium (V) and chromium (VI) by Shewanella loihica PV-4. Bioresour Technol 227:353–358. https://doi.org/10.1016/j.biortech.2016.12.070

Wang X, Zhang D, Pan X, Lee D-J, Al-Misned FA, Mortuza MG, Gadd GM (2017b) Aerobic and anaerobic biosynthesis of nano-selenium for remediation of mercury contaminated soil. Chemosphere 170:266–273. https://doi.org/10.1016/j.chemosphere.2016.12.020

Wen L, Lin Z, Gu P, Zhou J, Yao B, Chen G, Fu J (2009) Extracellular biosynthesis of monodispersed gold nanoparticles by a SAM capping route. J Nanopart Res 11:279–288. https://doi.org/10.1007/s11051-008-9378-z

Wu Y, Pang H, Liu Y, Wang X, Yu S, Fu D, Chen J, Wang X (2019) Environmental remediation of heavy metal ions by novel-nanomaterials: a review. Environ Pollut 246:608–620. https://doi.org/10.1016/j.envpol.2018.12.076

Yadav V, Sharma N, Prakash R, Raina KK, Bharadwaj LM, Prakash NT (2008) Generation of selenium containing Nano-structures by soil bacterium, Pseudomonas aeruginosa. Biotechnology 7:299–304. https://doi.org/10.3923/biotech.2008.299.304

Yadav K, Gupta N, Kumar V, Singh J (2017) Bioremediation of heavy metals from contaminated sites using potential species: a review. Indian J Environ Prot 37:65–84

Yeary LW, Moon J-W, Love L, Thompson JR, Rawn C, Phelps TJ (2006) Magnetic properties of biosynthesized magnetite nanoparticles. IEEE Trans Magn 41:4384–4389. https://doi.org/10.1109/TMAG.2005.857482

Yin Y, Yang X, Hu L, Tan Z, Zhao L, Zhang Z, Liu J-F, Jiang G (2016) Superoxide-mediated extracellular biosynthesis of silver nanoparticles by the fungus Fusarium oxysporum. Environ Sci Technol Lett 3. https://doi.org/10.1021/acs.estlett.6b00066

Zhang H, Hu X (2018) Biosynthesis of Pd and Au as nanoparticles by a marine bacterium Bacillus sp. GP and their enhanced catalytic performance using metal oxides for 4-nitrophenol reduction. Enzym Microb Technol 113:59–66. https://doi.org/10.1016/j.enzmictec.2018.03.002

Zhang H, Li Q, Lu Y, Sun D, Lin X, Deng X, He N, Zheng S (2005) Biosorption and bioreduction of diamine silver complex by *Corynebacterium*. J Chem Technol Biotechnol 80:285–290. https://doi.org/10.1002/jctb.1191

Chapter 14
Advanced Nanocomposite Materials for Remediation of Hazardous Pollutants

Abhinay Thakur and Ashish Kumar

Abstract Nanocomposite materials, with their distinctive physicochemical properties like high surface area, reactivity, and selectivity, have emerged as promising agents for remediating hazardous pollutants. Comprising nanoparticles ranging from 1 to 100 nm dispersed in organic, inorganic, or hybrid matrices, these materials require meticulous consideration of surface chemistry and interactions to optimize performance. Tailoring nanocomposites for specific applications involves adjusting composition, structure, and properties. Carbon-based, metal-based, and metal oxide-based nanocomposites have been developed and applied effectively in the environmental remediation of heavy metals, organic pollutants, and dyes. This chapter delves into the mechanisms underlying pollutant removal, addressing the challenges and opportunities in nanocomposite development for environmental remediation. The exploration extends to the intricate interplay between nanocomposite properties and pollutant removal mechanisms, particularly in the context of heavy metals, organic pollutants, and dyes. The chapter culminates by reflecting on the future prospects of nanocomposite materials, envisioning their role in advancing technologies for environmental remediation. The versatility and tunability of nanocomposites position them as pivotal contributors to the ongoing quest for sustainable solutions in mitigating environmental pollution.

Keywords Nanocomposite materials · Hazardous pollutants · Heavy metals · Organic pollutants · Sustainability

---

A. Thakur
Department of Chemistry, School of Chemical Engineering and Physical Sciences, Lovely Professional University, Phagwara, Punjab, India

A. Kumar (✉)
NCE, Bihar Engineering University, Department of Science, Technology and Technical Education, Government of Bihar, Chandi, Bihar, India

A. Guldhe et al. (eds.), *Biomaterials for Environmental Applications: Detection and Remediation*, Environmental Challenges and Solutions,
https://doi.org/10.1007/978-3-032-12701-3_14

14.1 Introduction

Nanocomposite materials, a fusion of nanoparticles within a matrix material, represent a paradigm shift in materials science. The inherent characteristics of nanoparticles, particularly their size ranging from 1 to 100 nm, contribute to the exceptional properties of nanocomposites (Lu et al. 2019; Elessawy et al. 2021; Javed et al. 2022). Nanocomposites are powerful instruments in the battle against pollution because of their extensive surface area, reactivity, and selectivity. As the introductory section unfolds, it navigates through the diverse landscape of nanocomposites, categorizing them into carbon-based, metal-based, and metal oxide-based nanocomposites. Each category brings forth distinct advantages, and the section provides insights into the synthesis methods, properties, and implementations of these nanocomposites in the context of environmental remediation. The significance of nanocomposite materials in environmental remediation is underscored by their ability to address a spectrum of pollutants effectively. This introduction delves into the multifaceted applications of nanocomposites, emphasizing their role in adsorption, catalysis, and photodegradation processes. Their tunability, or the capacity to modify the composition, structure, and properties of nanocomposites to suit certain applications, is a crucial component of their efficacy. The section explores the nuances of this tunability and offers a thorough explanation of how the overall performance of nanocomposite materials can be improved by carefully adjusting these parameters. Examining how well nanocomposites function in pollution removal mechanisms is essential to comprehending their applications. This introduction navigates through the intricate details of pollutant removal, focusing on heavy metals, organic compounds, and emerging contaminants (Alharthi et al. 2020; Chauhan et al. 2021; Yaseen et al. 2022). Through processes like adsorption, chelation, and precipitation, nanocomposites specifically target heavy metals, which are notorious for their toxicity and persistence. The conversation offers a thorough analysis of the relationships that exist between heavy metals and nanocomposites, illuminating the critical roles that surface chemistry and certain characteristics of nanoparticles play in their removal. The role of nanocomposites in addressing organic pollutants, arising from industrial discharges and anthropogenic activities, is also examined in detail. The catalytic functionalities of nanocomposites come to the forefront in degrading organic contaminants, and the section provides a nuanced understanding of the mechanisms involved in this degradation process. Furthermore, the dynamic landscape of emerging contaminants, whose impact may not yet be fully understood, is explored. Because of their diversity and adaptability, nanocomposites are becoming vital instruments for mitigating the environmental effects of these new contaminants. For instance, in an experiment, Mandal et al. (2023) introduced an innovative and eco-friendly approach to synthesizing multifunctional nanobiocomposites designed for the effective eradication of hazardous heavy metals, dyes, and disinfection of wastewater microorganisms. The chitosan (CS)-altered, wide-surface-region AC obtained via kenaf (KAC) was combined with photochemically produced silver nanoparticles (AgNPs) using a special self-activation technique to create the nanobiocomposites, known as KAC-CS-AgNPs. Scanning electron microscopy images demonstrated the even distribution of AgNPs, while Fourier transform infrared studies revealed major absorption peaks. Experimental results showcased the exceptional adsorption efficiency of KAC-CS-AgNPs for lead (Pb^{2+}), copper (Cu^{2+}), and Congo Red dye (CR), coupled with effective antibacterial attribute toward *Escherichia coli* and *Staphylococcus aureus*. The maximum adsorption capacities for Cu^{2+}, Pb^{2+}, and CR were 71.5, 72.3, and 75.9 mg/g, respectively, following Langmuir and Freundlich isotherm models and the second-order kinetic model ($R^2 > 0.99$). KAC-CS-AgNPs demonstrated effective reusability for up to four consecutive phases with fewer losses in adsorption capability. The adsorption mechanism appeared to be spontaneous and endothermic based on thermodynamic characteristics. Both *S. aureus* and *E. coli* showed significant bactericidal effects in bacterial inactivation assays, with MIC values of 44 and 32 μg/mL, correspondingly. The KAC-CS-AgNPs, which have been synthesized, offer a novel approach to efficient and sustainable wastewater treatment and have great commercial potential. Similarly, Vasu et al. (2021) presented a pioneering approach in synthesizing Cu/La/CeO_2/GO nanocomposites via hydrothermal techniques, where copper and lanthanum are incorporated into cerium oxide (CeO_2) loaded on graphene oxide (GO). The existence of dopant ions and a crystalline phase was verified by XRD examination, and the nanocomposites' rod-like structure was shown by FESEM pictures. The primary breakthrough is the in-situ synthesis of Cu/La/CeO_2/GO nanocomposites, which exhibit improved photodegradation of hazardous organic dyes such as Cibacron Red (CR), Rhodamine B (RhB), and Sunset Yellow (SY). The structure of the nanocomposites efficiently inhibits the quick integration of photoinduced electron-hole pairs, increasing the photocatalytic efficiency. Interestingly, the proportions of degradation for the dye mixtures containing RhB, SY, and CR were 81%, 60%, and 95%, correspondingly. The morphology and structural characteristics of Cu/La/CeO_2/GO were investigated through FE–SEM analysis, as depicted in Fig. 14.1a, b. The analysis highlighted the rod-like configuration of uniform size formed by the nanocomposites during the hydrothermal synthesis process. The TEM images of Cu/La/CeO_2/GO nanocomposites are presented in Fig. 14.1c, d, revealing aggregated nanocomposites with small particle sizes. Notably, these nanocomposites exhibited an irregular shape, contributing to uneven size distribution. Further insights into the elemental distribution were obtained through STEM–EDS elemental mapping, as illustrated in Fig. 14.1e–h. The mapping demonstrated the existence of all doping materials—La, Cu, Ce, and C—in Cu/La/CeO_2/GO nanocomposites. The overlapping of these elements was observed, indicating a cohesive integration of copper, lanthanum, cerium, and carbon within the nanocomposite structure. The comprehensive characterization of Cu/La/CeO_2/GO morphology and elemental composition enhances our understanding of the nanocomposites' structural features derived from the hydrothermal synthesis method.

Two further hazardous dyes, SY and CR, were used to compare the photocatalytic effectiveness of the nanocomposites that were created. The degradation of SY and CR organic dyes at wavelengths of 482 and 500 nm, respectively, was

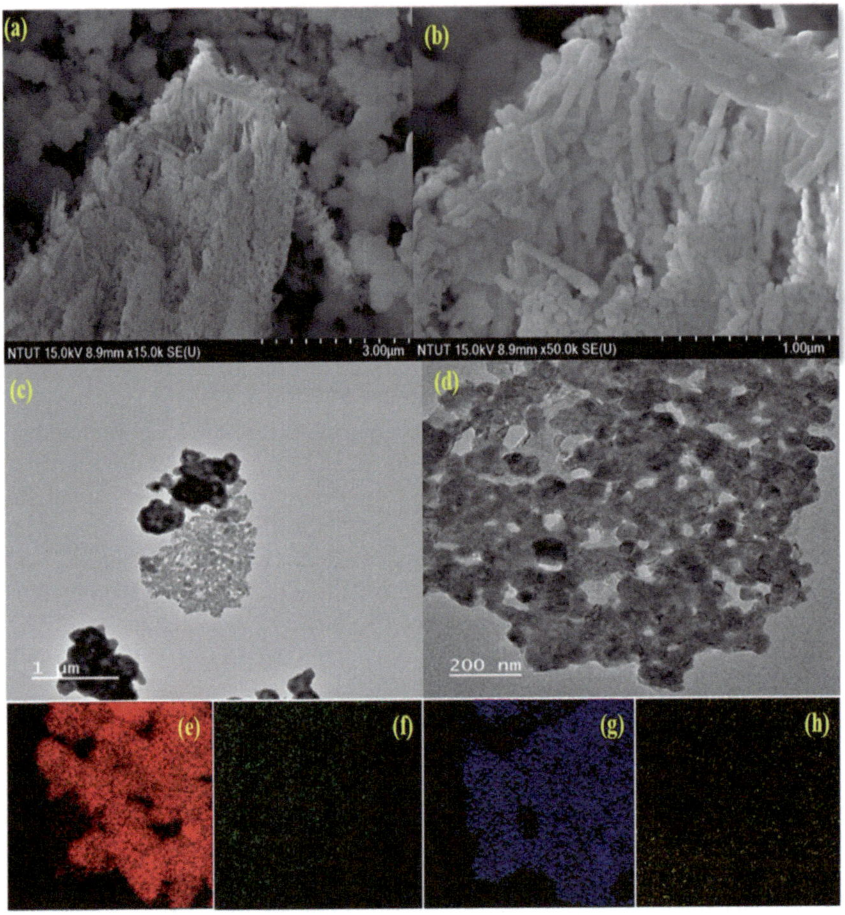

Fig. 14.1 Depiction of the surface morphology of Cu/La/CeO_2/GO nanocomposites in panels (**a**, **b**), showcases the rod-like structures formed during synthesis. Panels (**c**, **d**) present TEM images, offering a closer look at the nanocomposites, while panels (**e**–**h**) provide elemental mapping for Ce, Cu, La, and C within the nanocomposites. (Adapted from Vasu et al. (2021) under CCBY 4.0)

monitored using a UV–Visible spectrophotometer (Fig. 14.2a, c). One distinctive aspect of the photodegradation aided by the Cu/La/CeO_2/GO nanocomposite was the significantly reduced treatment time required to break down these dangerous organic dyes. When exposed to visible light, the produced nanocomposites showed exceptional efficacy in the destruction of harmful organic dyes (Fig. 14.2b, d). Notably, after just 50 min of treatment, the degradation of the CR dye showed its maximum mineralization. This faster deterioration is explained by the fact that the nanocomposites produce more electron–hole pairs, which in turn produce more superoxide and hydroxyl radicals. Reactive radicals were essential in causing organic contaminants to degrade in the presence of visible light.

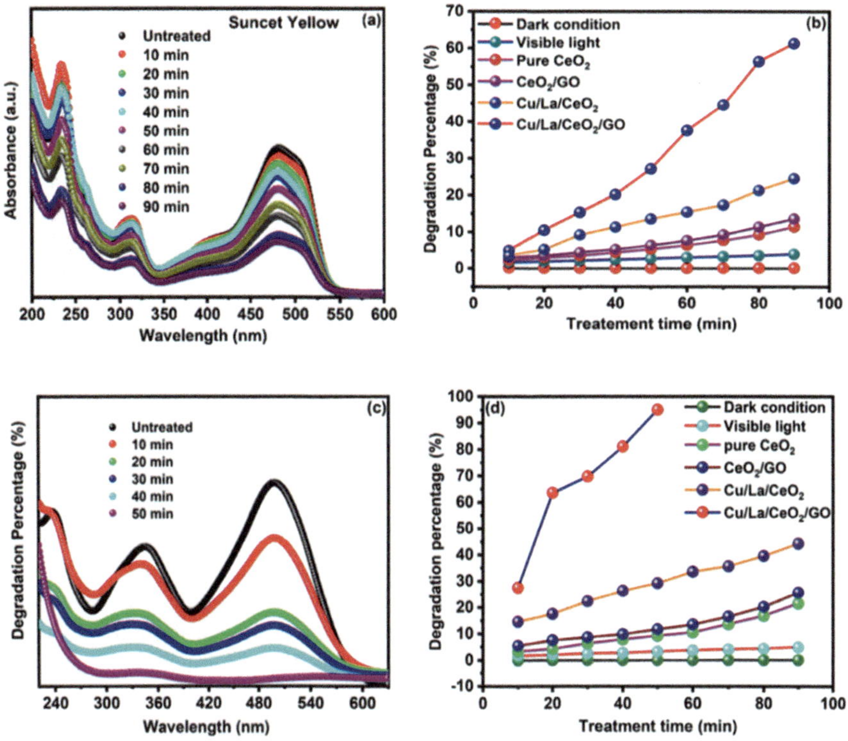

Fig. 14.2 Depiction of the photodegradation process of CR and SY organic dyes through UV–vis spectrum. Panels (**a**, **b**) depict the degradation and absorption % of SY, while panels (**c**, **d**) showcase the degradation and absorption % of CR. (Adapted from Vasu et al. (2021) under CCBY 4.0)

The chapter places a particular emphasis on the diverse applications of nanocomposites in environmental remediation. Adsorption processes, a cornerstone in pollutant removal, are explored in-depth, highlighting the unique adsorption mechanisms facilitated by nanocomposites. Catalytic applications, leveraging the catalytic properties of metal nanoparticles within nanocomposites, emerge as a powerful strategy for pollutant degradation. The section also explores the sustainable approach of photodegradation, where nanocomposites harness solar energy to drive the breakdown of pollutants.

14.2 Nanocomposite Properties

Nanocomposite materials, the fusion of nanoparticles within a matrix, possess intricate properties that underpin their efficacy in diverse applications, particularly in environmental remediation. This section delves into the physicochemical characteristics of nanocomposites, emphasizing the importance of surface chemistry and interactions, and explores how optimization strategies enhance their overall performance.

14.2.1 Physicochemical Characteristics

At the core of nanocomposite materials are their physicochemical characteristics, which stem from the unique attributes of nanoparticles embedded within the matrix. The size, shape, composition, and distribution of nanoparticles significantly influence the overall behavior of nanocomposites.

- Size and surface area: The high surface area of nanocomposites is a result of the presence of nanoparticles, which are usually between 1 and 100 nm in size. Because there are more active sites due to this increased surface area, their reactivity and adsorption capability are improved. Carbon-based nanocomposites, for example, capitalize on the vast surface area of carbon nanotubes or graphene, rendering them particularly effacious in environmental applications.
- Shape and distribution: The accessibility of active sites is greatly influenced by the morphology of the nanoparticles, including their distribution and shape within the matrix. Controlled dispersion of well-dispersed nanoparticles maximizes the use of their characteristics. For effective pollutant degradation, metal-based nanocomposites take advantage of the catalytic qualities of evenly distributed metal nanoparticles.
- Composition: The choice of materials for both nanoparticles and the matrix is crucial in defining the overall composition of nanocomposites. Carbon-based nanocomposites may utilize graphene or carbon nanotubes, while metal oxide-based nanocomposites combine the stability of metal oxides with the reactivity of nanoparticles (Sun et al. 2018; Thamer et al. 2021). The composition dictates the overall properties, influencing factors such as conductivity, stability, and reactivity.
- Optical properties: In some nanocomposites, particularly those with metal nanoparticles, optical properties become significant. Metal oxide-based nanocomposites, for instance, exhibit photocatalytic properties, allowing them to harness solar energy for environmental remediation through photodegradation processes. Understanding and manipulating these optical properties are key to enhancing their performance in such applications.
- Mechanical properties: The inclusion of nanoparticles could also influence the mechanical characteristics of nanocomposites. Carbon-based nanocomposites, known for their exceptional mechanical strength, find applications in pollutant adsorption and structural matrices. The section provides a comprehensive exploration of how these physicochemical characteristics collectively contribute to the unique properties of nanocomposite materials.

14.2.2 Surface Chemistry and Interactions

Surface chemistry is a defining aspect of nanocomposites, governing their interactions with pollutants, the matrix material, and the surrounding environment. The intricate interplay between nanoparticles and their surroundings determines the efficiency of nanocomposites in pollutant removal processes.

- Nanoparticle-Matrix Interactions

The success of nanocomposites in environmental remediation hinges on the nature of interactions between nanoparticles and the matrix material. These interactions are fundamental for determining the stability, durability, and overall performance of nanocomposites (Naveed et al. 2021; Chauhdary et al. 2022). The careful engineering of these relationships is pivotal in tailoring nanocomposites for specific applications. The choice of matrix material, whether organic, inorganic, or a hybrid combination, influences how nanoparticles are dispersed and how they interact within the composite structure. Understanding and manipulating nanoparticle-matrix interactions are essential steps in optimizing the design and functionality of nanocomposites for efficient pollutant removal.

- Surface Functionalization

An effective tactic to increase the propensity of nanocomposites for certain contaminants is surface functionalization. In order to modify the chemical properties of nanoparticle surfaces to better suit the target pollutants, particular functional groups are added to the surfaces. When it comes to carbon-based nanocomposites, surface functionalization is an effective technique (Wang and Shih 2021). The capacity for adsorption of carbon-based nanocomposites for organic contaminants might be greatly increased by adding particular functional groups.

- Adsorption Mechanisms

The adsorption mechanisms of nanocomposites are intricately linked to their surface chemistry. In the process of pollutant adsorption, the substantial surface area and particular interactions at the nanocomposite interface are essential. By using a variety of adsorption techniques, nanocomposites use their distinct physicochemical characteristics to draw in and absorb contaminants. Predicting and maximizing the adsorption capacity of nanocomposites requires an understanding of these mechanisms. The adsorption mechanisms determine how well nanocomposites can eliminate contaminants from the environment, whether by physical adsorption, chemical adsorption, or an amalgam of both.

- Electrochemical Interactions

In certain nanocomposites, particularly those incorporating metal nanoparticles, electrochemical interactions play a significant role in shaping their behavior. The electrochemical properties of nanocomposites become particularly relevant in processes such as catalysis. Metal-based nanocomposites leverage the catalytic capabilities of metal nanoparticles, and understanding the electrochemical interactions involved is crucial for optimizing their performance. As nanocomposites with metal components continue to find applications in environmental remediation, unraveling the electrochemical aspects becomes imperative for harnessing their full potential and ensuring their effectiveness in diverse environmental conditions.

14.2.3 Optimization for Performance

The optimization of nanocomposites is a critical aspect of harnessing their full potential in environmental remediation. This involves careful consideration of various parameters, including composition, structure, and properties, to tailor nanocomposites for specific applications and enhance their overall performance.

- Composition Optimization

Tailoring the composition of nanocomposites is a meticulous process involving the selection of the most suitable combination of nanoparticles and matrix materials. This section delves into the intricate relationship between the choice of materials and the overall properties of nanocomposites. Composition optimization is a crucial step in achieving desired outcomes in pollutant removal. Different types of nanoparticles, ranging from carbon-based to metal-based, and various matrix materials, whether organic, inorganic, or hybrid, offer distinct advantages. Understanding how the composition influences the stability, reactivity, and adsorption capacity of nanocomposites is essential for designing materials that are effective in specific environmental remediation scenarios (Guerra et al. 2018; Alvarez-Torrellas et al. 2019; Liu et al. 2019). Strategies for optimizing composition are explored, providing insights into the factors that contribute to the success of nanocomposites in pollutant removal applications.

- Structural Optimization

The structural arrangement of nanoparticles within the matrix material exhibits a vital role in determining the accessibility of active sites. Well-organized structures with controlled distribution enhance the stability and reactivity of nanocomposites. This section delves into the importance of structural optimization, emphasizing how the spatial arrangement of nanoparticles influences their exposure to pollutants. The controlled distribution of active sites ensures efficient interaction with contaminants, maximizing the efficacy of nanocomposites in environmental remediation. Strategies for achieving structural optimization, such as precise control during synthesis processes, are discussed.

- Property Optimization

A crucial component of optimization is adjusting the intrinsic characteristics of nanocomposites. The performance of nanocomposites can be improved by adjusting characteristics including reactivity, surface area, and particle size. In order to optimize properties, controlled synthesis methods and post-synthesis treatments are investigated. For example, to improve their adsorption capabilities, carbon-based nanocomposites can be heated or chemically treated. The understanding of how these properties influence pollutant removal efficiency is essential for designing nanocomposites with optimized performance.

- Functionalization Strategies

Surface functionalization emerges as a powerful strategy for optimizing the performance of nanocomposites. In order to improve the interaction of nanoparticles with specific contaminants, particular functional groups are introduced onto their surface. The section examines several functionalization techniques and how they affect the contaminants' affinity for nanocomposites. Nanocomposites can be fine-tuned to show improved adsorption capabilities and reactivity by functionalizing the surface chemistry. The versatility of functionalization strategies adds a layer of adaptability to nanocomposites, allowing them to address a broad array of pollutants in diverse environmental conditions.

- Adapting for Specific Applications

The adaptability of nanocomposites for specific applications is a key aspect of optimization. Tailoring nanocomposites to address particular pollutants or environmental conditions requires a nuanced understanding of the factors influencing their performance. This section explores how the optimization process extends beyond general properties to consider the specific requirements of environmental remediation projects (Shahabuddin et al. 2016; Liu et al. 2018; Cardoso et al. 2021). The adaptability of nanocomposites involves customizing their composition, structure, and properties to meet the challenges posed by different pollutants and environmental matrices. Strategies for adapting nanocomposites for specific applications are discussed, emphasizing the importance of a targeted and context-specific approach in optimizing their effectiveness.

14.3 Fabrication and Design

- The successful deployment of nanocomposite materials in environmental remediation relies heavily on meticulous fabrication and thoughtful design. This section embarks on a comprehensive exploration of the intricacies involved in the creation of nanocomposites, emphasizing composition adjustment, structural considerations, and property optimization. As we delve into the realm of fabrication and design, we uncover the innovative strategies that researchers employ to tailor nanocomposites for specific environmental challenges.

14.3.1 Composition Adjustment

Tailoring the composition of nanocomposites is a pivotal step in designing materials with targeted properties for effective pollutant removal. This subsection delves into the strategies and considerations involved in adjusting the composition, choosing appropriate nanoparticles and matrix materials, and achieving the desired synergy for optimal environmental remediation.

- Material Selection and Compatibility

The selection of nanoparticles and the matrix material is fundamental to the composition of nanocomposites (Singh et al. 2013; Modi et al. 2022). This chapter scrutinizes the characteristics of various nanoparticles, ranging from carbon-based materials like CNT and graphene to metal oxide-based nanoparticles. Understanding the compatibility and synergies between different nanoparticle and matrix combinations is paramount in crafting nanocomposites with enhanced stability and reactivity.

- Tailoring for Specific Pollutants

Environmental pollutants exhibit diverse characteristics, and tailoring nanocomposite compositions to specific pollutants requires precision in design (Bashir et al. 2019, 2020b; Kumar and Thakur 2020; Thakur et al. 2022a; Kaya et al. 2023a, b). For instance, the discussion unfolds how carbon-based nanocomposites can be tailored to adsorb organic pollutants, while metal-based nanocomposites leverage catalytic properties for the degradation of specific contaminants. The ability to fine-tune compositions allows researchers to address a spectrum of pollutants with unparalleled specificity.

- Synergistic Effects

The synergistic effects arising from the judicious combination of nanoparticles and matrix materials are explored in this subsection. Metal oxide-based nanocomposites, for example, capitalize on the stability of metal oxides combined with the reactivity of nanoparticles, showcasing how a harmonious interplay between components can elevate the overall performance of nanocomposites in environmental remediation.

- Responsive Nanocomposites

The dynamic nature of environmental conditions necessitates the development of responsive nanocomposites. These materials are more adaptable and efficient because they can change their characteristics in response to environmental changes (Seliem et al. 2020; Castro-Muñoz et al. 2021). The chapter delves into the design principles behind responsive nanocomposites, outlining how changes in factors such as pH, temperature, or pollutant concentration trigger adaptive responses, ensuring sustained effectiveness in diverse environmental contexts.

14.3.2 Structural Considerations

Beyond composition, the structural arrangement of nanoparticles within the matrix plays a crucial role in determining the accessibility of active sites and overall stability. This subsection explores various structural considerations, encompassing controlled synthesis methods and templated approaches, to optimize the architecture of nanocomposites for environmental applications.

- Controlled Synthesis Methods

Achieving a well-defined structure within nanocomposites requires precision in synthesis methods. The controlled synthesis of carbon-based nanocomposites, for example, can enhance their adsorption capacities by ensuring uniform dispersion and optimal structural arrangements.

- Templated Approaches

Templated approaches represent a strategic method to guide the assembly of nanoparticles within the matrix material. Templated approaches are instrumental in achieving specific structural architectures, influencing properties including surface area and porosity. Understanding the nuances of templated strategies allows for the deliberate design of nanocomposite structures tailored to desired applications in pollutant removal.

- Nanocomposite Morphologies

The morphology of nanocomposites, influenced by structural considerations, is a crucial factor in determining their functionality. Different morphologies, such as core-shell, hollow, or layered structures, offer unique advantages in pollutant removal processes. For instance, core-shell structures in metal-based nanocomposites can enhance catalytic efficiency, while layered structures in graphene-based nanocomposites provide increased surface area for adsorption.

- Hierarchical Nanocomposites

Hierarchical nanocomposites, featuring multiscale architectures, represent an advanced approach to structural design. Hierarchical arrangements enhance the accessibility of active sites, facilitate mass transfer, and improve overall stability.

14.3.3 Property Optimization

Optimizing the inherent properties of nanocomposites is imperative to fine-tune their performance for specific environmental challenges. This subsection delves into the strategies employed to optimize properties, including surface area, particle size, reactivity, and conductivity, ensuring that nanocomposites stand at the pinnacle of efficiency in environmental remediation.

- Manipulating Particle Size

Particle size is a critical parameter influencing the properties of nanocomposites. This section discusses how manipulating particle size, through techniques like ball milling or chemical vapor deposition, can impact the surface area and reactivity of nanoparticles (Fan et al. 2021; Shathy et al. 2022). Tailoring particle size becomes especially crucial in optimizing adsorption capacities and catalytic activities, enhancing the overall pollutant removal efficiency of nanocomposites.

- Enhancing Surface Area

The high surface area of nanocomposites is a key factor in their effectiveness. Strategies to enhance the surface area, such as creating porous structures or utilizing nanosheets, are explored in this subsection. Carbon-based nanocomposites, with their inherent porosity, exemplify how maximizing surface area contributes to superior adsorption capacities (Bashir et al. 2020a, c; Thakur et al. 2022b; Sharma et al. 2023; Thakur and Kumar 2023a, b).

- Tuning Reactivity

The reactivity of nanocomposites, particularly in catalytic and photocatalytic applications, is a critical property. Metal-based nanocomposites leverage the catalytic properties of nanoparticles, and metal oxide-based nanocomposites exhibit photocatalytic capabilities.

- Conductivity Optimization

In certain nanocomposites, especially those with carbon-based materials, conductivity is a crucial property. Enhanced conductivity facilitates efficient electron transfer, a key factor in catalytic reactions.

- Functionalization Strategies

Surface functionalization emerges as a versatile strategy for property optimization. By introducing specific functional groups to nanoparticle surfaces, researchers can tailor nanocomposites for targeted applications (Thamer et al. 2020; Kocijan et al. 2022). The ability to functionalize nanocomposites enhances their affinity for specific pollutants, expanding their applicability in diverse environmental contexts.

14.4 Types of Nanocomposites

The diverse landscape of nanocomposites unfolds through this exploration of various types specifically engineered for environmental remediation. Each category, such as carbon-based, metal-based, and metal oxide-based nanocomposites, brings forth unique properties, synthesis methods, and applications. This section traverses the intricacies of these types, delving into the nuances that make them indispensable tools in the pursuit of sustainable pollutant removal.

14.4.1 Carbon-Based Nanocomposites

Carbon-based nanocomposites represent a groundbreaking category with unparalleled versatility and unique physicochemical properties. The carbonaceous components, including graphene, carbon nanotubes, and activated carbon, serve as the cornerstone for their exceptional adsorption capacities and reactivity in environmental remediation. Graphene is a special two-dimensional nanomaterial with exceptional properties made up of a single sheet of carbon atoms arranged in a hexagonal lattice. The high surface area, excellent conductivity, and mechanical strength of graphene contribute to its effectiveness in adsorbing pollutants and catalyzing degradation reactions. Carbon nanotubes, cylindrical structures of rolled-up graphene sheets, possess unique electronic, mechanical, and adsorption properties. Delving into the synthesis techniques and applications of carbon nanotube-based nanocomposites, this subsection unravels how their inherent characteristics enhance pollutant adsorption and catalytic activities. Their tubular structure provides an extensive surface area for interactions with pollutants, making them formidable agents in environmental remediation. Activated carbon, characterized by a highly porous structure, emerges as a stalwart adsorbent in nanocomposites (Bekru et al. 2023; Shan et al. 2023). The porous nature of activated carbon facilitates pollutant adsorption through physical and chemical mechanisms. The subsection explores how tailored activation processes and hybridization with other materials enhance the pollutant removal capabilities of activated carbon-based nanocomposites.

- Challenges and Future Perspectives in Carbon-Based Nanocomposites

While carbon-based nanocomposites showcase immense potential, challenges such as aggregation, scalability, and cost hinder their widespread application. This section critically examines these challenges and envisions future research directions. Strategies to overcome aggregation issues, scale up production, and enhance cost-effectiveness are discussed, providing a comprehensive overview of the current landscape and the trajectory of carbon-based nanocomposites in environmental remediation.

14.4.2 Metal-Based Nanocomposites

Metal-based nanocomposites leverage the catalytic properties of metal nanoparticles to drive pollutant degradation reactions. Iron-based nanocomposites, particularly those incorporating zero-valent iron (ZVI) nanoparticles, exhibit remarkable reactivity in pollutant degradation. This subsection elucidates the synthesis techniques and applications of iron-based nanocomposites in environmental remediation. The redox properties of iron nanoparticles make them effective in transforming pollutants through reduction or oxidation reactions, providing a sustainable approach to remediation (Youssef et al. 2019; Kar et al. 2020). Silver nanoparticles, known for their antimicrobial properties, find applications in nanocomposites for environmental remediation. Their antimicrobial activities extend the scope of these nanocomposites to address microbial pollution in water bodies. Copper-based nanocomposites emerge as catalytic guardians in environmental remediation. This subsection unravels the synthesis techniques and applications of copper-based nanocomposites, emphasizing their catalytic activities in pollutant degradation. Copper nanoparticles facilitate the breakdown of pollutants through redox reactions and adsorption, showcasing their potential in diverse remediation scenarios.

- Challenges and Future Perspectives in Metal-Based Nanocomposites

Despite their catalytic prowess, metal-based nanocomposites face challenges related to stability, selectivity, and potential toxicity. This section critically examines these challenges and outlines future research directions. Strategies to elevate the stability of metal nanoparticles, improve selectivity in pollutant removal, and address potential toxicity concerns are discussed, providing insights into the evolving landscape of metal-based nanocomposites for environmental applications.

14.4.3 Metal Oxide-Based Nanocomposites

Metal oxide-based nanocomposites amalgamate the stability of metal oxides with the reactivity of nanoparticles, offering a versatile and synergistic approach to environmental remediation. Titanium dioxide (TiO_2) stands as a flagship in metal oxide-based nanocomposites, renowned for its photocatalytic properties (Chavan et al. 2015; Olteanu et al. 2019; Oyewo et al. 2020). The photocatalytic prowess of TiO_2 enhances its efficiency in breaking down pollutants and emerging contaminants, marking it as a sustainable solution for environmental remediation. Zinc oxide (ZnO), with its multifunctional properties, takes center stage in metal oxide-based nanocomposites. Because of its unique properties, including its broad bandgap and vast surface area, ZnO can be used to solve a number of environmental issues. Iron oxide-based nanocomposites, incorporating materials like magnetite (Fe_3O_4) and hematite (α-Fe_2O_3), bring magnetism into the realm of environmental remediation. This subsection elucidates the synthesis techniques and applications of iron oxide-based nanocomposites, emphasizing their magnetic properties for facile separation and recovery. The magnetic responsiveness of iron oxide nanoparticles enhances the recyclability and efficiency of these nanocomposites in pollutant removal.

- Challenges and Future Perspectives in Metal Oxide-Based Nanocomposites

While metal oxide-based nanocomposites offer a synergistic approach to remediation, challenges such as agglomeration, photocatalyst recovery, and scalability persist. This section critically assesses these challenges and envisions future directions for research. Strategies to mitigate agglomeration, improve photocatalyst recovery methods, and scale up production are discussed, providing a comprehensive outlook on the evolving landscape of metal oxide-based nanocomposites.

14.5 Applications in Environmental Remediation

Nanocomposites have transcended the realm of laboratory curiosities to become pivotal players in addressing the escalating environmental challenges. This section discusses the application of nanocomposites for the elimination of organic pollutants, heavy metals, and dyes.

14.5.1 Heavy Metal Remediation

When it comes to eliminating heavy metals from a variety of environmental matrices, nanocomposites are incredibly effective. They may bind heavy metal ions precisely and selectively because of their large surface area and particular surface chemistry. Ion exchange, chelation, and adsorption mechanisms in nanocomposites help remove metals like lead, cadmium, and mercury precisely. For instance, in an experiment, Dan et al. (2023) explored the adsorptive capabilities of a graphene oxide/SiO_2 nanocomposite-based adsorbent. The sorption of Cr (VI) and MB ions from wastewater was carried out using the graphene oxide and SiO_2 nanocomposite. Furthermore, the synthesized nanocomposites' antibacterial potential was examined. In their lab, adsorption studies were carried out with different variables, and the Langmuir isotherm model was used to evaluate the maximum capacities of the adsorbent. The highest adsorption capabilities for MB and Cr (VI) ions were determined to be 555.50 and 181.81 mg/g, correspondingly. The adsorption study findings showed that the graphene oxide/SiO_2 nanocomposite performed according to the order MB > Cr (VI). Furthermore, research on the nanocomposites' antibacterial activity against both gram-negative and gram-positive bacteria demonstrated how successful it was in getting rid of bacteria, highlighting its potential for application in multipurpose water purification procedures. Similarly, Sharma et al. (2014) Using the sol–gel method, polyaniline gel was combined with the inorganic precipitates of Th(IV) tungstomolybdophosphate (TWMP) to create Polyaniline Th(IV) TMTPP (PANI/TWMP). A comprehensive characterization was performed using FTIR, XRD, SEM, and TEM methods. The study demonstrated that PANI/TWMP had a higher ion exchange capacity (1.07 meq/g) than both its inorganic counterpart TWMP (0.78 meq/g) and organic counterpart PANI (0.30 meq/g). Additionally, PANI/TWMP showed thermal stability at 400 °C, retaining 53.27% of its ion exchange capacity. Remarkable antibacterial activity toward Staphylococcus aureus was shown by the nanocomposite. In order to separate Cu^{2+} and Pb^{2+} selectively, the paper also examined the possible uses of PANI/TWMP in the evaluation of dangerous heavy metals in the environment. The material's efficient ion exchange capacity, thermal stability, and antimicrobial properties make PANI/TWMP a versatile candidate for water remediation and environmental analysis.

Yaqoob et al. (2021) focused on the fabrication of a cellulose-derived graphene-polyaniline (GO-PANI) composite anode to enhance electron transfer rates

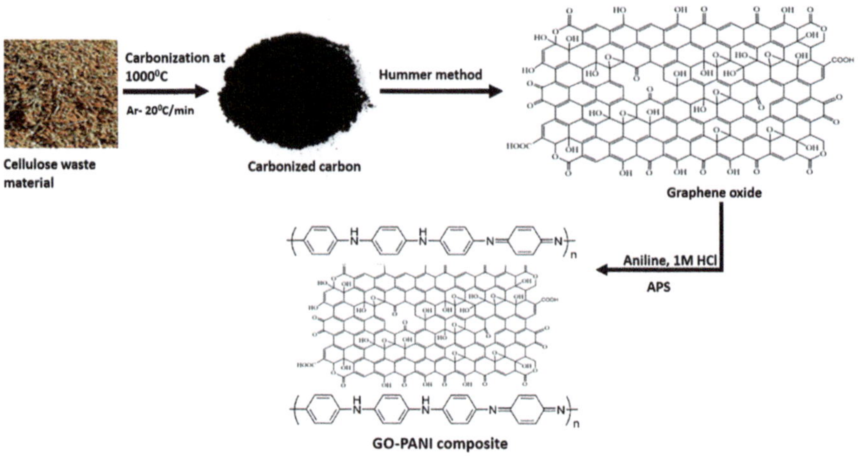

Fig. 14.3 The process of transforming biomass into graphene oxide (GO), ultimately results in the synthesis of the GO-polyaniline (GO-PANI) nanocomposite. (Adapted from Yaqoob et al. (2021) under CCBY 4.0)

(Fig. 14.3). The anode's efficiency in Benthic microbial fuel cells (BMFCs) is characterized by a range of electrochemical and physicochemical approaches. Using sweet potatoes as an organic substrate, the GO-PANI anode demonstrated a peak current density of 87.71 mA/m^2 during polarization behavior and 15.13 mA/m^2 in closed-circuit settings with an external resistance of 1000 Ω. The unaltered graphene anode performed four times as well as the enhanced one. Additionally, the GO-PANI anode demonstrated superior restoration effectiveness of 65.51% for Cd(II) and 60.33% for Pb(II) when compared to the unmodified graphene anode.

Remediation efficacy for Cd (II) reached 14% (GO) and 23% (GO-PANI) using sweet potato waste, while for Pb (II), it stood at 11% (GO) and 16% (GO-PANI). Figure 14.4 shows that the removal efficacy for Pb (II) in tapioca proved 7% (GO) and 11% (GO-PANI), while the remediation efficiency for Cd (II) was 10% (GO) and 16% (GO-PANI). In conclusion, tapioca proved less effective than sweet potato trash in both the cadmium and lead cleanup processes.

Chavan et al. (2015) utilized polydimethylsiloxane (PDMS) foams, which undergo appropriate modifications to render the hydrophilic state, facilitating the penetration of polluted water into their volume. The surface of the foams was adorned with ZnSe colloidal nanocrystals (NCs), serving as active components that capture the metal ions. When the nanocomposite foams were immersed in Pb^{2+} or Hg^{2+}-contaminated water, the heavy metal ions were effectively eliminated. The term "cation exchange reaction" refers to this process. The removal rate for Pb^{2+} ions surpasses 98% in the studied concentration range of 20–40 ppm, whereas the removal rate for Hg^{2+} ions is close to 100%. It is significant because over 95% of both metal ions are eliminated in the first hour of the reaction, even if it ends after 24 h. Furthermore, the color of the foams alters the way that heavy metal ions are trapped, making it simple to identify their existence in water. The produced foams

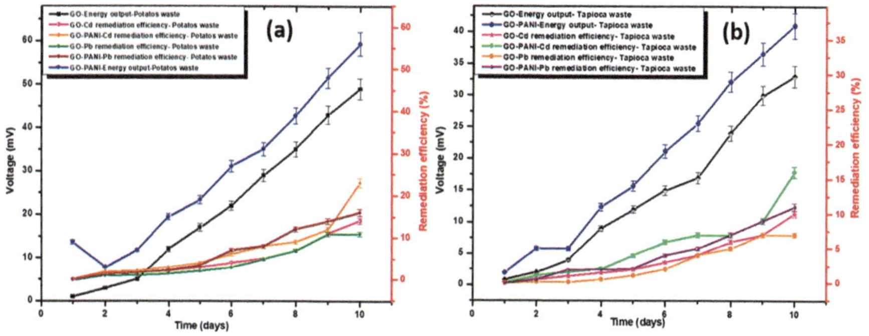

Fig. 14.4 Depiction of the impact of organic substrates, specifically (**a**) potato and (**b**) tapioca waste, over energy production and the efficacy of remedial through benthic microbial fuel cells (BMFCs). (Adapted from Yaqoob et al. (2021) under CCBY 4.0)

proved adaptable for application in various water remediation circumstances due to their good elastic characteristics and heat resistance. Similarly, Youssef et al. (2019) presented a novel approach for the preparation of cross-linked nanocomposite films comprising polyvinyl alcohol (PVA) and functionalized multiwalled carbon nanotubes (f-MWCNTs) at varying concentrations. The inherent limitations of multiwalled carbon nanotubes (MWCNTs), such as poor water solubility, inadequate surface chemistry, inflexibility, and incapacity to form film composites independently, hinder their widespread application. To overcome these drawbacks and enhance their suitability, the MWCNTs are modified and blended with PVA, leveraging the advantages of both materials. This process creates hydrophobic films that are highly recyclable and capable of efficiently removing bacteria, fungi, pesticides, and heavy metals from wastewater. Citric acid was used to crosslink the f-MWCNTs/PVA composite, resulting in a powder that was subsequently placed on Petri dishes and dried at 60 °C to form nanocomposite films. According to the experimental findings, the elimination efficiency of heavy metals rises as f-MWCNT concentrations rise, reaching an ideal efficiency of 8%. Moreover, higher pesticide capture is correlated with higher f-MWCNT concentration. Significant antibacterial activity is indicated by greater inhibition zones produced by increased loadings of f-MWCNTs in the nanocomposites, according to antimicrobial tests against a variety of bacterial and fungal species. These findings establish the potential of these nanocomposite films as a promising and economically feasible technology for wastewater treatment.

14.5.2 Organic Pollutant Removal

The catalytic functionalities of nanocomposites come to the forefront in the remediation of organic pollutants. Nanocomposites, especially those incorporating metal nanoparticles, catalyze the degradation of organic compounds through oxidation and reduction processes. Surface functionalization enhances their adsorption capacities, setting the stage for subsequent catalytic transformations. In an experiment, Akintunde et al. (2022) focused on the evaluation of visible-light-driven photocatalysts for the treatment of emerging contaminants in water and wastewater. The research particularly looked into the photocatalytic activities of nanocomposites, or GCN/ = Ag_2CrO_4, that result from the coupling of silver chromate (Ag_2CrO_4) and graphitic carbon nitride (GCN). The ability of the nanocomposites to degrade organic pollutants and inactivate microorganisms was assessed using an imperial blue light-emitting diode (LED). Methyl chlorophenoxy propionic acid (MCPP or Mecoprop-P) and 2,4-dichlorophenoxyacetic acid (2,4-D) were two of the targeted organic pollutants found in the herbicide KillexR. The bovine serum albumin (BSA) protein and the SARS-CoV-2 spike protein were also added. The results demonstrated that the GCN/Ag_2CrO_4 nanocomposites could degrade 2,4-D and Mecoprop-P by more than 85% in just 120 min. The BSA protein was completely broken down in 20 min, and 77.5% of the SARS-CoV-2 protein was destroyed in 30 min. Additionally, the nanocomposites decreased cellular ATP (cATP), total coliforms, and E. coli in wastewater treatment plants' (WWTPs') secondary effluent by more than one log after an hour of imperial blue LED irradiation. The investigation showed how efficiently the photocatalyst responded to light-induced degradation of specific pollutants. While the interaction strength of released metal ions of GCN/Ag_2CrO_4 to protein and ATP functional sites contributed to protein breakdown and cATP reduction, reactive oxygen species (ROS) assisted in disinfecting *E. coli* and total coliforms.

Balu et al. (2018) explored the photodegradation of MB dye in the existence of visible light utilizing SnS_2 (tin sulfide) produced via a facile hydrothermal methodology. To increase the photocatalytic efficacy, core-shell patterned $SiO_2@\alpha\text{-}Fe_2O_3$ nanocomposites were applied to the SnS_2 flowers. The work focused on the photocatalytic properties of the $SnS_2\text{-}SiO_2@\alpha\text{-}Fe_2O_3$ nanocomposites by monitoring the MB breakdown. The results showed that the photocatalytic efficiency was significantly improved, particularly for the 15-wt% $SiO_2@\alpha\text{-}Fe_2O_3$ nanocomposites on SnS_2 flowers. As semiconductor materials, such as SnS_2, have been shown to exhibit effective photocatalytic properties, they were helpful for the photodegradation of organic pollutants. $SiO_2@\alpha\text{-}Fe_2O_3$ nanocomposites' core-shell structure significantly improved the material's overall performance. Figure 14.5a illustrates the % of remaining MB dye in the media subjected to visible light irradiation for various nanocomposites, including Fe_2O_3, SnS_2, $SnS_2\text{-}SiO_2@\alpha\text{-}Fe_2O_3$, and $SiO_2@Fe_2O_3$ (SSF-15 wt%). Following 100 min, the dye degradation efficiencies were 19%, 42%, 48%, and 96%, correspondingly. Furthermore, specimens were taken at regular intervals throughout visible-light irradiation to determine the catalytic performance of the $SnS_2\text{-}SiO_2@\alpha\text{-}Fe_2O_3$ nanocomposite via phenol degradation (25 ppm). The excellent photocatalytic destruction of phenol was demonstrated by the same nanocomposite, as evidenced by the decrease of the characteristic phenol peak at 270 nm. This proves that $SnS_2\text{-}SiO_2@\alpha$-photocatalytic Fe_2O_3's activity, not the dye's sensitization, is responsible for the dye's deterioration. The

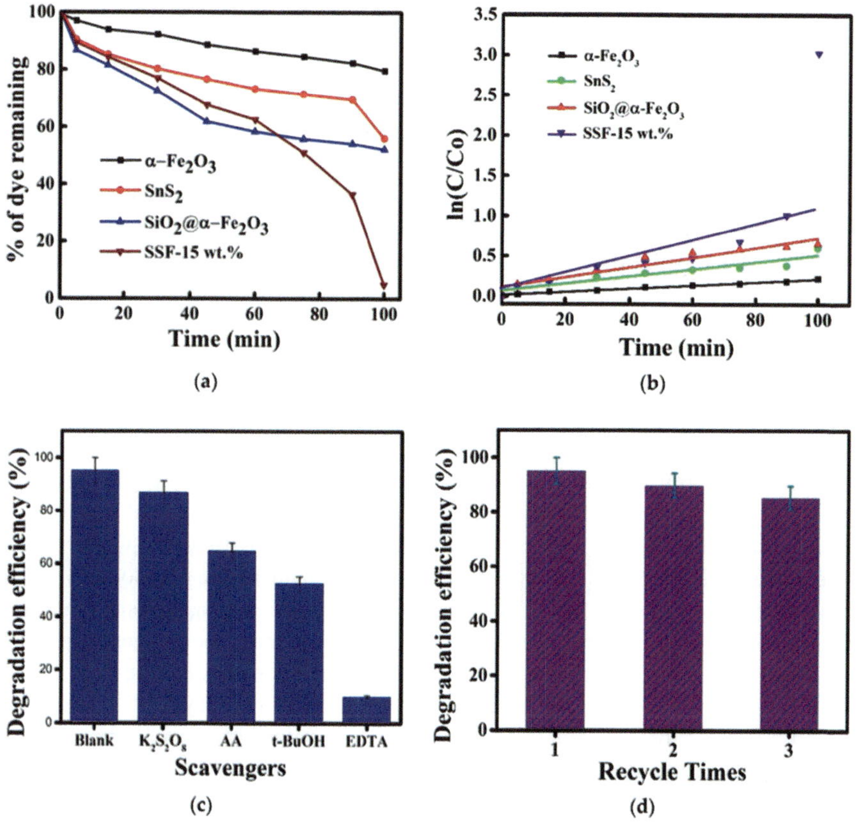

Fig. 14.5 (**a**) Degradation percentage and residual dye over various time intervals (5–100 min) under visible light irradiation for α-Fe_2O_3/SnS_2/SiO_2@α-Fe_2O_3/SnS_2-SiO_2@α-Fe_2O_3. (**b**) ln (C/Co) versus time plot. (**c**) Photodegradation efficacy of SSF-15 wt% with different active species (scavengers). (**d**) Reusability assessment of the SSF-15 wt% nanocomposite. (Adapted from Balu et al. (2018) under CCBY 4.0)

SnS_2-SiO_2@α-Fe_2O_3 nanocomposite has an excellent performance in photodegrading dye and phenol under visible light, as evidenced by the reported degradation efficiencies.

Shan et al. (2023) introduced a novel eco-benign Ag_2O/Fe_3O_4 magnetic nanocomposite, synthesized through a simple coprecipitation technique. This nanocomposite increases broad-spectrum responsiveness and efficiency by efficiently coupling photocatalysis and Fenton processes, in addition to establishing a type II heterojunction. The Ag_2O/Fe_3O_4 (10%) nanocomposite showed remarkable degradation ability against organic dyes when exposed to visible light, achieving a remarkable 99.5% breakdown of 10 mg/L methyl orange (MO) in less than 15 min. Its ability to degrade different dyes effectively while retaining high stability across several testing cycles demonstrates its adaptability. The Ag_2O/Fe_3O_4 (10%)

nanocomposite's ease of recovery with magnetic force is highlighted by magnetic testing. This novel method offers a viable plan for building high-efficiency photocatalytic systems to combat organic dye-related water pollution. Similarly, Bekru et al. (2023) demonstrated how to create (rGO-ZnO)/CuO nanocomposites (NCs) with the help of a microwave (MW) for effective (photo)catalytic uses. The results showed the formation of rod-like (rGO-ZnO)/CuO NCs having a nanoscale diameter. These NCs displayed well-matched band locations across CuO and rGO-ZnO as well as enhanced light absorption. Additionally, catalytic reduction studies utilizing 4-nitrophenol (4-NP) and photocatalytic breakdown studies utilizing MB produced outstanding results. The rate parameters for the breakdown of MB were 0.02213 min^{-1} during 105 min and 0.468 min^{-1} during 8 min for 4-NP elimination. Therefore, the deliberate insertion of CuO onto ZnO nanorods (NRs) at the (rGO-ZnO)/CuO NCs interface shows promise as a useful technique for creating extremely powerful photocatalysts.

Shabelskaya et al. (2023) deliberated a straightforward one-stage synthesis method for creating organic-inorganic composite materials featuring cobalt (II) and iron (III) oxide compounds exhibiting a developed substrate. The goal was to comprehend how the surface of biochar forms the cobalt(II) ferrite ($CoFe_2O_4$) structure. Specifically, rice husks, pea kernels, and sunflower husks are the sources of the biochar. Notably, this study marks the first-time production of composite materials, cobalt ferrite/biochar, using these agricultural byproducts. The suggested process for creating composite materials involved the chelate complex synthesis of transition component cations using citric acid over the outermost layer of biochar, followed by their subsequent heat degradation. The produced materials showed high adsorption effectiveness, particularly in terms of removing chromium (VI) ions in water solutions. Of all the substances tested, the sunflower husk biochar composite material based on $CoFe_2O_4$ demonstrated the highest adsorption capability, with a rate of adsorption of 6.98 mg/g. These findings demonstrate the enormous possibilities of biochar composites that utilize $CoFe_2O_4$ for practical industrial wastewater management applications. Furthermore, Soltani et al. (2023) examined the production of zinc oxide nanoparticles (ZnONPs) utilizing extract from the rind of a scallions plant and their potential uses as a photocatalyst to remove organic contaminants from wastewater and as a nanofertilizer for plants. According to macroscopic methods, the ZnONPs had a mix of round and rod-shaped shapes, including cylindrical particles measuring between 40 and 100 nm and rod-shaped particles larger than 200 nm. The existence of carbonyl and hydroxyl groups is shown by the distinctive bands of ZnONPs at 400–800 cm^{-1} that are visible in infrared spectroscopy. The diameter of roughly 200 nm is obtained by evaluating the hydrodynamic size using a particle size analyzer (PSA). The produced ZnONPs show promise as a nano adsorbent for the removal of methylene blue dye and antibiotics (ampicillin) from simulated wastewater, as well as a micronutrient for plant growth. ZnONP-treated seeds outgrow control seeds in terms of germination and seedling growth, demonstrating the benefits of ZnONPs as a nano-nutrient source in agriculture. Additionally, ZnONPs exhibit efficient ampicillin and MB dye cleanup in UV light in wastewater simulation conditions. This study identifies ZnONPs as a viable nano-based nutrient resource for agricultural uses and demonstrates their efficacy in the photocatalytic elimination of organic pollutants from wastewater.

14.5.3 Dye Remediation

In the realm of dye contamination, nanocomposites showcase their versatility through both adsorption and photodegradation mechanisms. The high surface area and specific interactions within nanocomposites facilitate the adsorption of dye molecules, providing an effective means of removal. Additionally, nanocomposites with photocatalytic properties harness solar energy to drive the degradation of dyes, offering a sustainable approach to remediation. In a research, Kocijan et al. (2021) presented a novel method for creating TiO_2/reduced graphene oxide (rGO) nanocomposites that may be used to photocatalytically break down dyes in water-based solutions. Ti–O–C bond formation enables the synergistic integration of TiO_2 and rGO, providing a rare chance to create nanocomposite photocatalysts with improved visible light absorption. This change makes it possible to activate photodegradation in both the visible and UV portions of natural sun energy. A hydrothermal process accompanied by a calcination process was used to create TiO_2@rGO nanocomposites containing various rGO contents (5 and 15 wt%). Using artificial solar-like radiation, the breakdown of MB and RhB dyes was used to evaluate the photocatalytic efficacy of these nanocomposites. In comparison to the produced TiO_2 photocatalyst, findings showed that TiO_2@rGO nanocomposites greatly improve the breakdown of RhB and MB. In particular, the TiO_2@rGO nanocomposite having a higher rGO content (15 wt%) works better after 2 h of exposure to solar-like light, achieving 99.8% destruction of RhB and 98.1% deterioration of MB. The enhancement was caused by the TiO_2@rGO nanocomposites' bandgap decreasing as the rGO concentration increased. Pristine TiO_2 particles exhibit a wider energy gap (Eg) of 3.15 eV, whereas the nanocomposites' Eg values were 2.71 and 3.03 eV, correspondingly. These findings demonstrate the potential of graphene-based TiO2 nanocomposites as incredibly efficient solar light-driven photocatalysts for water purification applications.

Heidari et al. (2019) explored the application of a photo-Fenton catalyst in the textile wastewater, specifically an AC/$CoFe_2O_4$ (AC/CFO), for the removal of dye. This study aims to assess the effectiveness of chemical oxygen demand (COD) and phenol removal by adjusting critical parameters like pH (from 3 to 11), reaction duration (from 1 to 50 min), quantity of persulfate (from 1 to 9 mM/L), and quantity of AC/CFO nanocomposite (from 0.1 to 0.9 g/L). The best eradicated rates of reactive red 198 and COD in actual wastewater were achieved at 98% and 100%, correspondingly, after ideal conditions were established. The ideal parameters were as follows: pH of 6.5; 0.3 g/L of AC/CFO; 25 min for the reaction; and 5 mM/L of persulfate under continuous (30 W) UV light irradiation at ambient temp. The outcomes demonstrate this system's excellent efficacy and feasibility as a wastewater remediation strategy for organic contaminants. This method's affordability, simplicity of use, use of resources that are readily available on Earth, and possibility for reuse are noteworthy features. This photo-Fenton-like catalyst system provides a viable and eco-friendly alternative to other sophisticated oxidation techniques for the removal of organic contaminants from textile effluent. Similarly, Kocijan et al. (2022) introduced the immobilization of titanium dioxide (TiO_2) nanoparticles (NPs) and reduced graphene oxide (rGO)-TiO_2 nanocomposite on glass sheets for the photocatalytic degradation of methylene blue (MB) under various radiation sources, including ultraviolet and simulated solar radiation. Utilizing titanium isopropoxide precursor and a simple hydrothermal process, the rGO-TiO_2 and TiO_2 NPs nanocomposite was created. Calcination was then used. Using the spin-coating technique, the produced photocatalysts were applied to glass sheets. To improve film adherence on the glass surface, ethylene glycol was added. The photocatalytic activity was investigated in both simulated solar radiation and UV light. Additionally, the effects of several water matrix (milli-Q, stream, reservoir, and seawater) and reactive elements (·OH, h^+, and e^-) on the immobilized rGO/TiO_2 nanocomposite's photocatalytic effectiveness were carefully evaluated. The relative results of photolysis, photocatalytic degradation, and methylene blue (MB) dye adsorption in a milli-Q water medium exposed to UV-A and simulated solar light irradiance are shown in Fig. 14.6a. Taking into account both UV and simulated solar irradiation, the photolytic and photocatalytic efficiency for MB degradation were assessed for the immobilized rGO/TiO_2 nanocomposite, bare TiO_2, and a control without a catalyst. The photolytic degradation of MB dye is illustrated in Fig. 14.6b, which shows that, in the exclusion of a catalyst, after 120 min of simulated solar irradiation, roughly 11.5% degradation takes place in a milli-Q water medium. Interestingly, no photolytic degradation of MB dye is observed under UV-A irradiation. This is likely due to the fact that UV-A light does not provide enough energy for breaking apart an intricate organic substance like MB dye. However, when photolysis takes place under simulated sun irradiation, MB dye only slightly degrades. This implies that the destruction of complex organic contaminants like MB is probably better suited for simulated solar radiation due to its increased energy density.

Abualnaja et al. (2021) examined the removal of methyl orange (MO) dye under varied circumstances using a polymeric composite, poly(acrylonitrile-co-styrene) P(AN-co-St), saturated with adsorbents, including sulfonated and multiwall carbon nanotubes (MWCNTs). MWCNTs were produced using a cost-effective chemical vapor deposition (CVD) technique, and P(AN-co-St) and SP(AN-co-St) were employed in a novel nanocomposite copolymer blend. The results of the experiment demonstrated that the efficacy of P(AN-co-St)/MWCNTs for MO removal increased at acidic pH circumstances (pH 3 and 5) at a stirring rate of 140 rpm, a sorbent weight of 0.01 g, and a mean dye concentration of 20 mg. According to the Langmuir model, the maximum sorption capacities for P(AN-co-St)/MWCNTs, SP(AN-co-St), and P(AN-co-St) composites were determined to be 121.95, 48.78, and 47.84 mg/g, respectively. Several different isotherm models, including Freundlich, Temkin, and Halsey, were used to assess the experimental results. The sorption mechanism was better explained by the pseudo-second-order model. Similarly,

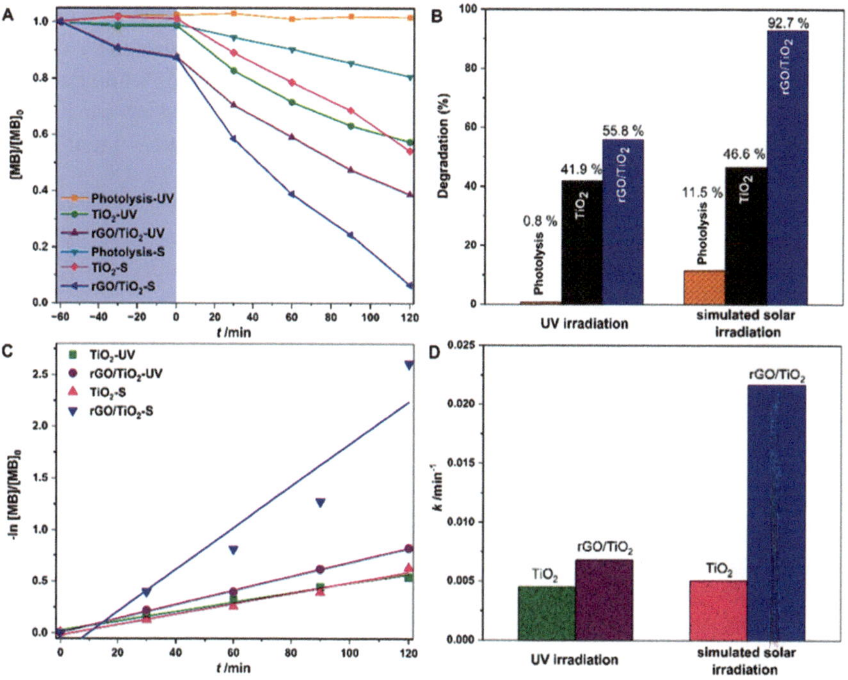

Fig. 14.6 (**a**) Assessment of MB absorption, photolytic degradation, and photocatalytic breakdown in milli-Q water, an aqueous media, as exposed to UV-A and simulated solar irradiation (S) for different periods of irradiation. (**b**) Following 120 min, the photodegradation yields MB. (**c**) The relationship between reaction time and –ln [MB]/[MB]0. The immobilized rGO/TiO_2 nanocomposite photocatalyst facilitates photocatalytic processes, as shown by the kinetic rate constants (**d**). (Adapted from Kocijan et al. (2022) under CCBY 4.0)

Thamer et al. (2020) To overcome the difficulties posed by the high swelling characteristics and difficult recovery of certain hydrogels employed as adsorbents for the removal of pollutants from wastewater, a novel porous nanocomposite hydrogel was created in situ. The found hydrogel nanocomposite, which relies on chitosan-grafted-polyacrylic acid/oxidized electrospun carbon nanofibers (CT-g-PAA/O-ECNFs), was made using an in situ grafting polymerization process. With this nanocomposite hydrogel, methylene blue (MB) could be cheaply and effectively removed from contaminated water thanks to its highly reusable adsorption characteristics. The effect of adding O-ECNFs on the hydrogel's ability to swell was investigated in both distilled water and MB solution at a pH of 7. The impact of many parameters on the adsorption process was examined, including the O-ECNFs ratio, temperature, beginning concentration, pH, contact duration, and investigation. The adsorption isotherm and kinetics were also examined in the study utilizing a variety of non-linear models. The results indicated that adding O-ECNFs to the hydrogel network improved its efficacy in removing MB dye while decreasing its capacity to swell. The adsorption process was pH dependent, and the Freundlich isotherm model and the pseudo-second-order model (PSO) both provided good explanations for the adsorption and kinetic findings. It is noteworthy that the recently created adsorbents demonstrated recyclability for a minimum of five cycles while retaining their adsorption capability. Their capacity for useful uses in water filtration is further demonstrated by the fact that they could be readily retrieved without losing their prior weight.

14.6 Pollutant Removal Mechanisms

The utilization of nanocomposites in environmental remediation signifies a transformative approach to addressing pollution challenges. This chapter delves into the intricacies of pollutant removal mechanisms, focusing on the versatile applications of nanocomposites in tackling heavy metals, organic compounds, and emerging contaminants. Heavy metals, notorious for their persistence and toxicity, find targeted solutions through nanocomposites, utilizing adsorption, chelation, precipitation, and ion exchange processes. The nuanced interactions between nanocomposites and heavy metals, governed by surface chemistry and nanoparticle attributes, are explored to comprehend efficient and selective removal strategies. In the realm of organic compounds, nanocomposites with catalytic functionalities demonstrate remarkable efficacy. Surface functionalization enhances adsorption capacities, setting the stage for catalytic transformations that break down organic pollutants through oxidation and reduction processes (Vasu et al. 2021; Dan et al. 2023). Figure 14.7 depicts the overarching procedure of photocatalytic eradication employed for organic pollutant removal in H_2O. The intricate process involves the utilization of photocatalysts to

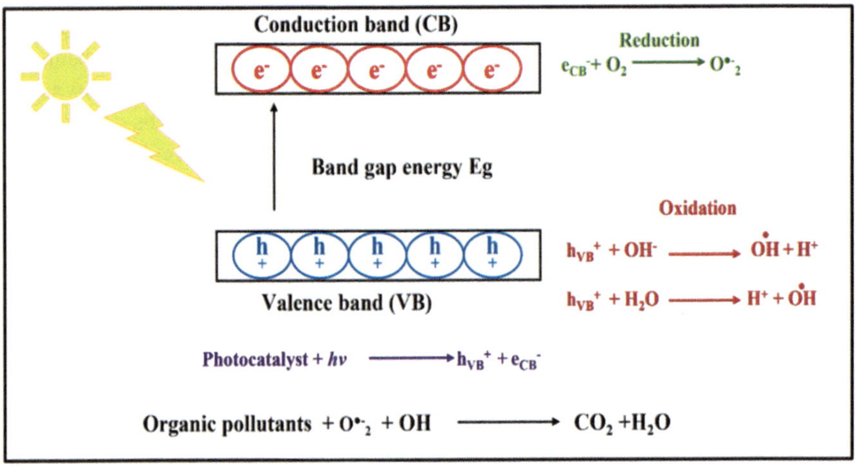

Fig. 14.7 Depiction of the fundamental process of photocatalytic degradation targeting organic pollutants in water. (Adapted from Alahmadi (2022) under CCBY 4.0)

initiate the breakdown of organic pollutants, harnessing the energy from incident light. This figure serves as an illustrative guide, offering a visual representation of the key steps and interactions within the photocatalytic degradation mechanism. In this process, photons from input light are absorbed by the photocatalyst, which is typically made of semiconductor materials like zinc oxide (ZnO) or titanium dioxide (TiO_2), creating electron-hole pairs. Following redox interactions with oxygen and water, these charge carriers generate reactive oxygen species (ROS), including superoxide radicals ($\cdot O_2^-$) and hydroxyl radicals (·OH). These ROS are essential for oxidizing and breaking down the organic contaminants in the water. The illustration continues on to outline the phases of pollutant deterioration, highlighting the role that the produced ROS plays in converting complex organic compounds into less hazardous byproducts. Additionally, it provides a visual narrative of the overall photocatalytic process, guiding the viewer through the sequential events that contribute to the effective remediation of organic pollutants in water.

Similarly, Shathy et al. (2022) described the mechanochemical–calcination process used to characterize and create a novel boron-doped ZnO/TiO_2 (B-ZnO/TiO_2) nanocomposite photocatalyst. During sunlight irradiation, the photocatalytic ability of B-ZnO/TiO_2 was assessed in comparison to conventional TiO_2, ZnO, B-ZnO, and ZnO/TiO_2 in the elimination of methylene blue (MB) dye. Approximately 95% of the MB was removed under ideal circumstances, which included a pH of 11, a B-ZnO/TiO_2 quantity of 30, a 15-min operating time, and an MB level of 15 mg/L. According to scavenger examinations, the active components in MB degradation were O_2^- and OH radicals. Figure 14.8 shows how B-ZnO/TiO_2

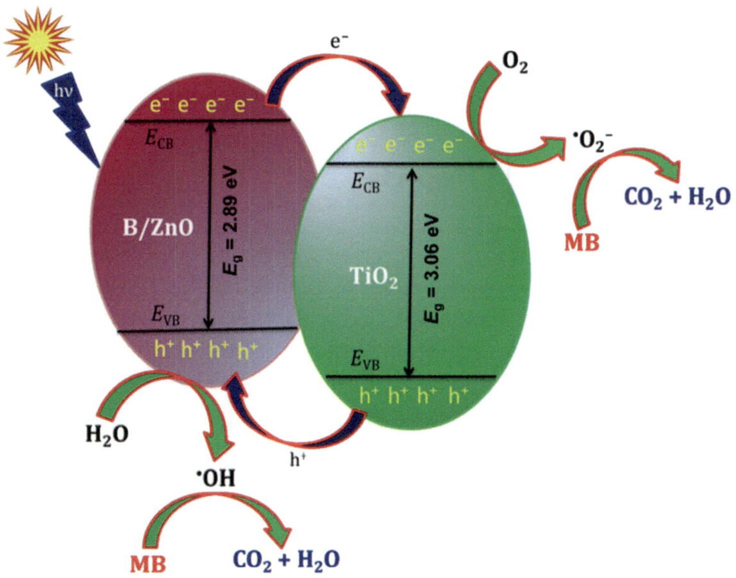

Fig. 14.8 Depiction of the schematic diagram depicting the photocatalytic degradation mechanism of B-ZnO/TiO_2. (Adapted from Shathy et al. (2022) under CCBY 4.0)

nanocomposites aid in the photocatalytic breakdown of MB. Both ZnO and TiO_2 underwent an alteration from the valence band (VB) to the conduction band (CB) when exposed to natural light, producing electrons (e^-) and holes (h^+). In comparison to TiO_2 (−0.14 eV), ZnO had a greater negative conduction band minimal (ECB) value (−0.16 eV). In order to improve the dispersion of charges, a small number of electrons (e^-) migrate from the CB of ZnO to the CB of TiO_2, and holes (h^+) migrate from the VB of TiO_2 to the VB of ZnO. When oxygen (O_2) and water (H_2O) are present in the MB solution, O_2 and H_2O are converted into $\cdot O_2^-$ and ·OH radicals, accordingly, by electrons (e^-) and holes (h^+). Scavenger investigations have confirmed that the most active species in charge of the photocatalytic breakdown of MB molecules are $\cdot O_2^-$ and ·OH radicals. These radicals react with MB dye because of their strong oxidation capabilities, producing carbon dioxide (CO_2), H_2O, and other environmentally safe byproducts.

Addressing the dynamic landscape of emerging contaminants requires adaptable remediation approaches. The variety of nanocomposites makes them indispensable resources for reducing the adverse effects of contaminants whose properties are still being determined. The chapter explores how nanocomposites can be tailored to adapt to emerging contaminants through adsorption, catalysis, and targeted functionalization strategies. By unraveling the pollutant removal mechanisms for emerging contaminants, the potential of nanocomposites to provide agile and effective solutions in the face of evolving environmental challenges becomes apparent (Falah and Mackenzie 2020; Pandis et al. 2022). The pollutant removal mechanisms employed by nanocomposites showcase their versatility in addressing diverse pollutants. The ability to tailor nanocomposites for specific applications through composition, structure, and functionalization strategies is a hallmark of their efficacy. Whether adsorbing heavy metals, catalyzing the degradation of organic compounds, or adapting to emerging contaminants, nanocomposites exhibit dynamic responses.

14.7 Challenges and Prospects

The development of nanocomposites for environmental remediation introduces a spectrum of challenges and opportunities, as explored in this chapter. In addressing hurdles faced in synthesis, application, and scalability, precision in nanocomposite synthesis without compromising scalability is a significant challenge. Achieving uniform dispersion of nanoparticles, controlling size and morphology, and ensuring reproducibility require meticulous design, balancing precision with scalability for real-world applications. Stability and durability challenges arise as nanocomposites face harsh environmental conditions, variable pH, and co-existing pollutants. Their long-term behavior and resilience under dynamic conditions necessitate comprehensive understanding and strategic development. Environmental and health concerns emerge as nanocomposites are deployed, raising questions about potential impacts, release of nanoparticles, persistence, and bioaccumulation. Navigating these unknowns demands a thorough examination of fate, transport, interactions, and effects on ecosystems and human health. Scalability and cost-effectiveness present formidable challenges, hindering large-scale adoption due to issues such as high production costs, specialized equipment needs, and limited scalability of effective laboratory synthesis methods.

Navigating uncharted waters in regulatory frameworks poses a challenge as nanocomposite technology evolves faster than regulatory adaptation. The lack of standardized testing protocols, risk assessment guidelines, and clear regulatory pathways hampers responsible and sustainable nanocomposite development. Bridging the gap between technological innovation and regulatory adaptation becomes crucial for navigating these challenges responsibly. Amidst challenges, opportunities for advancement in nanocomposite technology abound. The tunable design of nanocomposites for targeted applications is a key opportunity, enabling tailored composition, structure, and properties to address specific pollutants with enhanced efficiency. Integration with smart technologies offers real-time monitoring, adaptive strategies, and autonomous nanocomposites for dynamic environmental adjustments. Sustainable synthesis approaches align with green chemistry goals, utilizing eco-friendly precursors and reducing energy-intensive processes. Multifunctional nanocomposites provide synergistic solutions, combining adsorption, catalysis, and photocatalysis functionalities to address a broader spectrum of pollutants comprehensively. Interdisciplinary collaboration emerges as a crucial opportunity, fostering collaboration among diverse experts to accelerate understanding, establish standardized protocols, and inform responsible nanocomposite deployment. The chapter thus unveils a landscape where challenges necessitate strategic solutions, and opportunities offer avenues for transformative advancements in the field of nanocomposite-based environmental remediation.

14.8 Future Outlook

In envisioning the future of nanocomposite materials for environmental remediation, their prospects are rooted in precision and tailorability. As synthesis techniques advance and understanding of nanomaterial interactions deepens, the prospect of designing nanocomposites with unparalleled specificity for targeted environmental solutions emerges. This tailored approach opens avenues for highly efficient and customized remediation strategies. Moreover, the future sees nanocomposites evolving into multifunctional platforms, integrating various remediation mechanisms within a single material. This multifunctionality allows nanocomposites to simultaneously address a wide range of pollutants, providing comprehensive solutions to complex environmental challenges. The future of nanocomposites is largely dependent on sustainability, with green synthesis techniques becoming more and more popular. The development of environmentally sustainable nanocomposites aligns with global initiatives for responsible material development, reducing the environmental footprint associated with production. In the realm of smart and autonomous systems, the future envisions nanocomposites integrating sensors, responsive materials, and data-driven algorithms for real-time adaptability. By offering flexible and self-regulating solutions that increase productivity while reducing unexpected environmental effects, these smart technologies offer the ability to completely transform environmental remediation.

Furthermore, it is expected that nanocomposites would contribute to the circular economy, emphasizing reusability and recycling. Designing nanocomposites with efficient recovery and recycling strategies contributes to sustainable practices, aligning with the broader goal of transitioning towards a circular economy. Moving beyond theoretical potential, the future of nanocomposites involves real-world applications and field studies. Evaluating their performance in diverse environmental settings through field studies and pilot projects provides valuable insights, bridging the gap between theoretical potential and practical efficacy. Advancing environmental remediation technologies using nanocomposites also entails integration with existing methods. The prospect of combining nanocomposites with traditional remediation techniques creates synergistic approaches that enhance overall efficiency. Collaboration among researchers, industries, and regulatory bodies is pivotal for advancing environmental remediation technologies responsibly. Establishing standardized testing protocols, risk assessment methodologies, and regulatory frameworks contributes to the responsible development and deployment of nanocomposites. Furthermore, the future involves addressing emerging pollutants and staying ahead of environmental challenges. Anticipating and proactively addressing novel and emerging contaminants positions nanocomposites as adaptable tools in the ongoing battle against evolving pollutants. Public awareness and acceptance are integral to the future success of nanocomposite technologies in environmental remediation. Transparent communication, education, and engagement efforts are essential in building public understanding and trust, facilitating the responsible deployment of nanocomposite technologies. In conclusion, the future outlook of nanocomposite materials envisions transformative roles in precision, multifunctionality, sustainability, smart systems, circular economy practices, real-world applications, collaboration, addressing emerging pollutants, and fostering public trust. As these technologies continue to mature, nanocomposites are poised to become indispensable tools in the pursuit of a sustainable and resilient global environment.

14.9 Conclusion

In the evolving landscape of environmental remediation, nanocomposite materials stand as beacons of innovation and promise, offering transformative solutions to complex challenges. As this chapter delves into the multifaceted dimensions of nanocomposite development, from their unique physicochemical properties to their applications and challenges, it becomes evident that these materials hold the key to shaping the future of environmental remediation technologies. The exploration of nanocomposite properties, intricately tied to their physicochemical characteristics, surface chemistry, and optimization for performance, lays the foundation for understanding their prowess in pollutant removal. The design and fabrication of nanocomposites, as discussed in the context of composition adjustment, structural considerations, and property optimization, underscore the importance of precision engineering to unleash their full potential. Metal/oxide-based and carbon-based nanocomposites emerge as versatile categories, each with distinct advantages, synthesis methods, and applications in environmental remediation. As nanocomposites transition from laboratory experiments to real-world applications, the significance of their matrix becomes increasingly apparent. The stability and reactivity of nanocomposites are significantly influenced by the matrix material, which can be either organic, inorganic, or hybrid. The nuanced interactions between nanoparticles and the matrix material, influenced by surface chemistry, charge interactions, and compatibility, further highlight the intricacies of designing effective nanocomposite systems for environmental remediation. Applications of nanocomposites in environmental remediation, spanning adsorption processes, catalytic applications, and photodegradation, underscore their versatility in addressing a spectrum of pollutants. The mechanisms of pollutant removal, with a specific focus on heavy metals, organic compounds, and emerging contaminants, reveal the adaptability and innovation embedded in nanocomposite technologies. The synergy between nanocomposites and existing remediation methods creates a holistic approach, promising enhanced efficiency and effectiveness in tackling environmental pollution.

However, this journey is not without its challenges. The hurdles in nanocomposite development, ranging from synthesis precision and scalability to stability, environmental concerns, and regulatory frameworks, require careful consideration. Overcoming these challenges demands collaborative efforts, interdisciplinary research, and a commitment to responsible innovation. It necessitates the development of sustainable synthesis approaches, the integration of smart technologies for real-time monitoring, and the establishment of best practices for nanocomposite deployment. Yet, within these challenges lie opportunities for advancement. The tunable design of nanocomposites for targeted applications, the integration with smart technologies, the development of sustainable synthesis approaches, and the creation of multi-functional nanocomposites present avenues for overcoming current limitations. The circular economy prospects, real-world applications, and collaboration for standardization contribute to the maturation of nanocomposite technologies. Looking ahead, the future of nanocomposite materials in advancing environmental remediation technologies appears promising. Precision, tailorability, multifunctionality, sustainability, smart systems, circular economy practices, real-world applications, collaboration, addressing emerging pollutants, and fostering public trust collectively define the path forward. These materials are not merely agents of change; they are architects of a sustainable and resilient environmental future.

References

Abualnaja KM, Alprol AE, Abu-Saied MA et al (2021) Studying the adsorptive behavior of poly(Acrylonitrile-co-styrene) and carbon nanotubes (nanocomposites) impregnated with adsorbent materials towards methyl orange dye. Nanomaterials 11:1144. https://doi.org/10.3390/nano11051144

Akintunde OO, Yu L, Hu J et al (2022) Disinfection and photocatalytic degradation of organic contaminants using visible light-activated GCN/Ag_2CrO_4 nanocomposites. Catalysts 12:943. https://doi.org/10.3390/catal12090943

Alahmadi N (2022) Recent progress in photocatalytic removal of environmental pollution hazards in water using nanostructured materials. Separations 9:264. https://doi.org/10.3390/separations9100264

Alharthi FA, Alghamdi AA, Alanazi HS et al (2020) Photocatalytic degradation of the light sensitive organic dyes: methylene blue and rose bengal by using urea derived g-C_3N_4/ZnO nanocomposites. Catalysts 10:1–16. https://doi.org/10.3390/catal10121457

Alvarez-Torrellas S, Boutahala M, Boukhalfa N, Munoz M (2019) Effective adsorption of methylene blue dye onto magnetic nanocomposites. Modeling and reuse studies. Appl Sci 9:4563. https://doi.org/10.3390/app9214563

Balu S, Uma K, Pan GT et al (2018) Degradation of methylene blue dye in the presence of visible light using SiO_2@α-Fe_2O_3 nanocomposites deposited on SnS_2 flowers. Materials (Basel) 11:1030. https://doi.org/10.3390/ma11061030

Bashir S, Thakur A, Lgaz H et al (2019) Computational and experimental studies on phenylephrine as anti-corrosion substance of mild steel in acidic medium. J Mol Liq 293:111539. https://doi.org/10.1016/j.molliq.2019.111539

Bashir S, Lgaz H, Chung IM, Kumar A (2020a) Effective green corrosion inhibition of aluminium using analgin in acidic medium: an experimental and theoretical study. Chem Eng Commun 208:1–10. https://doi.org/10.1080/00986445.2020.1752680

Bashir S, Thakur A, Lgaz H et al (2020b) Corrosion inhibition efficiency of bronopol on aluminium in 0.5 M HCl solution: insights from experimental and quantum chemical studies. Surf Interfaces 20:100542. https://doi.org/10.1016/j.surfin.2020.100542

Bashir S, Thakur A, Lgaz H et al (2020c) Corrosion inhibition performance of acarbose on mild steel corrosion in acidic medium: an experimental and computational study. Arab J Sci Eng 45:4773–4783. https://doi.org/10.1007/s13369-020-04514-6

Bekru AG, Tufa LT, Zelekew OA et al (2023) Microwave-assisted synthesis of rGO-ZnO/CuO nanocomposites for photocatalytic degradation of organic pollutants. Crystals 13:133. https://doi.org/10.3390/cryst13010133

Cardoso IMF, Cardoso RMF, Esteves da Silva JCG (2021) Advanced oxidation processes coupled with nanomaterials for water treatment. Nanomaterials 11:2045. https://doi.org/10.3390/nano11082045

Castro-Muñoz R, González-Melgoza LL, García-Depraect O (2021) Ongoing progress on novel nanocomposite membranes for the separation of heavy metals from contaminated water. Chemosphere 270:129421. https://doi.org/10.1016/j.chemosphere.2020.129421

Chauhan HA, Rafatullah M, Ahmed Ali K et al (2021) Metal-based nanocomposite materials for efficient photocatalytic degradation of phenanthrene from aqueous solutions. Polymers (Basel) 13:2374. https://doi.org/10.3390/polym13142374

Chauhdary Y, Hanif MA, Rashid U et al (2022) Effective removal of reactive and direct dyes from colored. Water 14:1–20

Chavan AA, Li H, Scarpellini A et al (2015) Elastomeric nanocomposite foams for the removal of heavy metal ions from water. ACS Appl Mater Interfaces 7:14778–14784. https://doi.org/10.1021/acsami.5b03003

Dan S, Bagheri H, Shahidizadeh A, Hashemipour H (2023) Performance of graphene oxide/SiO_2 nanocomposite-based: antibacterial activity, dye and heavy metal removal. Arab J Chem 16:104450. https://doi.org/10.1016/j.arabjc.2022.104450

Elessawy NA, Gouda MH, Elnouby MS et al (2021) Novel sodium alginate/polyvinylpyrrolidone/TiO_2 nanocomposite for efficient removal of cationic dye from aqueous solution. Appl Sci 11:9186. https://doi.org/10.3390/app11199186

Falah M, Mackenzie KJD (2020) Photocatalytic nanocomposite materials based on inorganic polymers (geopolymers): a review. Catalysts 10:1–20. https://doi.org/10.3390/catal10101158

Fan B, Tan Y, Wang J et al (2021) Application of magnetic composites in removal of tetracycline through adsorption and advanced oxidation processes (AOPs): a review. Processes 9:1644. https://doi.org/10.3390/pr9091644

Guerra FD, Attia MF, Whitehead DC, Alexis F (2018) Nanotechnology for environmental remediation: materials and applications. Molecules 23:1–23. https://doi.org/10.3390/molecules23071760

Heidari MR, Varma RS, Ahmadian M et al (2019) Photo-Fenton like catalyst system: activated carbon/$CoFe_2O_4$ nanocomposite for reactive dye removal from textile wastewater. Appl Sci 9:963. https://doi.org/10.3390/app9050963

Javed I, Hanif MA, Rashid U et al (2022) Enhancing functionalities in nanocomposites for effective dye removal from wastewater: isothermal, kinetic and thermodynamic aspects. Water 14:2600. https://doi.org/10.3390/w14172600

Kar P, Jain P, Gupta RK, Tripathi KM (2020) Emerging carbon-based nanocomposites for remediation of heavy metals and organic pollutants from wastewater. In: Emerging carbon-based nanocomposites for environmental applications. Wiley, pp 1–29. https://doi.org/10.1002/9781119554882.ch1

Kaya S, Lgaz H, Thakkur A et al (2023a) Molecular insights into the corrosion inhibition mechanism of omeprazole and tinidazole: a theoretical investigation. Mol Simul 49:1–15. https://doi.org/10.1080/08927022.2023.2256888

Kaya S, Thakur A, Kumar A (2023b) The role of in silico/DFT investigations in analyzing dye molecules for enhanced solar cell efficiency and reduced toxicity. J Mol Graph Model 124:108536. https://doi.org/10.1016/j.jmgm.2023.108536

Kocijan M, Ćurković L, Ljubas D et al (2021) Graphene-based TiO_2 nanocomposite for photocatalytic degradation of dyes in aqueous solution under solar-like radiation. Appl Sci 11:1–15. https://doi.org/10.3390/app11093966

Kocijan M, Ćurković L, Bdikin I et al (2022) Immobilised rGO/TiO_2 nanocomposite for multicycle removal of methylene blue dye from an aqueous medium. Appl Sci 12:385. https://doi.org/10.3390/app12010385

Kumar A, Thakur A (2020) Encapsulated nanoparticles in organic polymers for corrosion inhibition. Elsevier Inc.

Liu Y, Hou C, Jiao T et al (2018) Self-assembled AgNP-containing nanocomposites constructed by electrospinning as efficient dye photocatalyst materials for wastewater treatment. Nanomaterials 8:35. https://doi.org/10.3390/nano8010035

Liu Y, Huang Y, Xiao A et al (2019) Preparation of magnetic Fe_3O_4/MIL-88A nanocomposite and its adsorption properties for bromophenol blue dye in aqueous solution. Nanomaterials 9:51. https://doi.org/10.3390/nano9010051

Lu H, Zhang L, Ma J et al (2019) Nano-cellulose/mof derived carbon doped CuO/$Fe3O_4$ nanocomposite as high efficient catalyst for organic pollutant remedy. Nanomaterials 9:277. https://doi.org/10.3390/nano9020277

Mandal S, Hwang S, Marpu SB et al (2023) Bioinspired synthesis of silver nanoparticles for the remediation of toxic pollutants and enhanced antibacterial activity. Biomolecules 13:1054. https://doi.org/10.3390/biom13071054

Modi S, Yadav VK, Gacem A et al (2022) Recent and emerging trends in remediation of methylene blue dye from wastewater by using zinc oxide nanoparticles. Water 14:1749. https://doi.org/10.3390/w14111749

Naveed AB, Riaz F, Mahmood A et al (2021) A facile synthesis of Bi_2O_3/$CoFe_2O_4$ nanocomposite with improved synergistic photocatalytic potential for dye degradation. Catalysts 11:1180. https://doi.org/10.3390/catal11101180

Olteanu M, Baraitaru A, Panait AM et al (2019) Advanced SiO_2 composite materials for heavy metal removal from wastewater. Water Air Soil Pollut 230:179. https://doi.org/10.1007/s11270-019-4225-7

Oyewo OA, Elemike EE, Onwudiwe DC, Onyango MS (2020) Metal oxide-cellulose nanocomposites for the removal of toxic metals and dyes from wastewater. Int J Biol Macromol 164:2477–2496. https://doi.org/10.1016/j.ijbiomac.2020.08.074

Pandis PK, Kalogirou C, Kanellou E et al (2022) Key points of advanced oxidation processes (AOPs) for wastewater, organic pollutants and pharmaceutical waste treatment: a mini review. ChemEngineering 6:8. https://doi.org/10.3390/chemengineering6010008

Seliem MK, Barczak M, Anastopoulos I, Giannakoudakis DA (2020) A novel nanocomposite of activated serpentine mineral decorated with magnetic nanoparticles for rapid and effective adsorption of hazardous cationic dyes: kinetics and equilibrium studies. Nanomaterials 10:1–12. https://doi.org/10.3390/nano10040684

Shabelskaya N, Egorova M, Radjabov A et al (2023) Formation of biochar nanocomposite materials based on $CoFe_2O_4$ for purification of aqueous solutions from chromium compounds (VI). Water 15:93. https://doi.org/10.3390/w15010093

Shahabuddin S, Sarih NM, Kamboh MA et al (2016) Synthesis of polyaniline-coated graphene oxide@$SrTiO_3$ nanocube nanocomposites for enhanced removal of carcinogenic dyes from aqueous solution. Polymers (Basel) 8:305. https://doi.org/10.3390/polym8090305

Shan C, Su Z, Liu Z et al (2023) One-step synthesis of Ag_2O/Fe_3O_4 magnetic photocatalyst for efficient organic pollutant removal via wide-spectral-response photocatalysis–Fenton coupling. Molecules 28:4155. https://doi.org/10.3390/molecules28104155

Sharma G, Pathania D, Naushad M, Kothiyal NC (2014) Fabrication, characterization and antimicrobial activity of polyaniline Th(IV) tungstomolybdophosphate nanocomposite material: efficient removal of toxic metal ions from water. Chem Eng J 251:413–421. https://doi.org/10.1016/j.cej.2014.04.074

Sharma D, Thakur A, Kumar M et al (2023) A convenient synthesis, electrochemical profiling, and morphological studies of a pyridine-based 1,3,4-oxadiazole hybrid: a promising study for corrosion mitigation of mild steel in strongly acidic environment. Inorg Chem Commun 158:111554. https://doi.org/10.1016/j.inoche.2023.111554

Shathy RA, Fahim SA, Sarker M et al (2022) Natural sunlight driven photocatalytic removal of toxic textile dyes in water using B-doped ZnO/TiO_2 nanocomposites. Catalysts 12:308. https://doi.org/10.3390/catal12030308

Singh S, Barick KC, Bahadur D (2013) Functional oxide nanomaterials and nanocomposites for the removal of heavy metals and dyes. Nanomater Nanotechnol 3:1–19. https://doi.org/10.5772/57237

Soltani S, Gacem A, Choudhary N et al (2023) Scallion peel mediated synthesis of zinc oxide nanoparticles and their applications as nano fertilizer and photocatalyst for removal of organic pollutants from wastewater. Water 15:1672. https://doi.org/10.3390/w15091672

Sun M, Senthil RA, Pan J et al (2018) A facile synthesis of visible-light driven rod-on-rod like α-FeOOH/α-$AgVO_3$ nanocomposite as greatly enhanced photocatalyst for degradation of rhodamine B. Catalysts 8:392. https://doi.org/10.3390/catal8090392

Thakur A, Kumar A (2023a) Computational insights into the corrosion inhibition potential of some pyridine derivatives: a DFT approach. Eur J Chem 14:246–253. https://doi.org/10.5155/eurjchem.14.2.246-253.2408

Thakur A, Kumar A (2023b) Exploring the potential of ionic liquid-based electrochemical biosensors for real-time biomolecule monitoring in pharmaceutical applications: from lab to life. Results Eng 20:101533. https://doi.org/10.1016/j.rineng.2023.101533

Thakur A, Kumar A, Kaya S et al (2022a) Suppressing inhibitory compounds by nanomaterials for highly efficient biofuel production: a review. Fuel 312:122934. https://doi.org/10.1016/j.fuel.2021.122934

Thakur A, Kumar A, Kaya S et al (2022b) Recent advancements in surface modification, characterization and functionalization for enhancing the biocompatibility and corrosion resistance of biomedical implants. Coatings 12:1459. https://doi.org/10.3390/coatings12101459
Thamer BM, Aldalbahi A, Meera MA, El-Newehy MH (2020) In situ preparation of novel porous nanocomposite hydrogel as effective adsorbent for the removal of cationic dyes from polluted water. Polymers (Basel) 12:1–18. https://doi.org/10.3390/polym12123002
Thamer BM, Aldalbahi A, Meera Moydeen A et al (2021) Modified electrospun polymeric nanofibers and their nanocomposites as nanoadsorbents for toxic dye removal from contaminated waters: a review. Polymers (Basel) 23:1–37. https://doi.org/10.3390/polym13010020
Vasu D, Fu Y, Keyan AK et al (2021) Environmental remediation of toxic organic pollutants using visible-light-activated Cu/La/CeO_2/GO nanocomposites. Materials (Basel) 14:6143. https://doi.org/10.3390/ma14206143
Wang ER, Shih KY (2021) Facile microwave hydrothermal synthesis of $ZnFe_2O4$/rGO nanocomposites and their ultra-fast adsorption of methylene blue dye. Materials (Basel) 14:5394. https://doi.org/10.3390/ma14185394
Yaqoob AA, Ibrahim MNM, Umar K et al (2021) Cellulose derived graphene/polyaniline nanocomposite anode for energy generation and bioremediation of toxic metals via benthic microbial fuel cells. Polymers (Basel) 13:1–26. https://doi.org/10.3390/polym13010135
Yaseen M, Humayun M, Khan A et al (2022) Photo-assisted removal of rhodamine B and nile blue dyes from water using CuO–SiO_2 composite. Molecules 27:1–17. https://doi.org/10.3390/molecules27165343
Youssef AM, El-Naggar ME, Malhat FM, El Sharkawi HM (2019) Efficient removal of pesticides and heavy metals from wastewater and the antimicrobial activity of f-MWCNTs/PVA nanocomposite film. J Clean Prod 206:315–325. https://doi.org/10.1016/j.jclepro.2018.09.163

GPSR Compliance

The European Union's (EU) General Product Safety Regulation (GPSR) is a set of rules that requires consumer products to be safe and our obligations to ensure this.

If you have any concerns about our products, you can contact us on ProductSafety@springernature.com

In case Publisher is established outside the EU, the EU authorized representative is:

Springer Nature Customer Service Center GmbH
Europaplatz 3
69115 Heidelberg, Germany

Batch number: 09550171

Printed by Printforce, the Netherlands